ON THE ORIGIN OF STORIES

ON THE ORIGIN OF STORIES

EVOLUTION, COGNITION, AND FICTION

Brian Boyd

The Belknap Press of Harvard University Press

CAMBRIDGE, MASSACHUSETTS · LONDON, ENGLAND · 2009

Library of Congress Cataloging-in-Publication Data

Boyd, Brian, 1952-
   On the origin of stories : evolution, cognition, and fiction /
Brian Boyd.
       p.   cm.
   Includes bibliographical references and index.
   ISBN 978-0-674-03357-3 (cloth : alk. paper)
   1. Fiction–History and criticism.   2. Fiction–Authorship.
I. Title.

   PN3451.B68 2009

   809.3—dc22   2009007642

TO BRONWEN

CONTENTS

Contents

ILLUSTRATIONS

ACKNOWLEDGMENTS

My first debt is to Art Spiegelman. The very day I read about, then promptly bought and read, *Maus*—before I had finished reading it, even—I decided to teach a new graduate course in narrative from the *Odyssey* and Genesis, through Giotto, Shakespeare, Hogarth, Austen, Victorian narrative painting, Tolstoy, Joyce, film of four continents, Nabokov, Australian Aboriginal painting of the Western desert, and *Maus*. I had no idea then that teaching this course would eventually goad me into searching for an evolutionary explanation for fiction, although preparation for the course introduced me to the work of David Bordwell. The theoretical acumen with which he applied a cognitive (and now also an evolutionary) approach to film was revelatory.

The invitations of a number of conference organizers helped precipitate some of this book's ideas. Mike Hanne asked me to give a keynote at his Narrative and Metaphor Conference at the University of Auckland in 1995, and prompted me to develop ideas emerging in the narrative course about the immemorial and still active role that other animals play in human story. Chris Price, director of the Writers and Readers Week at the Wellington International Festival of the Arts in 2000, invited me to join a panel on literature and science that returned me to thinking about evolution, art, and literature intensely enough to launch the writing of this book. At various stages of the project David Miall at the University of Alberta at Edmonton, George Gadanidis and Cornelia Hoogland at the University of Western Ontario, Anthony Uhlmann at the University of Western Sydney, and Frans Saris at the University of Leiden invited me to give keynotes that forced me to marshal ideas in new ways.

I thank the Royal Society of New Zealand for a James Cook Fellowship, which allowed me to begin this project; and Dr. John Hood, then Vice Chancellor of the University of Auckland, for the generous terms of a distinguished professorship that allowed me to complete it while continuing to publish in other areas.

Among the many astute readers of versions of this book or of articles or talks that prefigured it I would particularly like to thank Michael Corballis, Stephen Davies, Denis Dutton, Steven Pinker, Michelle Scalise Sugiyama, Ineke Sluiter, Blakey Vermeule, and Michael Wright, and above all David Bordwell, Joseph Carroll, Brett Cooke, Ellen Dissanayake, Jonathan Gottschall, Marcus Nordlund, Murray Smith, and David Sloan Wilson. David Bordwell, Joe Carroll, Brett Cooke, Jon Gottschall, and Marcus Nordlund all read and commented astutely on the whole manuscript. It has been a rare scholarly pleasure to work closely with Joe and Jon on my, their, and our projects, and to refresh my thinking in David Sloan Wilson's cascade of ideas. I am also grateful for comments from students in my "Narrative," "From Sonnets to Comics," and "Literature and Science" courses at the University of Auckland, especially Janet Hunt, Stephanie Miskell, Anne Ruthe, and Bruce Sheridan; and from audiences at the universities of Auckland, California at Berkeley, California at Los Angeles, Harvard, Leiden, Stanford, Texas A & M, and Wisconsin–Madison and at conferences of the Human Behavior and Evolution Society. My work has profited from discussions with Donna Rose Addis, William Benzon, Leda Cosmides, Fred Crews, Nancy Easterlin, Philip Fisher, Patrick Colm Hogan, Roger Horrocks, Tim Horvath, Anna Jackson, Mac Jackson, Edmund King, Stephen Kroon, Robert Nola, Zachary Norwood, Alan Richardson, Elaine Scarry, C. K. (Karl) Stead, John Tooby, Arie Verhagen, Julian Young, and Lisa Zunshine.

Some material has been published before, usually in different form and contexts, in *Philosophy and Literature* (Boyd 2001, 2004, 2005b, 2007a) and in *American Scholar* (Boyd 2006b, 2008c, 2009b).

Ann Hawthorne has been an ideally sensitive and responsive editor, and Bronwen Nicholson, as always, has been an eager first reader, an alert editor, and an unfailing support in work and play.

We are absurdly accustomed to the miracle of a few written signs being able to contain immortal imagery, involutions of thought, new worlds with live people, speaking, weeping, laughing. We take it for granted so simply that in a sense, by the very act of brutish routine acceptance, we undo the work of the ages, the history of the gradual elaboration of poetical description and construction, from the tree-man to Browning, from the caveman to Keats.

Vladimir Nabokov, *Pale Fire* (1962)

INTRODUCTION: ANIMAL, HUMAN, ART, STORY

SOME PEOPLE EAT WITH CHOPSTICKS, some with knives and forks, and some with neither. Are these just discrete cultural inventions, or do they have common roots? What about kabuki, Western opera, and Aboriginal dance ceremonies: are they local inventions with nothing in common, or reflections of something we share as humans?

All people use their hands to convey food to their mouths—in fact all *primates* do—and all peoples modify some food before eating it. Feeding with hands is a primatewide biological adaptation, a trait shaped by natural selection because it offers advantages in terms of survival and reproduction. Modifying food before eating—cutting, cooking, or more—is a specifically human adaptation. I will suggest that despite its many forms, art, too, is a specifically human adaptation, biologically part of our species. It offers tangible advantages for human survival and reproduction, and it derives from play, itself an adaptation widespread among animals with flexible behaviors. But I will focus most on the art of storytelling.

We can tell stories to explain things, from a child's or a country's pouty "*They* started it" to why the world is as it is, according to myth or science. We also tell stories just because we cannot stop, because they fascinate and engage us even if we know they are untrue. What links these different kinds of stories, observed, discovered, or invented, and why has the richest explanatory story of all, the theory of evolution by natural selection, been so little used to explain why and how stories matter?

I recall a colleague asking, as academics do: "What are you working on?" "I'm trying to figure out," I answered, "an evolutionary—Darwinian—approach to fiction." Not waiting to hear more, he shut down his face and the conversation: "That must be very reductive." "No, not reductive, but expansive," I might otherwise have answered: extending the historical context from decades to millions of years, and increasing the historical precision, from decades down to the moment of choice.[1]

An evolutionary understanding of human nature has begun to reshape psychology, anthropology, sociology, philosophy, economics, history, political studies, linguistics, law, and religion.[2] Can it also help explain even art, even human minds at their freest and most inventive?

In art, as in so much else we had thought uniquely human, like tool-using or tool-making, counting or culture, we have begun to find precursors elsewhere in nature.[3] But can evolution account even for the one human art with no known precedent, the art of fiction? Can it show why, in a world of necessity, we choose to spend so much time caught up in stories that both teller and told know never happened and never will? I want to show that it can, in ways far less reductive than much recent literary scholarship, in ways both wider in scope and finer in detail.[4]

In literary studies, and in the humanities in general, a biological approach to the human has been anathematized for the last four decades by the recently dominant paradigm that calls itself "Theory" or "Critique." But after announcing decades ago first the death of the author and then the death of the "subject" (the individual), Theory has recently raised the question of its own death, and there has been a widespread cry in literary studies for a return to texts.[5]

A biocultural approach to literature invites a return to the richness of texts and the many-sidedness of the human nature that texts evoke. But it also implies that we cannot simply go back to literary texts without assimilating what science has discovered about human nature, minds, and behavior over the last half-century, and considering what these discoveries can offer for a first truly comprehensive literary theory.

Even some who accept evolution as the most powerful explanation of living things insist that it can say little about human nature and behavior.[6] To concede that natural selection has shaped the structure and function not only of our bodies but also of our minds, they fear, would impinge on our freedom or our capacity to transform ourselves and our world. But as we shall see, their fears are misplaced: evolution can explain the bases not only of human behavior, from mating to murder, but also of culture and freedom.[7]

Art shows imaginations at their freest, shaping the world on their own terms, at the furthest remove from biological necessity. If evolution can help to account even for art, it can surely contribute to *any* explanation of human behavior. Of course, evolution alone cannot explain every feature of an art or an artist: not even Shakespeare had genes for writing *Hamlet*. But without considering fiction's origins we cannot follow its full story; indeed, we start almost at the end of the story. Without a biocultural perspective we cannot appreciate how deeply surprising fiction is, and how deeply natural.

In answering the question why humans in all societies have such a fascination for art, and for the art of fiction, we can appreciate not only why art began, but also why we feel compelled to tell and listen to stories, why we can understand them so readily, why they are formed as they are, why they treat what they do in human nature, and why they continue to break new ground.

Dolphins breathe air and blow out bubbles as they exhale. They can use these bubbles almost like nets to herd fish together before closing in for the kill. Untrained Amazon River dolphins sometimes play with the necklaces of bubbles trailing from their mouths by turning to swim through them or bite them. Dolphins in several marine species have been observed in more elaborate play, releasing air from their blowholes to form underwater rings that hover and hold their form for several seconds as they expand. But like humans blowing smoke-rings, dolphins must practice to master such a quirky skill.

In a marine park on the coast of Hawaii's Oahu in the 1990s a small pod of bottlenose dolphins turned these bubble-rings into their own art form. Watched but not prompted or rewarded by the scientists at Sea Life Park, half of the dolphins now engage in elaborate air-bubble play. They take their cue from others, practice the rings until they become stable, inspect their own performance, explore new possibilities, and intently monitor others' efforts. Some dive through the rings as they expand. Others create vortices with their tail flukes, and release the rings into the swirling current so that they travel not upward but sideways or even downward in the water. An adult male, Kaiko'o, can emit two controlled bubble-rings, one after the other, which he then nudges together with his rostrum to form a single large ring. A young female, Tinkerbell, has developed several unique techniques, such as creating a vortex with her dorsal fin as she swims across the tank, then retracing her path and releasing into the vortex a stream of air that shoots out in a helical pattern in front of her.[8]

Is this behavior art? Would its closest human equivalent be smoke-rings, which we might call play or display but not art? Or rhythmic gymnastics, which, like figure skating, loops and arches across the boundary between sport and art? Or dance, which we would agree *can* be art, or definitely *is*, by the time we get to ballet or Bali? In many ways these dolphin air-rings *do* approximate incipient art. But for all their playfulness dolphins seem to have no specieswide predispositions to "artistic" behavior. Dolphin air art blurs the boundary between play and art, and it is from play, widespread across the animal world, that I suggest art evolved in humans.

Like some human arts, dolphin air art involves design but not representation. Without representation, fiction—and indeed much song and dance or painting and sculpture—would be impossible. Can animals other than ourselves represent things in ways more optional and open than the honeybee waggle that tells hive-mates the direction and distance of flowers? Can they do this in nature in ways that serve no immediate function of reproduction or survival?

Dian Fossey described two groups of gorillas meeting on the slopes of Mount Visoke in Rwanda. The male leaders in each group strutted defiantly to warn the other group against attack. But when on one side the inexperienced leading male performed clumsily, the youngsters in his group followed behind him, "exaggeratedly mocking his awkward displays of bravado."[9] Like dolphin air art, these youngsters' mockery is play, but play that now involves rudimentary representation. Behavioral imitation occurs widely in social species, since individuals can learn from others' successes. But these young gorillas imitate not out of would-be bravado, but in play, to hold up the model for their mimicry to derision and amusement.

The gorilla anecdote records a brief moment of mimetic mockery, perhaps dependent on an immediate model. But another great ape even closer to us can engage in deliberate and sustained play. In the Kibale Forest of Uganda, primatologist Richard Wrangham watched an eight-year-old chimpanzee, Kakama, "pick up and keep a small log for two days, hugging it, carrying it in every possible fashion, lying on his back in his nest and balancing it on his feet as mothers often do with their babies. He made a little day-nest and laid the log in it while he sat beside. He retrieved the log when it fell." Wrangham eventually lost the chimp's—and the log's—trail when he had to circumvent snorting bush-pigs. Wrangham concluded:

> My intuition suggested a possibility that I was reluctant, as a professionally skeptical scientist, to accept on the basis of a single observation: that I had just watched a young male chimpanzee invent and then play with a doll . . . A doll! The concept was novel enough that I simply filed away my notes without saying much about it to anyone else, and left Uganda the following week. Four months later, two field assistants at Kibale, Elisha Karwani and Peter Tuhairwe, happened to be following Kabarole and Kakama. Neither Karwani nor Tuhairwe knew of my observation. Yet for three hours they watched Kakama carry a log—not the same one as before, surely—taking it with him wherever he fed. This time they saw him leave it. Once they were certain Kakama had disappeared, they collected it, brought it to camp, and stapled to it a label that described their own straight-forward interpretation of the object's meaning: "Kakama's Toy Baby."[10]

We would not call Kakama's actions art, but they do suggest a capacity for sustained imaginative imitation and for representation that we see in the pretend play of human children. We would not call pretend play art, either, but it can easily shade into mime, theater, or storytelling. Imitation or representation, common in some arts (painting or fiction, for instance), may be absent in others (music or decorative, abstract, or conceptual art), but two things distinguish Kakama's behavior from art. Whereas play comes naturally to chimpanzees, the kind of elaborate scenario-building observed in Kakama seems no normal part of a natural specieswide behavior, unlike childhood pretend play. And unlike human art, it seems never to be carried out with others. Human children, by contrast, construct imaginary scenarios perfectly "naturally," without training, alone *or* in company.

How close can we get to the origins of art in our own species? In a purely temporal dimension, not very. But we can recognize something close to the start of art in scenes not much more elaborate than the ones played out above by gorillas and chimpanzees. The following incident happened among Mbuti in the Ituri Forest of central Africa, but it might have occurred at any time over the last 100,000 or even million years, anywhere there were humans. In an apparently unstaged *National Geographic* special a father, talking with his infant son, swings him from his back to his chest, which upsets the boy. The father distracts him by telling the child that

> this is the way monkey babies are carried. The father has an idea. He begins to walk like a great ape, making soft hooting noises as he does. At first he seems to be doing this for the benefit of his son, but a group of women sitting in front of a nearby shelter soon become an audience, laughing and shouting encouragement to the father in such a way that it is obvious that his portrayal is meant for them as well. Indeed, the cause of the appreciation is obvious: the father is doing a very good imitation of a mature ape with its young.[11]

Here imitative play shades into art, into drama. The father knows his son will recognize his actions as imitation, even without a model nearby, and presumably, like parents everywhere, he knows his son is fascinated by animals. He knows his son will recognize the movements of the ape and also recognize the imitation as playful pretense. Hearing the response of an audience wider than he intended, the father enjoys evoking a still richer appreciation of his play, his pretense, his artistry.

As we will see, unique aspects of human parent-child interaction, a special instance of our species' singular capacity to share attention, hold a major key to the origin of art. Here the father engages his son's attention to change his mood. He thereby affects the mood of others, whose appreciation in turn alters his own mood. The feedback of action, attention, reaction, and the refinement of action to shape further attention and reaction provide an exclusively human basis for art.

In other species we can recognize the first impulses of art but no more. Other animals can engage in exploratory behavior that seems designed to appeal to the mind, but only in isolated and incidental fashion. But in our own species the impulse to art develops reliably in all normal individuals. The isolated sparks in other species have become the steady current of human art. What could have caused the change?

Evidence for the origins of human art necessarily remains extremely patchy. As in exploring the origins of language, we have to deduce much from the way individual development often roughly replays the development of the species. But although spoken language does not fossilize, and written language stretches back only a few millennia, full-fledged works of art survive from tens of thousands of years ago. The 1994 discovery of the drawings in France's Chauvet Cave pushed back the earliest date for cave art from about 17,000 years ago to 32,000 years ago, a leap large enough to suggest how much we still have to discover.[12]

We do not know all that the drawings in Chauvet Cave meant for those who drew them. Like the Mbuti father's play, they represent animals. They may well have had magical significance or conveyed some sense of power over the powerful prey they depict. But we can be sure that here we have art, that the people who drew these figures knew that they were drawing, and drawing animals, and doing so with skill, and that without both the likeness and their sense of power in achieving it any further meaning or magic would have been impossible.

Chauvet seems a sudden leap forward, presumably not because

art had just been invented there but because only there had it become so impressive that these drawings seemed worth executing in a site difficult of access but sure of preservation. The wall markings were hardly the casual doodles of idle afternoons. The grotto at Chauvet was no dwelling place, and the drawings were no stone-age wallpaper. This remote cave, deep underground, accessible only by the light of a burning brand or a tallow candle, seems to have been selected precisely for its remoteness from disturbance, whether by weather, plant, or animal, expressly to preserve the art of particularly awe-inspiring craftsmen. We, too, prize and preserve our art, but who would bet that the most treasured possessions of our greatest galleries, or the galleries themselves, will be intact in another 30,000 years?

No other species comes near this achievement. Apes and elephants prove ready enough to slap paint on paper when offered the means. The chimpanzees whom Desmond Morris provided with paper, brushes, and paint in the 1950s even became committed to painting, for no reward but the pleasure of the activity, and daubed with fierce concentration and individual flair. But for all their engagement in the process while it lasted, the apes lost interest in the results immediately afterward and never attempted anything remotely like representation.[13]

In some ways the art of Chauvet resembles dolphin air art or chimpanzee painting: not required for immediate survival, or even for practicing survival skills, it seems designed especially to appeal to the eyes and minds of the artists and others. Of course there are differences. Unlike dolphin air art, the cave drawings endure in a permanent medium and present a likeness recognizable many millennia later. And perhaps because of the strong cognitive appeal of the likeness, because it moved the spirits of those who watch, it could then be imagined to have power over other unseen forces.

But unlike dolphin air art or chimpanzee painting, the drawings at Chauvet reflect a human universal. Almost all psychically normal humans will have tried to represent animals, on a surface, or with a stick, or with some mud or clay, or through a mime or dance, or by imitation of sounds, or through a story, or in many of these ways. Even though both the animals depicted and the culture that encouraged their depiction no longer exist, viewers from any modern society can feel a shock of recognition at the Chauvet images, whatever other effects the artists may have intended.

By the time of the Chauvet drawings art would seem to have become characteristically human. We can only presume that works as elaborate as this bespeak a long prior process of practice and experiment on surfaces like bark or skins that have not survived. The intact and monumental example of Chauvet Cave, along with a few other isolated but much more modest relics over the preceding 30,000 years or so, implies extensive traditions of art, as something humanly shared, incorporating sophisticated representation, involving expert practitioners, earning the ready comprehension and eliciting the admiration of others around them.

Dolphins and humans last shared a common ancestor about 100 million years ago. Chimpanzees and humans diverged around 6 million years ago. Chauvet Cave's drawings date back over 30,000 years. Spray paint on a concrete wall brings us right into our own time.

A 1981 New Zealand graffiti calendar includes a photograph of an Auckland city scene. A man jogging on the grass indicates the scale of the retaining wall behind him. On it someone has spray-painted in letters four feet high, in a message almost forty feet across: "Ralph, come back, it was only a Rash."

In eight words, in nine syllables, less than a line of Shakespearean verse, unfolds a story, a drama, a tragedy, a joke, a wry reflection on the very medium of graffiti. If you really have lost your lover because he thinks you have caught a sexually transmitted disease that brands you unfaithful, but that turns out to be no more than an innocuous rash, you may want to call him back, you may not know where to reach him, but you won't broadcast the most intimate details of your emotional and physical life in a message forty feet long. This tragedy of jealousy, abandonment, and loss, with a poignantly desperate attempt at recovery, collapses in a moment into a self-conscious joke on the disparity between the intimacy of the message and the publicity of the medium.

Writing is much younger than cave painting, but fictional storytelling is far older, and a human universal. Think of this graffiti-writer's behavior in terms of evolutionary costs and benefits, and it appears to make no sense: the cost in materials; the cost in time, to invent and refine the message, to spray it on that scale; the cost in risk, in the chance of being caught and prosecuted for defacing public property. And for what benefits? The graffito remained unsigned and anonymous until it was removed altogether. Perhaps, like New York's Keith Haring or London's Banksy, the perpetrator won respect from friends and others, but she or he could equally have acted solo and in secret. Regardless of how circumscribed the glory that the graffito earned, the effort must have seemed worthwhile to the artist, simply for the pleasure of the idea and the pleasure of an audience, even an anonymous audience who did not know whom to thank. Someone was so taken by the notion that viewers could reconstruct a story, an implicit tragedy, *and* the joke that undermines it, from just these few words that that alone made the graffito worth the time, energy, and risk.

That is what I want to explain in evolutionary terms: our impulse to appeal to our own minds and reach out to others for the sheer pleasure of sensing what we can share even in an unprecedented new move.

We respond with ridiculous ease to a joke that leaps right out at us, but how do we explain *why* we see the story and the joke almost in an instant, when we have so little to go on, not even a cue that this is fiction or funny? The mind is not inductive, as Shakespeare's contemporary Francis Bacon suggested thinking should be. We do not patiently wait for all available evidence before advancing as short an additional distance as possible to our conclusion. Instead, we hastily construct inferences that reach well beyond what we find and that nevertheless, as in this case, thanks to the writer's skill, hit home.

Storytelling lies at the heart of literature, yet literary studies all too rarely explore our ability as readers to construct a story on meager hints, to fill gaps and infer situations. We take the process for granted. Sometimes nothing proves more difficult than seeing what's just under our noses.

We tend to consider vision immediate and transparent, but when artificial intelligence researchers tried to program visual comprehension into computers, they found how extraordinarily complex it is.

Narrative comprehension has proved still more difficult to program, even using the flattest of stories.[14] I'll believe that computers can think not when they can beat a Kasparov at chess, with its rapidly proliferating but after all calculable permutations, but when they can be fed something as unexpected as the photograph of "Ralph, come back, it was only a Rash" daubed on a wall, and can read the words, deduce the story, then laugh at the joke they have recognized for themselves.

An evolutionary explanation of human behavior does not entail genetic determinism. We have not evolved for spray-painting. Yet our impulse to engage in and respond to art, verbal, visual, musical, kinetic, exists across human cultures and develops in all normal children without special teaching. And it depends on the capacity of those around us to understand what we have done, even when we are absent or have devised something utterly new. How did a behavior so complex, often so costly in terms of time and even resources, and of so little apparent benefit in a competitive struggle for existence, ever become established throughout humankind?

On the Origin of Stories has two main aims: first, to offer an account of fiction (and of art in general) that takes in our widest context for explaining life, evolution; and to offer a way beyond the errors of thought and practice in much modern academic literary study, which over the last few decades has often stifled—and has even sought to stifle—the *pleasure,* the *life,* and the *art* of literature.

After introducing evolution and human nature in Part 1, I explain first art in general (Part 2) and the art of fiction in particular (Part 3) as biological adaptations. In Parts 4 and 5 I offer two detailed examples, from different kinds of "origins" of stories: the historical origin, in Homer's *Odyssey,* the most successful of ancient stories; and the individual origin, in *Horton Hears a Who!,* one of the finest works of the greatest storyteller for the very young. The Conclusion and Afterword suggest the implications of a biocultural perspective for understanding literature and life.

BOOK I

EVOLUTION, ART, AND FICTION

Everything is what it is because it got that way.

D'Arcy Thompson, *On Growth and Form* (1917)

In the distant future I see open fields for far more important researches. Psychology will be based on a new foundation, that of the necessary acquirement of each mental power and capacity by gradation. Light will be thrown on the origin of man and his history.

Charles Darwin, *On the Origin of Species* (1859)

IN BOOK 1 we pick up the long-distance lenses of evolution to focus in turn on human nature, on art as a human attribute, and on fiction as a mode of art.

Part 1 considers human nature in the light of evolution. Despite disagreements about particular perspectives, Chapter 1 shows, a wide consensus has emerged that we need to see human nature, like the rest of life, within the framework of evolution. We are as we are because we got that way, and we have nothing to fear and much to gain—including a better sense of how we can change—from understanding ourselves better.

Chapter 2 explains the evolutionary concept of an adaptation, a feature of a species' form or behavior especially designed by natural selection for the benefits it offers, on average, in terms of survival and reproduction. Chapters 3 and 4 focus on the evolution of intelligence and of cooperation, two key human adaptive complexes that therefore naturally play a key role in art, including in the stories we explore in Book II.

Part 2 offers an evolutionary account of art in general. Could art be an adaptation in its own right, Chapter 5 asks, or a product of sexual selection, or a byproduct of adapted minds, or an orderly series or an untidy assortment of independent local cultural inventions allowing little scope for evolutionary explanation, or perhaps even an insufficiently coherent category to permit any single description or explanation?

I propose in Chapter 6 that art is a human adaptation that derives from play, a behavior widespread across animal classes and perhaps universal in mammals. Play evolved through the advantages of flexibility; the amount of play in a species correlates with its flexibility of action. Behaviors like escape and pursuit, attack and defense, and social give-and-take can make life-or-death differences. Creatures with more motivation to practice such behaviors in situations of low urgency can fare better at moments of high urgency. Animals that play repeatedly and exuberantly refine skills, extend repertoires, and sharpen sensitivities. Play therefore has evolved to be highly self-rewarding. Because it is compulsive, animals engage in it again and again, incrementally altering muscle tone and neural wiring, strengthening and increasing speed in synaptic pathways, improving their capacity and performance.

Humans uniquely inhabit "the cognitive niche":[1] we gain most of our advantages from intelligence. We therefore have an appetite for information, and especially for pattern, information that falls into meaningful arrays from which we can make rich inferences. Information can be costly to obtain and analyze, but because it offers an invaluable basis for action, nature evolves senses and minds to gather and process information appropriate to particular modes of life. Like other species, humans can assimilate information through the rapid processing that specialized pattern recognition allows, but unlike other species we also seek, shape, and share information in an open-ended way. Since pattern makes data swiftly intelligible, we actively pursue patterns, especially those that yield the richest inferences to our minds, in our most valuable information systems, the senses of sight and sound, and in our most crucial domain, social information.

We can define art as cognitive play with pattern. Just as play refines behavioral options over time by being self-rewarding, so art increases cognitive skills, repertoires, and sensitivities. A work of art acts like a playground for the mind, a swing or a slide or a merry-go-round of visual or aural or social pattern. Like play, art succeeds by engaging and rewarding attention, since the more frequent and intense our response, the more powerful the neural consequences. Art's appeal to our preferences for pattern ensures that we expose ourselves to high concentrations of humanly appropriate information eagerly enough that over time we strengthen the neural pathways that process key patterns in open-ended ways.

Much play is social, and much of the most intense play is most social. Biologists have termed humans ultrasocial.[2] Our extreme sociality amplifies our predilection for the cognitive play of art, through both competitive and cooperative processes, especially through our unique inclination to share and direct the attention of others. Chapters 7 and 8 show how the individual and social benefits of art create a positive feedback process that amplifies individual and collective invention. Art generates a confidence that we can transform the world to suit our own preferences, that we need not accept the given but can work to modify it in ways we choose; and it supplies skills and models we can refine and recombine to ensure our ongoing cumulative creativity.

Part 3 presents a naturalistic account of fiction as a human behavior related to but also moving radically beyond what other animals can do in understanding one another (Chapter 10), communicating with one another (Chapter 11), and playing with one another (Chapter 12).

Chapter 9 establishes that true stories need little special explanation: many animals share information, and language allows humans to share information about the past. But the pseudo-information of fiction poses an evolutionary puzzle: why do we not prefer only true information?

Evolutionary adaptation involves *design* for specific *functions*, for specific benefits in terms of survival and reproduction. Chapters 10–12 show the evidence for the biological origin and design of fiction. Chapter 10 investigates first the origins of our capacities for understanding events in the evolutionary past and in human infancy, and especially in the uniquely human levels of theory of mind, our capacity to understand one another in terms of beliefs as well as in terms of desires and intentions. It then explores our capacities for recalling events. Recent research strongly suggests that our memories of past experience saturate our present thought: cognition partially reactivates, almost simulates, the multimodal nature of our prior experience. Still more importantly, the apparent weakness of memory, in reconstructing rather than passively recording experience, seems an evolved design that allows us to recombine freely our past experience so that we can imagine or presimulate our future. Chapter 11 shows how the unique human capacity for narrative emerges from animal capacities for representing events but adds specifically human capacities for joint attention, imitation, and language. It also shows how narrative as strategy reflects the tension between cooperation and competition in social life. Chapter 12 explains fiction as an art, a form of cognitive play with patterns of social information, especially in terms of the emergence of pretend play in human infancy.

Chapter 13 then considers the biologically adaptive *functions* of fiction, including the complex interrelation between fictions recognized as such and the often more powerful fictions of religion.

Athiest

PART 1

EVOLUTION AND NATURE

Our ancestors have been human for a very long time. If a normal baby girl born forty thousand years ago were kidnapped by a time traveler and raised in a normal family in New York, she would be ready for college in eighteen years. She would learn English (along with—who knows?—Spanish or Chinese), understand trigonometry, follow baseball and pop music; she would probably want a pierced tongue and a couple of tattoos. And she would be unrecognizably different from the brothers and sisters she left behind.

Kwame Anthony Appiah, *Cosmopolitanism* (2006)

1

EVOLUTION AND *HUMAN* NATURE?

HAMLET TELLS THE ACTORS at Elsinore not to overact, for doing so would pervert the "purpose of playing, whose end, both at the
first and now, was and is to hold as 'twere the mirror up to nature"—
by which he means human nature. As he says, "both at the first and
now" people have thought that literature reflects nature, especially
human nature.[1] "At the first," at least from Plato's unease at literature's power to mimic human life and Aristotle's admiration for literature's mimetic power, to Stendhal's image of a novel as a mirror
passing along a roadway, and beyond.

But not "now," at least not in university literature departments.
There, many have denounced the notion of a human nature as "essentialism," a belief in a human essence. For many in the modern
humanities and social sciences, there is no human nature, only the
constructions of local culture, and to think otherwise can only en-
danger hopes for changing what we are and do.

This position is confused. Even to deny a universal human nature
and insist only on local cultural difference already constitutes a claim
about human nature: that the minds and behavior of all humans,
and only humans, depend solely on culture.[2] This happens to be
false: our minds and behavior are *always* shaped by the interaction of nature and nurture, or genes and environment, including the
cultural environment. And false, too, about other species, many of
which have culture, and at least one other, chimpanzees, have a different culture in every group observed in the wild and could not survive without their local culture.[3] Even on an everyday level we could
not engage with other humans without an implicit theory of human

nature. We do not respond to other people as if they were dogs, mice, or chimpanzees. Bared teeth on a Rottweiler mean one thing but something very different on a smiling human face.

If many in literature departments reject the idea of a human nature, let alone of evolution's power to explain it, many in other fields from anthropology to religion and sociology have recently recognized that the deep past that shaped our species can help to explain our present and our recent past.[4] Evolutionary aspects of mind may not be as accessible in the fossil record as we would like, but evidence has converged from many sources across species, cultures, eons, and life-stages—from evolutionary theory; from observation and experiment with animals and human infants, children, and adults; from game theory, artificial intelligence, and computer simulations; from clinical and cognitive psychology and neuroimaging—that many aspects of our minds and behavior have been configured over evolutionary time. By nature we have much in common before and even after local culture shapes us.

No one who has open-mindedly considered the evidence doubts that evolution produced humans as it did our near relatives, chimpanzees, bonobos, and other great apes. Yet many who accept the fact of our physical evolution nevertheless resist seeing human minds and behavior as shaped in any way by our deep past. Let me try to remove misunderstandings and allay misgivings behind this resistance.

Human Differences

Although an evolutionary view of human nature will often focus on "universals," on common features of our brains and behavior, it does not ignore or deny the enormous cultural differences between peoples.

Human differences have been exaggerated in contrasting ways over the century and a half since the publication of *On the Origin of Species,* first in a racist, then in an antiracist direction. In the late nineteenth century, in the heyday of Western imperialist expansion and laissez-faire capitalism, Darwinism and especially the notion of "the survival of the fittest" (not initially Darwin's term) were seized on by Social Darwinists to justify the idea that the strong deserve to win, since their success in competition "proves" their superiority. They and others used this to rationalize a "science" of human racial difference—and, unsurprisingly for the times, of European racial superiority. The result was "the mismeasure of man":[5] a program to measure "racial" differences to "prove" the superiority of some human "races" and the inferiority of others.

Modern evolutionary approaches to human nature have nothing in common with this program. Science itself shows that despite differences strikingly visible to us, humans are genetically an unusually uniform species (three humans selected from around the world will differ genetically much less on average than three chimpanzees selected from any of their shrinking African habitats) and that there is more variation *within* a local ethnically homogeneous population than between one ethnic population and another.[6] Aware of the ghastly consequences of racism in the twentieth century, modern evolutionary psychologists stress "the psychic unity of humankind"[7] and focus more on what human minds have in common than on differences.

The nineteenth-century Western assumption that the West embodied the standards by which all peoples should be judged caused a reaction among early twentieth-century cultural anthropologists. To reject the idea that observed differences in behavior arise from biological differences and to show that they do not prove superiority or inferiority, cultural anthropologists and sociologists stressed the malleability of human nature under the pressure of culture. Early twentieth-century anthropologists insisted on the power and variety of culture, but although they claimed an almost limitless diversity of human behaviors, the behaviors they observed actually remained within narrow boundaries. No human culture, for instance, "even begins to compare with the social system" of any of our closest primate relatives.[8] Physiology renders visible the huge differences between human and great-ape sociosexual systems: all adult male chimpanzees try to couple with all adult females in their group, their massive testicles testifying to the intensity of male sexual competition; among bonobos, the other species closest to us, fervent and frequent female-female couplings result not only in enlarged clitorises

[handwritten marginal note:] does not make sense of avg. of a people to another people is less than variation w/in a people

but also in female alliances holding the balance of power; among gorillas no other male dares attempt to contest the dominant silverback's sexual control of all the adult females, and in the absence of direct male sexual competition, these huge apes have tiny testicles.

In the late 1960s the anthropologists' insistence on human difference began to permeate the humanities as well as the human sciences, and to seem like a moral crusade, a rejection of the hegemony of a Western sense of human nature. Roland Barthes, for example, criticized "the mystification which transforms petit-bourgeois culture into a universal nature."[9] Many in the humanities and social sciences began to deny human nature and to excoriate "essentialism."[10]

But as some anthropologists realized late in the century, their stress on human diversity had led them to overlook human universals.[11] On the receiving end, a Samoan scholar bemoaned Margaret Mead's enormously influential depiction of her people: she "took away our oneness with other human beings . . . We are no different from you."[12] To insist only on difference and deny commonality actually frustrates the commendable motives that led to the rejection of narrowly Western standards of human nature. Without any sense of all that we share in what the Ghanaian-American philosopher Kwame Anthony Appiah calls "the one race to which we all belong,"[13] what can be the basis for our special concern for human beings, for human justice?

A critique of unquestioned Western assumptions about human nature was needed, but the anthropological critique extended by Barthes and Michel Foucault in the 1960s and dogmatized over the following decades was shortsighted. It did not look nearly far enough. While structuralism and its more or less rebellious offspring continued in the direction set by early twentieth-century social sciences, the natural sciences moved in the opposite direction, toward the first comprehensive scientific attempt to understand human nature in the context of evolution and human, animal, and artificial cognition and behavior. The best way to critique Western bourgeois assumptions about human nature is not to deny that human nature exists, but to apply the hard tests of science, examining humans *against other species; in many cultures,* from hunter-gatherer bands to modern indus-

a big "human nature" ⟵

trialized states; *in many ages,* in their history *and* their arts, especially the arts of orature (oral literature) and literature: to investigate human universals *and* human particulars, similarities *and* differences.

Far from denying cultural difference, an investigation of human nature that takes into account our evolutionary past makes it possible to explain cultural difference in a way that insisting that humans are completely "culturally constructed" cannot. Constructed out of what, in any case?[14] The argument that people are culturally different because of their culture, or because humans are shaped just by culture, is merely circular.[15] Moreover, explanations of differences in, say, male and female behavior in terms of "our culture" or "their culture" look empty when we realize that similar systematic differences exist not only across human cultures but also in hundreds of animal species.[16]

An evolutionary account of human nature can show the advantages of sociality, and of social learning, in many species, culminating in the unique human susceptibility to culture. The extent of human cultural differences has been *made possible* by the evolution of the mind. Without the complex shared architecture of the mind, culture could not exist. *Because* of that shared design, there are many universals across cultures: there *is* a human nature.

BIOLOGICAL OR GENETIC DETERMINISM

Many fear that accepting the idea of an evolved human nature means accepting that our minds and actions are biologically or genetically determined. This fear is so mistaken and muddled that I hardly know where to start dispelling it.

You could call it "genetic determinism" that you were born a human being and not a creature of some other species. Does this fact restrict your freedom? Again, you could call it "genetic determinism" that you were born the unique individual you are, like and unlike your siblings and parents, and unprecedented in the history of the universe. Is *this* unwelcome? The notion that genes shape us is *less* deterministic than the notion that we are the product of our environment, since the complexity and randomness of genetic recombination in sexual reproduction means that we are each the result of an unpredictably generated variation unique to each of us rather than of anything imposed from without.

We should see genes less as constraints than as enablers.[17] A single mutation in a regulatory gene compels our neurons to continue the process of cell division several times more than in chimpanzees, so that our brains end up with several times the cell count of chimpanzee brains.[18] Mirror neurons in the primate line proliferate to such an extent in humans that in our relations with others we have "a deeply felt mirroring that moves people closer to each other and makes emotional connectedness possible."[19]

Our genes determine that, unlike other mammals, we walk on two legs, not four. We cannot slither like snakes or fly like swallows, but does the fact that our genes determine *that* we walk dictate where we walk, or how? Because we walk rather than gallop, over time our forelimbs became free to knap stone tools and hence, eventually, even to construct and play computer games. And the fact that we walk does not stop us from using two legs to skip, surf, or ski.

Richard Dawkins called it "a simple lie" to claim that an evolved human nature entails inevitable genetic determinism.[20] Genes do not dictate "always do this." Even in organisms without minds, even in single-celled organisms, they build in sets of if-then rules sensitive to context. All the more so in creatures with even simple minds, let alone in creatures with intelligence. We should see genes not as deniers of the role of the environment but as devices for extracting information *from* the environment.[21]

Dawkins notes that our sexual desires presumably derive from our genes, yet we usually manage to curb them when doing so is socially necessary. We can readily accept "that genes exert a statistical influence on human behavior while at the same time believing that this influence can be modified, overridden or reversed by other influences"[22]—including other genetic capacities, like the widespread capacity for empathy or the human capacity to inhibit and consider. Janet Radcliffe Richards explains: "Evolutionary psychology claims that various aspects of our character are deep in our genes, but it does not suggest for a moment that any particular emotion is overwhelming to the extent of preventing self-control or rational judge-

ment. Its claims are about the *origins and depth* of particular tendencies, not about pathological force."[23]

A biological view of human nature stresses that individuals are free agents endowed with the flexibility that evolution provides and active strategic choosers rather than passive products of their place and time.

NATURE VERSUS CULTURE?

A biocultural view of human nature does not exclude or slight the social or the cultural. It makes no sense to set biology in opposition to society or culture. Sociality occurs only within living species, and hence within the biological realm, through genes that encourage social animals to associate. Culture occurs only within the social and therefore, again, the biological realm.[24] And far from being unique to humans, culture—the nongenetic transmission of behavior, including local customs and even fashions—has been discovered over the last few decades in many social species, in birds as well as mammals.

Evolution has allowed humans to develop our singular capacity for culture because culture helps us track changes in the environment more rapidly than genes do.[25] Genetic change normally takes many generations to pervade a population; culture can enable advantageous options to spread rapidly in a single generation and to be passed on to successive generations. I therefore use "biocultural" and "evolutionary" almost interchangeably to characterize an approach to human nature that takes full note of biology *and* of the culture that evolution has made possible.

NATURE VERSUS NURTURE?

That our minds reflect evolution's design does not mean that all is nature and not nurture, that genes are everything and environment nothing. In any sophisticated biological thinking these oppositions have been thoroughly discredited.[26] Biologists stress that *phenotypes* (whole organisms) are not determined solely by *genotypes* (the genetic recipe in each cell's DNA) but are always inextricably *co*determined by the interacting of genes and the environment. Nature "ver-

nat and nurt work together

sus" nurture is not a zero-sum game, in which nature's x percent is 100 minus nurture's y percent. Rather, it is a product, x times y, nature *activated* at each stage according to its input from nurture.

The Natural as Right

To argue that biology provides a base for human life does not mean that it must impose a model for human morality. There is no reason why an origin must predetermine a particular end; biology in any case offers a vast array of different models; and culture is itself a part of biology and has the power to generate a cascade of new possibilities.

Nature as Selfish

Genes are "selfish" only in the sense that they prosper according to what benefits them in successive reproductive rounds, and not necessarily according to what benefits the organisms in which they reside, let alone other organisms with which they happen to interact. Yet most genes benefit from the health of the whole organism, or even from the success of a whole group of individuals, many or all of which carry those genes. Dawkins points out that he could with equal validity, though with less impact, have called his famous first book not *The Selfish Gene* but *The Cooperative Gene*.[27]
 Those uneasy about applying evolution to human behavior often assume that doing so must require stressing selfishness and competition at the expense of altruism and cooperation. In fact sociobiology's central preoccupation has been cooperation, or more precisely the complex mix of cooperation and competition in any society, and evolutionary psychology and evolutionary economics have placed far *more* emphasis on generosity, trust, and fairness than *non*evolutionary psychology or economics ever had.

Nature versus Freedom

An evolutionary view of human nature, far from threatening freedom, offers a reason to resist the molding of our minds by those who think they know best for us. The cultural constructionist's view of the mind as a blank slate is "a dictator's dream."[28] If we were entirely socially constructed, our "society" could mold us into slaves and masters, and there would be no reason to object, since those would henceforth be our socially constructed natures.

NATURE AS FIXED AND UNCHANGEABLE

Many who resist applying evolution to human nature associate it, curiously, with the fixed and inalterable, even though the theory shows species in constant change. Darwin's and Wallace's core discovery was that species are not rigid but in continuous transformation, slow if conditions are stable, swifter if they alter. And evolution has itself evolved a series of ways for organisms to respond more and more diversely to rapid change: sexual reproduction, nervous systems, flexible intelligence, social learning, culture. Biologist David Sloan Wilson argues that because of the unique role of culture in humans, "We have not escaped evolution, as so commonly assumed. We experience evolution in hyperdrive."[29]

Many suppose that to see people as the product of nurture alone offers more hope for changing them, as if they can be readily changed by cultural means. But "some environmentally caused characteristics are quite impossible to undo. Nobody can unbake a baked potato."[30] Seeing the world as discourse, as text, as many in the humanities convinced themselves they did in the late twentieth century, encouraged the idea—comforting for professional readers of texts—that they could transform the world just by reading it differently. But no, there is a world outside language, and that world does call out for substantive social change.

An evolutionary view allows for *informed* social change. Understanding biological conditions, what causes lead to what effects, makes it *more* possible to change conditions to achieve desired effects. When we discover which "if"-conditions produce which "then"-outcomes, we can look for ways to institute the conditions likeliest to lead to our preferred outcomes.[31] Owen Jones compares the law to a lever to change human behavior, and an informed knowledge of human nature to the fulcrum the lever needs to exert its force.[32]

NATURE AND POWER

Some think that the idea of "Nature, red in tooth and claw" must valorize power and class. In fact evolutionary anthropology and biology increasingly stress that a major difference between humans and other mammals is that they have found ways to control the urge for dominance, by collaborating to resist being dominated, and that this capacity has unleashed the unique power of human cooperation.[33] Multilevel selection theory (see Chapter 4) stresses that *any* group can compete more effectively against other groups by minimizing within-group fitness differences.[34]

NATURE VERSUS HUMAN SINGULARITY?

Humans have language, advanced technology, and high culture. Doesn't trying to see human lives in terms of evolution ignore all that differentiates us from the rest of the biological world?

No. Evolution can explain both our substantial continuity with other forms of life, in things as important as our social emotions, and our substantial difference. Evolution has passed through a number of major transitions, from inorganic chemistry to life, from nonnucleated (bacterial) cells to nucleated (eukaryotic: animal, plant, and fungal) cells, from single-celled to multicellular organisms, from individual organisms to societies. Each has involved new forms of cooperation at one level, making possible radically new and more complex possibilities at a higher level. Biologists now see the cooperation that makes human culture possible as the latest major transition in evolution.[35]

NATURE AND KNOWLEDGE

Many in the humanities reject an evolutionary account of human nature partly because they think it must presuppose a naïve empiricism, a view of knowledge that assumes we see just what's out there. Most in the humanities, by contrast, consider knowledge as not objective but culturally constructed, apparently "true" only within the terms of a particular culture.[36]

Vision can serve as a shorthand example to explain not only why a commonsense account of human knowledge fails but why the cultural constructivist rejection of that account also fails. In everyday experience the mind seems transparent, something we can understand just by looking into ourselves; it seems all of a piece, seamless, putting us into direct touch with reality, when in fact it integrates many built-in subroutines that we remain unaware of because they deliver results to us so fast and automatically. We look out and effortlessly see a world of space and objects and our place in it; our eyes seem to offer immediate access to the world.

But constructivists challenge that account, saying: "What we see is what we, through cultural experience, have learnt to see."[37] Since some peoples have only two words for color (the equivalent of light and dark or white and black), and since most cultures did not develop perspectival drawing, color and perspective, for instance, have been claimed as arbitrary and conventional.[38] In the mid-twentieth century, tribesmen not previously exposed to Western culture were said to be unable to interpret photographs because of their Western "conventions" (perspective, black-and-white)—a typically insulting and offensive consequence of the denial of human nature.[39] In fact even pigeons can read photographs and recognize human individuals from them.

Whatever their language, people see the same things, agree on the same colors as central to particular hues, and so on.[40] But vision does not transfer all the space, movement, shape, surface, and color out there into the mind in any immediate or transparent way, nor is human vision finer, as we had often smugly assumed, than many other kinds. We supposed for a long time that we were unique in having color vision; in fact we cannot see ultraviolet light as bees do, infrared light as pit vipers do, colors at night as some moths do, polarized light as many birds, insects, and even plants can; we have trichromatic vision, better than the dichromatic of many species, but pigeons have tetrachromatic vision. But although human vision misses out on much potential information, it effortlessly does far more than we realize, unconsciously integrating some fifty different brain areas to make complex, real-time sense of visual input. Specialized cells respond to edges, or speed of movement, or direction of movement, or animate but not inanimate movement; others respond to not just human faces but only faces at a particular angle, or a particular face that they then give an automatic emotional weighting to via the brain's emotional router, the amygdala. We have different day and night visual systems, different *where* and *what* visual systems, different unconscious early-warning visual systems and conscious visual systems.

As in so many other aspects of the human mind and human nature, such as our social cognition and social emotions, we err in supposing that we know what we see, think, or are, *either* through immediate introspection *or* through the mediation of local culture. Nature has built far more into vision and into the mind in general than we can know without science, and it has done so in ways that are the same in normal human minds of whatever ethnic or cultural origin.

2

EVOLUTION, ADAPTATION, AND ADAPTED MINDS

EVOLUTION BY NATURAL selection is a simple principle with staggeringly complex and unpredictable results. At its simplest, the principle says only that what works well enough in life gets passed on, and what works best gets passed on most frequently.[1] Yet, observes Richard Dawkins, "Never were so many facts explained by so few assumptions."[2]

Since distinguished biologists, like Dawkins himself in *The Blind Watchmaker* and *Climbing Mount Improbable,* have written excellent, accessible introductions to evolution, I will not attempt to cover their ground, but will merely prepare a path toward an evolutionary account of art and the art of fiction in particular. I outline here the basic principle of *evolution,* introduce the notion of biological *adaptation,* then show how human minds can be explained as adapted to the kinds of problems they have recurrently had to solve.

Everything here will prove relevant in chapters to come. Not only have our bodies evolved from simpler forms; so, too, have our minds and behaviors. Since evolution challenges deeply held intuitions about our special place in the world and about how complexity arises, since the controversies have been sharp and the confusions and misrepresentations profuse, we have to tread carefully.

As I will argue in Parts 2 and 3, our predispositions to engage in art of all kinds, including the art of fiction, have evolved and now constitute human psychological and behavioral adaptations. We need to know what that term means, what criteria have been proposed to identify adaptations, and what tests can decide whether we should explain a feature as a biological adaptation, or as a byproduct of other adaptations, or primarily in cultural terms.

EVOLUTION

Evolution by natural selection revolves like a perpetual-motion machine through cycle after cycle of reproduction. Darwin could arrive at the principle of evolution by reapplying the insights of Charles Lyell, the founder of modern geology. The mountains and plains around us seem almost to define solidity and stability, but geology shows them to be in flux and to contain the detailed record of their own past, if we can dig deep enough and look closely enough. So, too, with species. Seemingly stable species change form and function dramatically over hundreds, thousands, or millions of generations. Just as seabeds can become mountaintops, the descendants of fish can become Komodo dragons or hummingbirds.

The modern theory of evolution (essentially Darwin's theory of natural selection plus a modern understanding of genes) is as well supported as the theory of gravity and, like it, simple in outline but complicated in detail: (1) *if* there is inheritance, and we know there is (a child resembles its parents more than a random stranger, let alone a treefrog or a toadstool); (2) *if* there is variation, and we know there is (a child never exactly resembles its parents, and indeed by definition one child could never be exactly like two different parents); and (3) *if* some variations are more successful than others, and we know they are (the ability to run a little faster or think your way out of difficulties a little more often, for instance), *then* in a world of limited resources and competing interests, not all will be equally successful in producing offspring that themselves produce reproductively viable offspring. Over time those who leave more offspring will come to predominate within their populations, and the variants that they embody will therefore become either fixed, like two legs in humans, or polymorphic, one of a stable set of options, like the colors of human eyes or hair.

In one sense the principle of evolution seems almost tautological: the features of those that can outreproduce others will come to predominate within their species. Yet in another sense evolution meant revolution. For until Darwin and Wallace, scientists assumed that the complexity of organisms in the living world could be explained only by deliberate design. The explanation Darwin and Wallace offered showed that even intricately elaborate structures could emerge without conscious design, without any foreplanning, merely as a consequence of repeated generation, selection, and regeneration. Evolution accounts for the living world as "generated from the ground up rather than from the heavens down."[3]

Enormous complexity, Darwin showed, could happen through a process that appears almost random. But although chance often enters the living world, evolution's power does *not* derive from chance alone. Which genes recombine with which others in meiosis, the division of the genome in fertilization, *does* depend on pure chance, and allows for an astronomical number of new combinations in each generation—in the human case, trillions of options even from the same parents. But selection criteria exert consistent pressures that allow favorable combinations to accumulate. As Dawkins insists, "Darwinism is *not* a theory of random chance. It is a theory of random mutation plus *non-random* cumulative natural selection."[4]

Selective retention accumulates success, a little more, on average, in each feature in each cycle. It accretes even minute advantages, so that in the next cycle the most advantageous features in the previous cycle will become the basis for new recombinations. If the selection pressures remain relatively stable, success compounds, and although the process has no forethought it can rapidly arrive at highly effective design. Eyes, a decided advantage in negotiating the world, have evolved independently perhaps over sixty times.[5] Their complexity and efficiency often seem a trump to those who think the cards are stacked against evolution: "What use is half an eye?" In his brilliant rejoinder, Dawkins not only shows how even the feeblest capacity to register light could have been of benefit, but also describes a computer simulation demonstrating that a fish-eye lens could evolve from a simple structure of three cell layers, transparent over photosensitive over opaque, in as little as 364,000 generations, less than half a million years for organisms that reproduce at a year old.[6]

In complex real-world environments, selection pressures never remain the same forever. As one organ or organism grows larger and more effective at its role, it competes for energy with others, within a body, a species, or an ecosystem. Even without atmospheric or topographical change, the competitive environment shifts in many small systematic directions over generations. Predators and prey, parasites and hosts, sexual pursuers and the sexually pursued continually evolve in a series of arms races that push evolutionary design to its furthest extremes.

ADAPTATION

Not all evolution is adaptation—there are other factors, such as nonbiological "chance" events (asteroids, volcanoes, climate or geological change) or genetic drift—but only natural selection can produce complex design for biologically advantageous functions.[7] Catastrophes may cause the extinction of whole taxa and have a massive impact on the biota that endure, but they are too severe and rare to shape species except by eliminating their competition or reconfiguring their environment. Genetic drift—chance changes in frequency in genes not under strong selection pressure—can have consequences, especially in small, isolated populations, but drift is directionless, as much back or sideways as forward, unlike the steady compounding of advantages under selection pressure. Adaptation is not *all* there is to evolution, but for good reason it is the core of modern evolutionary biology. Since only natural selection can give rise to adaptation, to functional biological design, "adapted" is often used as almost equivalent to "evolved."

An adaptation is any trait modified by natural selection that enhances fitness, the capacity to survive and produce viable offspring.[8] An adaptive trait need not be perfectly adapted to its environment; it need only have given organisms that possess it some advantage over their kindred that lack this trait to the same degree.

The concept of adaptation drives so much research in modern evolutionary biology and psychology because it prompts us to ask "Why?" Rather than merely accepting a trait, even an apparently insignificant or decorative one, as part of an organism, we can ask: "Is it an adaptation? Why does the organism have this feature, with this value, and not some other in its stead? How does it help the organism to survive or reproduce better than its competitors without these specifications?"

Often traits assumed to be accidental or trivial, mere grace notes in the species design, have been properly understood for the first time only once researchers asked whether they had some fitness-enhancing function. A. J. Cain singled out as "the most remarkable functional interpretation of a 'trivial' character . . . [Sidnie] Manton's work on the diplopod *Polyxenus*, in which she has shown that a character formerly described as an 'ornament' (and what could sound more useless?) is almost literally the pivot of the animal's life."[9] Religion, an elaborate set of beliefs in and behaviors toward beings with no reality outside the minds of particular human believers, and art, an elaborate set of practices designed to appeal only to human preferences, might equally seem "ornaments" in *our* species. Others have suggested that religion is an adaptation;[10] I will try to show that art in general and storytelling in particular are also adaptations in our species.[11] Far from being ornaments, they too, often, become a pivot of human lives.

When Karl von Frisch proposed that honeybees dance to communicate to their hive-mates the location of nectar-bearing flowers in the vicinity, some biologists challenged him, while others demonstrated not only that the bees dance to express information about food, direction, and distance, but also that their hive-mates *act* on that information. As Dawkins comments: "An adaptationist could not have rested happy with the idea of animals performing such a time-consuming, and above all complex and statistically improbable, activity for nothing."[12] Religious and artistic practices, from the Owerri Igbo of Nigeria's constructing *mbari* houses over years but leaving them immediately to decay as soon as the construction has ended,[13] to Lorenzo Ghiberti's half-century of work on Florence's Baptistery doors, still there for all to admire more than a half millennium later, are nothing if not complex and improbable—and also, I suggest, part of behaviors as adaptive to humans as the waggle-dance is to honeybees. But this is to anticipate.

George Williams, who clarified the modern concept of biological adaptation, sees it as a powerful but strict and demanding notion, not to be used without warrant. Nevertheless he asks: "Is it not reasonable to anticipate that our understanding of the human mind would be greatly aided by knowing the purpose for which it was designed?"[14]

Applying the principle of adaptation to human minds and behavior arouses strong reactions. Even those who not merely accept but actively champion the adaptationist program in human psychology can doubt that it applies to activities like religion or art.[15] Others doubt that the adaptationist program applies to human psychology at all, and still others that adaptation plays as wide a role anywhere in biology as has been proposed.

Stephen Jay Gould led the attack on adaptationism. He charged adaptationists with erroneously supposing that all organic features are adaptive. In fact they suppose nothing of the sort. There are inevitably many organic traits that no one will claim to be adaptive: that daisies float, that elephants have fewer eyes than legs, that human flesh can be tasty to lions or even other humans.[16] Many properties of biologically adaptive features, like the redness of blood or the whiteness of bone, are not themselves adaptive but *byproducts* of features especially selected for.

Nor do adaptationists assume, as critics charge, that adaptation will show perfection of design, and for three reasons. First, the function that an adaptation serves cannot be achieved from scratch but only from whatever the organism has available. Speech, an invaluable human adaptation, had to start from the mammalian respiratory tract. Humans therefore have a jerry-built larynx that must simultaneously serve the ends of digestion, respiration, and speech, with the result that we choke more and can swallow less than other species.[17] Indeed the mix of "astonishing adaptive sophistication and botched improvisation" in biological design—including human voices, eyes, and minds—offers conclusive evidence *against* so-called Intelligent Design and *for* evolution by natural selection.[18]

Second, an adaptation need not be perfect to establish itself; it needs only to perform better on average than the available competition. If further variation can generate a still better refinement, this will in turn predominate, so that design may continue to improve. But any design improvement must not only start with what is already at hand but also be advantageous at every step. "No future use-fulness is ever relevant," as Williams points out,[19] because although on a drawing board this potential could lead to more elegant design, natural selection has no foreknowledge and records only what out-performs the existing competition one round at a time.

Third, adaptationists do not assume the *current* adaptiveness of any adaptation. Evolution built into dung beetles a craving for the smell and taste of dung. That still serves them well. Likewise it built into humans cravings for sweet and fat, preferences adaptive when high-energy food sources were hard to obtain but worth the effort, but now, with candy stores and burger bars so temptingly close, mal-adaptive, raising the risk of obesity, heart attack, and diabetes.

The key elements of adaptation are *design* for some *function*. The function need not be a single one, especially since evolution often turns existing traits into the basis for continued selection: as biologists say, "one ancestral, many derived functions."[20] Subsequent functions may modify the design features of the original adaptation, not usually erasing the initial function but adding an ever-wider range of uses.[21] Eyes evolved for vision, but we also use them for communication: hence our contrastive white sclera, which highlight the direction of another's gaze, and our highly refined capacities for registering and inferring attention and intention from others' eye direction.[22] That we can now intimidate others with our stare does not refute the fact that eyes evolved for vision. Art and storytelling may be adaptations, we will see, but their functions in the video age need not all have developed in the ice age.

Research into adaptations can work in two directions, from design to function (reverse engineering) or from function to design.

Given a particular *design,* what could its function be? In the case of physical features, these may be easily manipulated experimentally, a marking obliterated or highlighted, a tail cut shorter or artificially lengthened, and the effect on fitness—on the number of offspring—tabulated. One of the breakthroughs in twentieth-century biology was the realization that a behavior, too, can be an adaptation. A particular bird species lays *n* eggs per clutch. Why? By experimental manipulation, by removing eggs from some nests and adding them to others, it can be shown that a larger clutch size would strain the parental resources and lead to too few chicks' surviving, while a smaller clutch size would let parental food-gathering capacity go to waste. The precise mechanisms by which the bird lays the right number of eggs for *its* species may remain to be discovered, but *this* clutch size can be shown to be optimal, and therefore adaptive, for *this* species in its normal conditions.

The complementary question proves no less powerful. Given this apparent *function*, what design could underlie such behavior? For instance, given that as a highly social species we can benefit from social exchange, what design features would human minds need to enable this exchange and ensure that on average it was beneficial?

But perhaps the human mind is not a hierarchy of adaptive functions? Gould argued that many purportedly adaptive features of our minds are mere byproducts of our large brains, and that attempts to explain the adaptive function of this or that aspect of mind or behavior sink to the level of "Just-So" stories. Others counter that Gould's claims that mental or behavioral traits are mere "spandrels" (unintended side effects) of big brains are "mostly opaque 'it-just-happened' stories" with no explanatory power.[23]

The human brain exacts high costs, constituting only 2 percent of our body weight but demanding roughly 20 percent of our resting energy.[24] Natural selection will always seek to cut costs that provide no benefits. It has reduced or eliminated sight in cave-dwelling and burrowing animals and wings in many New Zealand birds that evolved without mammalian predators. It dispenses with the brains of sea-squirts once they assume their fixed adult position and shrinks the brains of domesticated animals that no longer need fend for themselves. Had every part of our brain's design not served an adaptive function, evolution would have reduced our surplus brain mass —an observation that will prove crucial for explaining art.

EVOLUTIONARY PSYCHOLOGY
AND EVOLUTIONARY PSYCHOLOGY

Darwin was the first to propose, in *On the Origin of Species,* that we should seek to understand human nature by taking evolution seriously into account, and the first to undertake such research, especially in *The Descent of Man* and *The Expression of the Emotions in Man and Other Animals.* For a long time little came of his hopes, but over the last half-century many research programs have begun to investigate human nature in light of our origins.

The evidence comes from many fields, at first tilled separately, but now often cross-fertilizing: from evolutionary theory, ethology, linguistics, artificial intelligence, game theory, evolutionary anthropology, evolutionary economics, neurophysiology, analytic and experimental philosophy, evolutionary epistemology, and many branches of psychology—clinical, comparative, developmental, evolutionary, personality, and social. The broad research program to which all these more focused programs contribute centrally or peripherally can be called evolutionary psychology, which I, like others, will distinguish from the narrower school of Evolutionary Psychology (EP).

Evolutionary psychology, in both broad and narrow senses, stresses that the mind is not just a large, powerful, general-purpose information mechanism that can apply standard content-neutral reasoning and local cultural reasons to any problem. It takes seriously the *frame problem* in artificial intelligence, the difficulty of defining what information or inferences may be relevant or irrelevant to a problem. In order for us to do what we do, our minds must have been prepared before birth to learn the information specifically relevant to human problems. Henry Plotkin explains the difficulty in general terms:

> the world can be partitioned, described and learned about in an infinitely large number of ways. If you let loose in the world a truly general-purpose learning device that learns entirely without constraint, it will, by chance, set off acquiring information in what is an infinitely large search space along any one of an infinitely large number of search paths. The chances of such a device learning anything that is biologically useful within a single lifetime is vanishingly small . . .
>
> There is really only one way to solve . . . [the] almost intractable [frame] problem. This is indeed to have vast reaches of time in which to search and the ability to conserve selec-

tively knowledge gained over such massive search spaces. And that is what evolution does. It gains knowledge of the world across countless generations of organisms, it conserves it selectively relative to criteria of need, and that collective knowledge is then held within the gene pools of species. Such collective knowledge is doled out to individuals, who come into the world with innate ideas and predispositions to learn only certain things in specific ways. Every human, every learner of any species, begins its life knowing what it has to learn and be intelligent about—we all come into the world with the search space that we have to work in quite narrowly defined. There is empirical evidence to support this view in a whole range of species.[25]

Plotkin adds that if this were not the case, if we all were blank slates at birth, "we would all of us enter into some individual and probably unique part of the search space and hence have little shared knowledge in common" and little chance of "understanding that there are experiences and minds other than our own."[26] This outcome could hardly be more different from actual human experience. Under experimental conditions human infants have been shown to imitate adult smiles within an hour of birth, a task that involves singling out the adult's face and the smile in particular, the motivation to imitate or respond to that smile, and the knowledge of which muscles to call on and in what way to produce a smile on its own face, even before it has ever seen it itself.[27] We come into this world prepared especially to learn from and share with each other—and that, as we will see, offers a powerful start for art.

Many of the most telling criticisms of narrow EP come from biologists. They note that EP hypothesizes adaptations on the basis of theoretical information-processing problems confronting human minds in an abstract Pleistocene and does not pay enough heed to real animals in real and testable situations; or that it remains too close to artificial intelligence and underrates the natural intelligence in other species and the gradation between human capacities and others.[28]

Biologists accept that any full biological explanation of a behavior needs to answer four questions: *why?* (fitness: what is its ultimate function?); *how?* (mechanism: how does it operate, what stimuli cause it to occur?); *whence?* (phylogeny: what are its evolutionary predecessors, what did it evolve from?); and *when?* (ontogeny: when does the adaptation develop and change in the individual?).[29] As some have noted, EP has often focused too exclusively on the first question, and needs to take the other three seriously—as indeed its best work often does, and as I hope to do in accounting for storytelling. Evolutionary Psychologists are themselves aware that proposing a functional adaptation in the mind is not an endpoint but an opening gambit, an invitation to test the proposal by any means available, such as cross-cultural and cross-species comparisons, clinical studies of selective mental deficits, and neuroimaging of normal and damaged minds.

Yet even most of the biologists, psychologists, and philosophers who have critiqued EP in the strict sense accept the necessity of studying human nature in the context of evolution. Though a critic of EP, primatologist Frans de Waal has no doubts about the future of evolutionary psychology: there is "no way around an evolutionary approach to human behavior," he declares, predicting that in another half-century all psychology departments will have a portrait of Darwin on their wall.[30] As psychologists and others note, psychology and the social sciences are full of unconnected minitheories or empirical findings with no theoretical framework, which an evolutionary psychology could supply.[31] Evolutionary psychology can ask *why* or *how* questions all too often ignored in earlier psychology. Why do we have this mental capacity or bias, or this behavior? Why has it been advantageous enough to become established in the species? How were we able to establish this aspect of human behavior? What information-processing or motivational capacities would it have required?

Some of the answers proposed in an evolutionary explanation of human nature may be premature, but they will be tested, sifted, and refined in due course. But incorporating deep time into our knowledge of the species adds a dimension whose absence had distorted all our thinking.

Literature has been the great repository of our detailed knowledge of human nature in the past. It will be illuminated by, and it will illuminate, our knowledge of an even deeper past.

Why so much about Evolutionary Psychology Very weak connection

It tells us our past and helps evo. psychology

Weak

41

3

THE EVOLUTION OF INTELLIGENCE

IF HUMAN MINDS AND BEHAVIOR have been adapted by evolution, *how* are they adapted?[1] Here we have room to consider only two major lines of development: *mind* in terms of the evolution of *intelligence,* and in the next chapter *behavior* in terms of the evolution of *cooperation.* We later follow both lines in our two case studies from Homer and Dr. Seuss, both because intelligence and cooperation matter so much in life and literature, and because some doubt that evolution can explain either.

Often those wary of Darwinism suppose that the more we see evolution as shaping minds, the less room we leave for flexibility. But flexible behavior requires *more,* not less, genetic underpinning, just as a computer can do more, not less, the more software it has installed.[2]

The core argument over evolutionary psychology is about how much evolution has shaped human minds.[3] Have our minds, as Gould claimed, evolved beyond evolution, becoming so large and powerful that we can now cope with the world through culture and language in almost endlessly flexible ways?[4] Has general-purpose conscious reasoning overridden the more specific functions of animal minds and set us free?

Or is the mind, as Leda Cosmides and John Tooby, the leading theorists of Evolutionary Psychology, claim, like a Swiss army knife, full of particular mechanisms that evolved to solve different adaptive problems in different domains, like choosing mates, avoiding dangers, maintaining status, or seeking food?[5] Nature has never before evolved an organism with massive spare capacity or a general-purpose organism; there are no general organs because there are no general problems.[6] Our minds, like every other organ, have surely evolved in response to specific problems: we do not seek information of the same sort, or respond to it in the same ways, when we are hungry, or frightened, or sexually aroused.

Are all or most human mental processes shaped by evolution to deal with specific problems, so that the human mind operates mostly through domain-specific mechanisms or "modules"? Or does it function in a primarily domain-general way?

David Geary's *The Origin of Mind: Evolution of Brain, Cognition, and General Intelligence* (2005) offers a resolution of this debate that allows room for both perspectives. Geary integrates neuroscience and evolution to show how the earliest-evolving "hard" modular systems were complemented first by "soft" modular systems modifiable by experience and then, especially in humans, by conscious and deliberate decision-making in novel situations.[7]

Brains evolved to allow organisms to respond to relevant features in their environments. These features span a spectrum from phenomena unvarying across time, like three-dimensional space or the motion of light and sound through air or water, to the more changeable, like biological kinds, local flora and fauna, to still more volatile phenomena, like the intentions and reactions of others: predators, prey, and conspecifics like allies or rivals, mates or offspring. As brains evolved to track these more changeable features, they included mechanisms that became less automatic, more complexly integrated for specialized combinations of functions, and more likely, as pressure for behavioral novelty increased, to feed into more deliberate systems of control.

Senses evolved first to react to comparatively invariant features of the environment, like space, light, or sound. The neural systems interpreting low-level visual or other sensory information can be automatic and unconscious, operating rapidly through massively parallel networks. We need not be aware how our minds interpret incoming light signals to cope with changing illumination, horizontal or vertical motion, or gradients in intensity and color that signify surfaces or edges and therefore objects. Such systems are self-enclosed, impervious to introspection, "modular" in the classic sense.[8]

Light and sound operate similarly everywhere, but each local environment has its unique flora and fauna. Geary proposes that to cope with such variability, brains evolved a more flexible modularity that permits plasticity (modifiability) in brain development.[9] We can conceive of modularity as having a hard exoskeleton, like a crab's, for those cognitive systems that process invariant patterns, and a soft internal structure of systems modifiable by experience, accommodating local differences within the constraints of the exoskeleton.[10] Within the exoskeleton each species will be predisposed to assign biologically relevant information into categories according to simple rules. To the degree that species can learn, brains automatically reallocate costly mental resources, expanding or shrinking neural space in response to the inputs of experience.

Let us focus again on vision, the clearest example, and on questions of survival, where the most intense selection pressures occur. Even before they have seen a real hawk, chicks fear a bird-shaped kite that experimenters tow one way, as if with wings near the head (a schematic bird of prey), but not when towed in the reverse direction, as if with wings toward the tail (a schematic duck or goose). If chicks feared all large bird shapes flying above, they would waste energy running for cover and time for foraging. But if they had to wait for experience to distinguish between a predator and a grazing bird, they might never survive their first lesson. Over time evolution therefore selected for chicks that could from the first distinguish one broad category of bird from another, the potentially harmful from the harmless, and react appropriately. Even this hard-wired system complexly integrates visual processing, emotional response, and motor reaction.

Although evolution can partly prepare us for whatever may help or hinder *our* particular needs, it cannot build into different kinds of minds instructions precise enough to cope with all they may encounter. Humans have a wide range of habitats, predators, and prey, so we need to learn the species of most relevance, and as children's early fascination with animal kinds shows, we have a strong disposition to do so. But as for young birds, so for children, certain animal kinds have particular evolutionary significance and emotional value. Even in modern urban environments children still fear lions, tigers, and monsters—an open category of large threatening creatures—without ever having encountered them.

Being highly social, we have a strong innate predisposition to attend to one particular species: *Homo sapiens.* We respond with decided emotional biases: infants smile in answer to smiles and show fear if they meet unflinching stares or unresponsive faces. But within our innate schema for human faces, we need to learn individual faces. We categorize faces as human very rapidly, in 100 milliseconds, and recognize individual faces still rapidly, through even more specialized mechanisms, but a little slower, in 170 milliseconds, through neurons trained to respond to particular faces.[11] And for faces we know well, a positive or negative emotional loading automatically accompanies each face we recognize.

For social species like dolphins, chimpanzees, or humans, the social world fluctuates far more subtly than the rest of the biological world. A chimpanzee will have much the same reaction to any colobus monkey or any leopard, but its reaction to another chimpanzee varies according to the other's size, sex, age, personality, status, alliances, and situation. Relating to other conspecifics places heavy demands on mental flexibility.

Most researchers accept the social intelligence hypothesis: that the greatest pressures for advanced intelligence arise from the need to track the identities, status, powers, and intentions of conspecifics and to respond to them to best advantage.[12] Animals like dogs, dolphins, and primates have to *cooperate* with conspecifics subtly enough to earn the resources obtainable only together. But they also need to *compete* with them to maximize their share of socially earned resources without risking prospects for future cooperation. This shifting balance of competition and cooperation exacts high computational demands. Individuals must track other individuals and their predispositions and relations to themselves and others, a complex task in species with changeable hierarchies and fluid alliances.

In human evolution, social selection pressures on intelligence have been unusually intense. Both humans and chimpanzees coordinate flexibly in ways that make them effective hunters—presumably a trait of their common ancestor. After hominids became fully bipedal, some time after the split from the common ancestor, and before their rapid growth in brain size, they learned to wield sticks, stones, and fire. They gained such environmental dominance that they became superpredators, eradicating much of the megafauna in new regions into which they ventured before these animals had time to evolve wariness of these not especially powerful-looking new neighbors.[13]

Hominid brain size then experienced the kind of rapid growth that usually results from an evolutionary arms race, a competitive feedback cycle either between species (as antelopes and cheetahs increased in speed to outrun each other) or within species. Once the advantages of belonging to a hominid band had become overwhelming, the main pressure for our forebears became to secure their share of the resources obtained through the band's efforts. They benefited by maximizing their own gains without alienating others, by allying with them strategically, outwitting them strategically, and resisting being outwitted in return. The complexity of such pressures seems to have caused the rapid growth in hominid brain size over the last two million years, and especially the development, as we will see, of an advanced theory of mind, a capacity to infer the beliefs, desires, and intentions of others, and a self-awareness that allows us to understand how others might infer our motives and react to our moves.

Geary explains the neurological consequences of the increasing brain size needed to cope with such demands. The area that expanded most in the hominid brain, the neocortex, lies on the outside of the brain. The neocortex could not double in size merely by doubling the size of its neurons, since to maintain conduction properties, like the speed of transmission, any *dendrites* (the short signal-*receiving* branch fibers of neurons) that doubled in length would have to quadruple in diameter. The expanded regions in the human neocortex are therefore less densely packed with neurons (cell bodies, gray matter), but they have a higher proportion of *axons* (white matter), the signal-*transmitting* nerve fibers, long enough to loop into other brain regions. Each neuron in the neocortex therefore communicates with proportionately fewer neighboring neurons than before the expansion. The reduced density of neurons resulted in the "increased specialization of interconnected clusters of neurons. Stated somewhat differently, the microarchitecture of expanded regions of neocortex necessarily becomes more modularized and specialized for processing finer-grained pieces of information."[14]

But because of the higher proportion of axons relative to neurons, and the increased folding of the new outer lobes of the brain, the neocortex in humans is more highly interconnected with other more *distant* brain regions than in other species. This structure permits more integration of information and more top-down control of the rest of the brain than in other species, more capacity to inhibit automatic response and attend to and manipulate information in search of novel responses.[15]

The combination of expansion of the neocortex and the altered proportion of axons to neurons therefore allows for the emergence of relatively new brain systems or the modification of older ones. In particular, our brains can maintain better attentional focus and control and stop "task-irrelevant information from entering conscious awareness,"[16] while new subroutines can engage in controlled problem-solving and generate mental models to cope with new or conflicting information.

Geary contrasts the neural responses possible at the invariant and variant ends of the information continuum. At the invariant end, processing information from the world of physics, responses can move rapidly and automatically along parallel channels. Toward the variant end, especially in subtle interactions with other conspecifics, ready-made rules no longer suffice to handle complex and unpredictable situations. A quite different response is required.

When a problem cannot be solved automatically, the output of the modular networks can be bumped up into consciousness, to be addressed in working memory, the space within the mind that consciously attends to and manipulates information, the active window, as it were, on the mind's computer screen. There the brain's central executive maintains attentional control. Deliberate attention *amplifies* information relevant to the problem and *inhibits* the irrelevant. In the severely limited space of working memory, we process information not unconsciously or implicitly, but consciously or "explicitly," and not in parallel but in series, through representations—perceptions, memories, projections—in one of our central executive's slave systems, via images (in our mind's eye), words (in our mind's ear), or episodic memories.[17] The ability to use working memory for controlled problem-solving in response to novel or complex situations correlates strongly with measures of general intelligence.

Another example from vision can show the kind of problem we or our ancestors have often had to face, and the relation between visual and other cognitive subsystems. We are born primed, through the exoskeleton of these mental modules, to attend to human faces and voices and to respond in particular ways to the gross features of either. Over our lifetimes, we learn subtler refinements of facial and vocal expression. If, say, we detect a mismatch between the words or tone of a person's speech and facial expressions or eye movements, we may suspect deception.[18] Attention and emotion will alert us to search for relevant information: we may look for other signs in the present, or recall our sense of this person in the past, or particular past encounters that now throw new light on his or her character, or other situations relevantly similar to this. We can sketch possible courses of action *if* we think we can confirm deception—such as either revealing or concealing our suspicions—and the other's likely response to our actions in either case. Our emotional reaction to these sketchily projected scenes will often guide our next move.

To sum up: First, organisms evolve simple neural systems as they become sensitive to relatively constant sources of environmental information. Second, animals develop minds that can respond to the less constant, more diverse features of the biological world in rough-and-ready ways, through networks of more or less modifiable modules, innate or developed through experience. Third, advanced, socially honed minds, especially human minds, access an even larger network of modules, but in problematic or novel situations consciously attend to information and run inferences and scenarios that they assess via evolved emotional systems. In mind, as throughout evolution, the more complex does not supplant the simpler but builds on and integrates it in new ways that allow remarkable new functions to develop.[19]

Geary explicitly associates the invariant-variant continuum, from (a) light or sound at one end through (b) plants and animals in the middle to (c) other conspecifics at the other end, with what development psychologists call (a) folk physics, (b) folk biology, and (c) folk psychology. As we will see in Part 3, these evolved mental systems prove essential to our capacity to understand events and therefore narrative.

As experience makes us more familiar with situations, we can respond to them in ways that require less of the effortful processing of conscious attention operating within working memory.[20] With increasing familiarity, situations can be processed in older, more posterior regions of the brain in faster, parallel, and less effortful ways, leaving more room in working memory to handle real novelty. I will suggest that this is one function of storytelling: that it makes us more expert in social situations, speeding up our capacity to process patterns of social information, to make inferences from other minds and from situations fraught with difficult or subtle choices or to run complex scenarios. Childhood play and storytelling for all ages engage our attention so compulsively through our interest in event comprehension and social monitoring that over time their concentrated information patterns develop our facility for complex situational thought.[21]

In order to assess novel or problematic situations human minds can draw not only on our individual present and our species' past, as all minds can do, but also on their individual pasts, even on particular episodes, and can consider projected futures, as they turn ideas around through a possibility space enlarged by the dimensions of the hypothetical and the counterfactual. Thanks to theory of mind, which we will explore in Part 3, we can also infer the beliefs as well as the desires and intentions of others, and have enough self-awareness to run scenarios in which we can test our own possible courses of action against the possible reactions of others.

For the great bulk of the 600-million-year evolution of mind on Earth, this ability to think in sustained fashion beyond the here and now has not been available to *any* species. But humans not only have this ability; we also have a compulsion to tell and listen to stories with no relation to the here and now or even to any real past. As we will see, our compulsion for story improves our capacity to think in

49

the evolutionarily novel, complex, and strategically invaluable way sketched above. By developing our ability to think beyond the here and now, storytelling helps us not to *override* the given, but to be less restricted by it, to cope with it more flexibly and on something more like our own terms.

4

THE EVOLUTION OF COOPERATION

HALFWAY THROUGH HIGH SCHOOL I discovered *Cracked* com-ics, edgier than the better-known *Mad*. I painstakingly copied onto my school briefcase, in light green house paint, an image I found there: Albert the Alligator from *Pogo* (as I retrospectively realize) with a two-yard-long screw skewering his midriff, framed in the cap-tion: "Do Unto Others . . . Then Cut Out." Forty years on I can't res-urrect the impulse that made this image so irresistible. I can only suppose that a standing alligator blithely sporting a gigantic screw tickled my fancy, and that at fifteen and chafing at adult proprieties I liked even more the parodic affront to the Golden Rule, "Do unto others as you would have them do unto you," the implicit "Screw you," in slang new to me. I must have copied out image and text in what I thought a protest against conformity. But not for a moment did I call into question the Golden Rule itself. Precisely because it made such universal good sense, pretending to subvert it seemed to my teenage self a perfect provocation.

Why does the Golden Rule make such sense to us all? Why do we cooperate, even when we might also itch to rebel?

Cooperation has been a prime focus of sociobiology and its off-shoot evolutionary psychology. In the mid-1960s William Hamilton and George Williams transformed evolutionary theory by turning sociality from an accepted fact into a scientific problem.[1] Darwin, always aware of difficulties his theory needed to confront, had been troubled by eusociality in ants and bees, in whose colonies most renounce their reproductive rights. But before Hamilton and Wil-liams, others less inclined to confront difficulties had been able to think that individuals could easily evolve to act for the good of their group or species, a position now known as "naïve group selectionism." Hamilton and Williams made clear how hard it was to explain the cooperation necessary for social life. If one animal behaves for the good of others—draws attention to predators or fights them off, for instance, at risk to itself, or forgoes a share of food because others are hungry—but another does not, the second will be more likely to survive. Its more selfish genes will be more likely to pass to the next generation than the genes of the more selfless. Over time, cooperative genes should therefore die out and uncooperative genes come to dominate in any population.

Biology rules out invariably unstinting altruism. Any creature that never, regardless of context, behaved first for itself could be easily exploited by others. It would be unlikely to last long enough to reach the age of reproduction, and if it lasted that far, it would get no further. The ultimate unconditional altruist would be a male who refrained from sex in favor of other males, to help *their* genes, rather than his own, into the next generation.[2] He could have no heirs. Yet if unconditional altruism is impossible, conditional cooperation exists. But how could it establish itself?

The challenge to explain rather than merely to accept sociality, and to do so in terms of strict genetic accounting, led to the new disciplines of sociobiology in the 1970s and, in the 1980s, evolutionary psychology, which has particularly probed the psychological mechanisms enabling sociality. Much of the energy of early sociobiology drove research programs (Hamilton's inclusive fitness theory, Dawkins' selfish gene theory, John Maynard Smith's evolutionary game theory) proposed as alternatives to group selectionism. Recently however these programs and their leading proponents have come to accept the need for *multilevel selection theory,* which rests on one key postulate: *"Selfishness beats altruism within single groups. Altruistic groups beat selfish groups."*[3] Other things being equal, selfishness—reaping a benefit without contributing a full share of the cost—pays most in the short term. But in the long term cooperation can often yield more, enabling a group to achieve more than the sum of what its members could achieve individually, whether that be building hives or dams, for bees or beavers, or hunting large prey, for hyenas or humans, and thereby to outcompete other groups.

Multilevel selection theory retains the competitiveness that drives natural selection, and like early sociobiology rejects naïve "for-the-good-of-the-group" thinking.[4] It explains how cooperation, like other traits impossible to establish at a given level, can in principle establish itself, can earn its competitive keep, by offering advantages in competition between higher-level units. *In principle:* but it also requires that any adaptive trait be assessed on a case-by-case basis, to see which levels of selection have been most powerful in shaping that particular trait: the individual or some higher-level group.

Multilevel selection theory also makes it more possible to see the history of life as a series of major transitions in which cooperation has been established (or competition suppressed) at one level, enabling greater cooperation and complexity at another level: the coordination of complex molecules to produce the first self-replicating organisms, the first life; the symbiosis of two bacterial cells, without nuclei, into the eukaryotic or nucleated cell; the integration of cells into multicellular organisms; the cooperation of individuals in social groups; and the special cases of eusociality—ant colonies, beehives, termite mounds, and the like, which function in effect as superorganisms—and human ultrasociality.[5] Human cooperation enables millions of individuals and myriad groups with distinct purposes to cooperate on a more-than-local and sometimes a global scale.

Our increased cooperativeness, many researchers now believe, may also be the key to the runaway growth of human intelligence. Through greater motivation to share emotions, attention, intentions, and information, we have learned to understand other perspectives and to detach thought from the immediate; we have developed language and what we could call human ultraculture.[6] Culture easily amplifies group differences, and hence the force of group selection, since although genetic mutations can spread only over generations, cultural changes with significant effects on the relative fitness of groups can spread within a single generation.[7]

But how could cooperation begin, if evolution operates on only immediate rather than future advantages, and if in the short term selfishness offers greater rewards than cooperation does? Cooperation poses a difficult engineering problem,[8] but since its advantages can be so substantial, and since nature has had so much time for trial and error, natural selection has found a suite of partial but surprisingly effective solutions.[9]

Biologists call the first solution *mutualism:* individuals help one another merely as each pursues its own interests.[10] If I watch for predators as I forage for food, I can benefit from the vigilance, and be alerted by the alarm, of my neighbors if they are close enough to face the same threat. In prey species natural selection gradually favored those most inclined to stay together. To restate in design terms: evolution engineered a first step to sociality by inculcating in animals with few natural defenses a sense of comfort in the presence of conspecifics and of unease at straying far from them. An isolated monkey will repeatedly pull a lever with no other reward than to glimpse another.[11] We, too, are "intrinsically a group-living ape" and dread the thought of being excluded from our group.[12]

The sense of empathy with others of one's kind, the positive feeling of shared purposes, underlies all sociality, but it can readily be overridden by the need to compete for resources like space and food that become scarcer with group living. How, then, can sociality develop still further?

The next step, *active* cooperation, could come as natural selection favored those whose behavior helped promote their genes not only in themselves but in others. In the most obvious case, parents predisposed to care for their offspring improve the chances that their own genes will survive and reproduce. Their offspring will inherit a high proportion of their genes, and therefore also their predisposition to promote *their* survival and reproduction. That genes for parental care should be powerfully selected for seems a no-brainer to us precisely because evolution has designed parental care so well that we take it for granted.

Hamilton generalized the principle at work here as *inclusive fitness* (or *kin selection*), a sort of selfishness slightly beyond the self, not only for ourselves but also for copies of our genes in others. This, the "central theorem of selfish gene theory," has been confirmed in hundreds of species.[13] If the cost to me of producing a benefit to a relative is less than the benefit to the other divided by its relation to me (if $C_a < B_b/r$), then evolution should favor my offering the benefit, since it will have a net beneficial effect on *my* genes, albeit within the bodies of others.

By now caring for kin seems not quite such a no-brainer. It does not mean however that I have to calculate the formula, whether I happen to be a tree shrew or a human hunter-gatherer; rather it means that natural selection has run the calculation over many generations and built it into my emotions: we "execute evolutionary logic not via conscious calculation, but by following [our] feelings, which were designed as logic executers."[14] Natural selection, by favoring us if we favor our close kin, thereby also selects for the genes that *incline* us to such helping behavior, via, for instance, neurotransmitters like oxytocin, which in mammals fosters social recognition and bonding. On average we have been designed by evolution to feel strongly attached to our parents, offspring, and siblings (to each of whom we are related by, on average, 50 percent of our variable genes),[15] less strongly attached to grandparents or grandchildren (25 percent), and still less to aunts, uncles, nephews, nieces (12.5 percent), and should on average feel inclined to help each according to the coefficient of relatedness.

The theory of *parental investment* among other things explains family *conflict* as well as cooperation in terms of genetic relatedness.[16] Parental investment includes any expenditure of resources on one offspring that would take from the resources (time, energy, food, and so on) needed to rear others. Parents can produce many children, and it will be in their interest to do so unless intense investment in fewer children would produce better results in terms of survival and eventual reproduction, as in mammalian rather than reptilian reproductive systems.

In diploid species such as mammals, parents are genetically related by 50 percent to each of their children, but each child is related 100 percent to itself and only 50 percent to its actual or potential siblings. Each child should therefore prefer parents to invest in *it*, unless the same investment would benefit at least two other actual or potential siblings by at least the same amount or benefit one other by at least twice as much. Each offspring should therefore prefer to concentrate

55

its parents' investment in itself—in mammals, for instance, through continued breast-feeding—while its parents should prefer to maximize their genetic transmission to the next generation by offering, on average, equal investment in each of their present and future offspring. Here we see why even in the most cooperative of relationships competition is inevitable, and why the powerful emotions engendered by family loyalty and conflict saturate stories from Genesis to *The Sopranos.*

If cooperation is fairly easy to establish, yet already fraught, among close kin, how can it extend even among non-kin? Game theory studies interactions in which one party's success depends to some extent on moves made by others with competing interests. In a zero-sum game like chess or tennis, your gain is my loss. In a non-zero-sum game, both of us can gain or lose different amounts. The prisoner's dilemma is the classic non-zero-sum model of the complications of cooperation.[17]

Imagine, Peter Singer asks, that you and another prisoner are sitting in separate cells of a Ruritanian prison, charged with plotting against the state. There is too little evidence to convict either of you. If you confess and the other prisoner does not, he will get twenty years and you will go free, and vice versa. If both of you confess, each of you will get ten years. If neither confesses, both of you will be held for six months under emergency regulations.

> The prisoner's dilemma is whether to confess or not. Assuming that the prisoner wishes to spend the minimum time in prison, it would seem rational to confess. No matter what the other prisoner does, that will be to the advantage of the confessing prisoner. But each prisoner faces the same dilemma, and if they are both to follow their own individual interests and confess, they will end up serving ten years, when they could have been out in six months!
>
> There is no solution to this dilemma. It shows that the outcome of rational, self-interested choices by two or more individuals can make all of them worse off than they would have been had they not pursued their own short-term self-interest.[18]

If each cooperates with the other, if each *trusts* the other enough, both can save years in prison, but if either assumes that the other can be trusted, but is wrong (the other "defects"), he or she gets the sucker's payoff, twenty years. Cooperation may be best, but how can the trust on which it depends be established securely enough to make the risks worthwhile? Singer offers an everyday modern example. If we all took buses to work, the roads would be unclogged, and we would get there faster. But since we do not trust that enough others will opt for buses, we continue to drive our cars, since if the roads remain clogged, cars will at least be faster than buses.

To explain how cooperation in non-kin evolves, we need to combine Hamilton's gene-centered view of evolution with game theory to generate *reciprocal altruism:* I help you in the expectation that you may help me later.[19] Such cooperation can extend far beyond relatives, and therefore bring many more of the benefits of cooperation, but like the prisoner's dilemma, it faces the risks of defection by the party that owes the return favor.

In a single prisoner's-dilemma situation, it always pays to defect. But in indefinitely repeated encounters, players may see that both will lose unless they trust each other. If you and I live in close proximity and are likely to encounter each other repeatedly, we have reason to develop the trust that can earn us the benefits of active cooperation. Reciprocal altruism can most readily begin, therefore, among individuals living in small, settled groups—where many may also be related.[20]

Even so, there still lurks the danger of defection, of failing to repay. Cheaters will thrive in exchanges with altruists unless altruists discriminate against—refuse further exchange with, or actively punish —cheaters. If they discriminate against cheaters, altruists will exchange only with other altruists and benefit from their cooperation, while cheaters will be left to deal with other cheaters and reap no net benefit. On this condition altruists can outcompete cheaters and, in evolutionary terms, ultimately outreproduce them.

For altruism to work robustly a whole suite of motivations has to be in place: sympathy, so that I am inclined to help another; trust, so that I can offer help now and expect it will be somehow repaid later; gratitude, to incline me, when I have been helped, to return the favor; shame, to prompt me to repay when I still owe a debt; a sense of fairness, so that I can intuitively gauge an adequate share or repayment; indignation, to spur me to break off cooperation with or even inflict punishment on a cheat; and guilt, a displeasure at myself and fear of exposure and reprisal to deter me from seeking the short-term advantages of cheating. We can reverse-engineer the social and moral emotions so central to our engagement with others in life and in story.[21] Rather than merely taking these emotions as givens, we can account for them as natural selection's way of motivating widespread cooperation in highly social species.[22]

Reciprocal altruism among non-kin has since been documented in vampire bats, which share surplus blood with other sharers but withhold from nonsharers,[23] and has been tested experimentally among primates. Capuchin monkeys will share more with partners who have helped them with food or grooming.[24] Baboons offer social support.[25] Chimpanzees and bonobos tally services they have received, return favors (food, sex, grooming, support, or nonintervention in power struggles), and can even show indignation when an ally who owes them support fails to deliver.[26]

Apart from theory, observation, and behavioral experiments, those studying the evolution of cooperation have found another way to clarify how cooperation could emerge. Computer simulations using weighted costs and benefits in iterated prisoner's-dilemma encounters make it possible to test multiple strategies against one another, over hundreds or thousands of cycles, to see which fares best in competition against whatever other strategies, nice or nasty, simple or complex, programmers can devise.

Academics from around the world have devised computer-programmed strategies to test the constraints on cooperation and pitted them against one another in computer tournaments.[27] Surprisingly, the simplest strategy, Tit for Tat, won the first such tournament: cooperate on the first move, then copy your partner's last move. If the partner cooperates, cooperate; if it defects, defect. In a second tournament even strategies designed specifically to counter Tit for Tat's success could not do so, although variations on this basic design (like Contrite Tit for Tat, Generous Tit for Tat, Suspicious Tit for Tat) have since been shown to fare even better under some conditions:

> as we carefully develop these models, adding complications one by one, a profile of traits emerges that many would regard as uniquely human—niceness, retaliation, forgiveness, contrition, generosity, commitment . . . and self-destructive revenge. It is amazing that such a human profile of traits emerges so naturally from such simple models based entirely on the survival of the fittest.
>
> . . . The capacity to forgive should extend far beyond us and our primate ancestors to any creature with sufficient brainpower to employ TFT-like rules. Even guppies have been shown to qualify . . . They recognize their social partners as individuals, behave according to their own dispositions at the start of relationships, and become less altruistic in response to selfish partners. They try to associate with altruistic partners regardless of their own degree of altruism.[28]

Building on the work of Hamilton and others, psychologist Leda Cosmides realized that if humans are to benefit from social exchange, we must always be alert for cheating. She saw that this requirement might offer a test of how our minds work. Do we assess information in a content-independent, general-purpose way, or do we assess it, when facing problems recurrent throughout our past, in ways shaped by evolution to deal with these particular situations?

Cosmides showed that a common test of purely logical reasoning, the Wason Selection Task, normally difficult for most to solve, turns out to be very easy to pass when it takes the form of a problem involving cheating in social exchange.[29] In the Wason test we are asked which of four cards we must turn over to test whether a proposition is valid. For instance, if the proposition is "If a person is a Leo, then that person is brave" and there are four cards with an astrological sign on one side and a rating of courage on the other, the face-up sides being "Leo," "brave," "Aries," and "coward," then most of us realize we have to turn over the "Leo" card, many think we also have to turn over the "brave" card, and considerably fewer realize that in fact we need to turn over both the "Leo" and the "coward" cards and no more. If we had special circuits for logical reasoning, we would see that *If P then Q* needs to be tested against both *P* and *not-Q*, but usually "fewer than 10% of subjects spontaneously realize this."[30]

When the content of the question changes so that it requires us to look for cheaters in social exchange, the performance jumps dramatically, to 70–90 percent, the highest ever recorded for the test, even when the situation described is culturally unfamiliar and deliberately bizarre (for instance: "If a man eats cassava root, then he must have a tattoo on his face"). Success remains high even with these unfamiliar scenarios, but only if they are set up as social contracts, not if they are presented as noncontractual situations based on the same bizarre condition. Cosmides then "tested unfamiliar social contracts that were *switched*" so that the correct social-contractual result was now the wrong logical response. Seventy-one percent still answered with "the 'look for cheaters' response, *not-P and Q,* . . . even though this response is illogical according to the propositional calculus."[31]

People who cannot usually detect violations of "if-then" rules can do so easily when the violation involves cheating in social exchange.[32] No one had thought, until evolutionary theory showed the constraints on evolving social exchange, to expect minds prepared to identify cheating, and colleagues thought Cosmides crazy even to investigate.[33]

Cosmides and her husband and coworker John Tooby ascribe the results to a cheater-detection module. I would prefer to call it a subroutine and to define its role as emotional highlighting rather than detection. We are indeed wary of cheating, and discrepancies between voice tone and facial expression or posture may place us on guard, but we have no sure way of *detecting* concealed cheating. Evolution has not made us Sherlock Holmeses, but when we *recognize* behavior that crosses a social rule, it springs out to our attention.

That quibble aside, Cosmides' findings are so striking that they have been repeatedly challenged and tested by those inside and outside evolutionary psychology. The debate still continues, as it does at the forefront of any science, but in every case so far Cosmides and Tooby and their colleagues have rerun or extended the tests, from logic students to hunter-gatherers, and have shown that proposed alternative explanations cannot account for their results.[34] Neuroimaging experiments also show activation in brain areas used for reasoning about social exchange rather than in areas used for reasoning about logically similar behavioral rules.[35] It really does seem that even on what might appear to be higher-level and domain-general problems, even on the logic of the Wason test, even in highly conscious mental problem-solving, our thoughts can be strongly guided by evolved predispositions.

Not only three-year-old humans, but even animals with tiny brains, like guppies or cleaner fish, seem to be able to solve such problems—obviously not via Wason tests, but as observed by biologists. This predisposition seems an element of social behavior that stretches far back into social exchange between unrelated individuals, and for that reason this highly focused system works more efficiently than our general-purpose logical reasoning: the answers pop out to our attention.

Across human cultures and in many other species individuals incline to punish others for cheating. In our case at least, our emotions not only alert us to register unfairness but also motivate us to punish it, because in the long run doing so improves the chances of our benefiting from cooperation.

Recent work in evolutionary economics has shown that in other respects, too, we, like other species, work less by reason, the economists' assumption of "rational individuals," than through emotions primed by evolution. In neoclassical economics, individuals are assumed to behave "rationally," to seek to maximize their gains. This assumption would predict that if we are offered real money for nothing, we will not turn it down. But in experiments people repeatedly reject profit in order to satisfy their sense of justice.[36]

In one experiment, the dictator game, two strangers play for (usually real) money, say $100. I, as dictator, must offer you a share. If you accept the division, both of us keep our agreed portions. If you reject the offer, neither of us receives anything. In terms of strict economic rationality, an offer of a dollar, even a cent, would leave the second participant better off, and should therefore be accepted. But even as you read this, I expect you recoil at the thought of such measly offers. Participants in the experiment behave the same way. No dictator proposes such token sums but tends to offer amounts closer to $50. Often if the sum offered is only a little under $40, the respondent rejects it. A sense of fairness in social exchange overrides the rational calculation of gain. We have evolved not to be "rational individuals," profit maximizers, but social animals, holding others to fair dealings even at our own cost.

Culture inflects the results of these experiments. In fifteen different societies, the average amount the dictators offered varied between 15 percent (in a low-trust society) and 58 percent (in a society sustained by the intense cooperation necessary for dangerous communal deep-sea fishing), but in ten out of fifteen societies the average was 50 percent. Local ecological and cultural conditions can modify the parameters, yet everywhere we act on a sense of fairness rather than on a "rational" pursuit of pure profit.[37]

The same sort of "irrationality" exists for the same reasons in creatures with much smaller brains than ours. Capuchin monkeys are known to be good cooperators. Experimenters gave each monkey a token, then, with hand outstretched, palm up, solicited the token in return for a slice of cucumber. The monkeys happily exchanged a token each time for cucumber, although it is not their favorite food. But then unfairness was introduced. In sight of one monkey, another was given for its token not a slice of cucumber but a much more appealing juicy big grape. When the other was then offered a slice of cucumber for *its* token, it reacted angrily in 40 percent of trials. When one monkey received a grape without even needing to pay a token, the other, four times out of five, refused to hand over its token or to take the proffered cucumber unless to toss it away in disgust.[38]

Humans have far more complex cooperative relationships than capuchins or chimpanzees. We cooperate enough to live in cities of millions. How have we managed to extend cooperation not only beyond nonrelatives but even to complete strangers? How is this possible, especially, when the best conditions for the emergence of reciprocal altruism are indefinitely repeated encounters between individuals settled close to one another?

Of course human cooperation does often break down. Crime simmers or seethes in all large cities. Nevertheless we have achieved remarkable results through building on our evolved dispositions, both our positive and our punitive feelings, our desire to punish cheats. But free-riding, taking benefits without paying the full cost, persists as the fundamental problem of social life.[39] Yet a real if imperfect solution has come through social control—in terms of multilevel selection theory, suppressing within-group selection—rather than through highly self-sacrificial altruism: through second-order punishment, discriminating against and punishing not only cheaters but also those who fail to discriminate against or punish them.[40]

Detecting and punishing others incurs real and substantial costs, as taxpayers paying for police or prisons know. This was the case even before there were detectives and detention centers. Discovering a transgression when someone aims to hide it or confronting someone who has cheated can be costly in time, energy, and risk. Just as we face a first-order temptation to take the benefits of cooperation without the costs, so we can face a second-order temptation to hang back, to reap the benefits of the cooperativeness that others maintain by enforcing the detection and punishment of cheaters. But if I fail to notice or punish cheaters but am noticed and punished for my failure, the advantages of my second-order free-riding rapidly diminish. Soon few will avoid playing their part in detecting and punishing cheaters. Second-order punishment in turn rapidly discourages first-order cheating and makes the second-order cost of monitoring it relatively light, since everybody contributes. So long as people share similar notions of what constitutes cheating, the need for vigilance, and ways it should be dealt with, few will dare step out of line. The social monitoring already intense elsewhere in the primate line becomes still more intense for humans—and a powerful prompt for storytelling.[41]

To extend cooperation through still larger and looser populations, however, we need additional measures, especially cultural ones like shared norms and institutions. You and I need not only to share norms but also to *know* we share them, so that we feel the pressure not only to resist the temptation to cheat but also to resist the temptation not to slacken in dealing with others who cheat.

In small-scale societies, uncooperative acts were often punished through personal revenge, motivated by an evolved sense of outrage but often leading to destructive cycles of vengeance. Especially in larger societies, better means were needed. Centralized systems of justice, and eventually a police force, could detect transgressions, assess charges, and administer punishment. Depersonalizing justice could dampen incendiary emotions and diminish vendettas.

But even before such cultural systems were invented, other ways of motivating cooperation emerged. One was through story. Stories arose, as we will see, out of our intense interest in social monitoring. They succeed by riveting our attention to social information, whether in the form of gossip—indirect but real and relevant social information—or fiction—admittedly invented and heightened versions of the behaviors we naturally monitor. Modern hunter-gatherer societies preserve their strong egalitarianism by gossip, sharing reports of anyone seeking status, and by admonitory stories warning against violating egalitarian norms.[42] And in societies of any size, stories involving agents with unusual powers capture attention and commandeer memory, and stories with unseen agents who can monitor our behavior and administer punishment or reward—the stories we call religion—permeate and persist partly because they offer such powerful ways of motivating and apparently monitoring cooperative behavior. Religious stories establish a secret spirit police.[43]

In both factual and fictional forms, stories can consolidate and communicate norms, providing us with memorable and shared models of cooperation that stir our social emotions, our desire to associate with altruists (like Dr. Seuss's Horton), and our desire to dissociate ourselves from cheats and freeloaders (like the suitors whom the *Odyssey* repudiates and Odysseus routs). Such memorable images of pro- and antisocial characters and actions common to whole communities can not only define and communicate shared standards but ensure that all know what others know of these standards.[44] As we will see, stories have multiple origins and functions, but among them not least is that they have so often discouraged defection and aided cooperation. *Graphing Jane Austen,* a recent internet study of readers' responses to the characters of nineteenth-century British novels, confirms our folk sense of the polarizing power of the goodie-baddie axis and its centrality in our responses to fiction.[45]

Our continually refined methods of prevention, detection, conviction, and punishment allow us to coexist in societies of many millions. Let us hope we will continue to find even better solutions. But we could not have started on this path without evolved dispositions for cooperation or have advanced as far as we have without elaborating them through culture.

Research of the last few decades in evolutionary theory and game theory, in biological field study and experiment, and in evolutionary psychology and economics shows how cooperation can evolve in creatures with different genes and, to that extent, different interests. It emerges from shared needs, in *mutualism;* from shared genes, in *inclusive fitness;* from shared impulses, in *empathy;* from common interests over time, in *reciprocal altruism;* and from a shared recognition of the benefits of cooperation, even on a large scale, incorporated by nature into our *emotions,* our sense of fairness, our wariness against cheating, our indignation and readiness to punish cheaters, and incorporated by culture into our *norms, institutions,* and *narratives.*

Some who look at human nature through evolutionary lenses argue that our intelligence allows us to overcome—to say no to—our genes. It does: we can inhibit ourselves through conscious reflection, through cultural norms, through the fear of social sanctions. Others argue that our genes themselves give rise to the suite of moral emotions that we partly share with other social animals and motivate us to *want* to inhibit our more selfish impulses.[46] They, too, are right. Although our genes compete with one another, many of our evolved emotions also point us toward cooperation, because its benefits are so substantial, and we can use our intelligence and our sociality—also aspects of our evolved nature—to devise still better solutions to the problems of cooperation.

Evolution offers a much more complex and nuanced view of the social world than the artificial model of the rational individual of economics, or the romantic idea, common since Rousseau, of good people perverted by evil systems, or the paranoid Nietzschean or Foucauldian suspicion that all moral claims mask a lust for power. An evolutionary view of cooperation allows us to look at the social world without inordinate hopes, but with real confidence that we can continue to find better solutions, even to the new problems that the very successes of our cooperation create.

PART 2

EVOLUTION AND ART

But perhaps the most remarkable functional interpretation of a "trivial" character is given by [Sidnie] Manton's work on the diplopod *Polyxenus,* in which she has shown that a character formerly described as an "ornament" (and what could sound more useless?) is almost literally the pivot of the animal's life.

A. J. Cain, "The Perfection of Animals" (1964)

5

ART AS ADAPTATION?

EVOLUTION MAY HELP EXPLAIN copulation and even coopera-tion, but can it account for the creative side of human life? Can it explain art? I will suggest that it can—and that an evolutionary ac-count of art, far from being reductive or deterministic, can do more than any other to explain art's force and freedom.

On the Origin of Stories focuses on one art in particular, the art of fiction. Why can we not discuss fiction solely in terms of narrative, without considering art as a general behavior? An evolutionary anal-ysis of narrative, of *true* stories (see Chapter 11), will prove straight-forward, precisely because it does not require art at all. But the more we gain from sharing the information in true stories, the less need we would seem to have for the false information of fiction. How can we explain our human compulsion to invent or enjoy stories we know to be *un*true?

To account for the art of fiction, we need first to consider the hu-man tendency to engage in any art. Although each of the arts has unique features and functions, there are also features common to all forms of art that need a common explanation.

Not everyone agrees. Let us consider some alternative positions.

> *Position 1: Art is not even a meaningful category.* Examples
> of "art" ranging from "Hickory dickory dock" to the Great
> Sphinx at Giza do not form a coherent class.

But while art is indeed a fuzzy category, so is much else that mat-ters in life, like love, which there is also reason to think has a biological origin, mechanism, and function.[1] An evolutionary explanation of art needs to account for art from the infant to the expert or from ocher face-painting to O'Keeffe.

> *Position 2: Art is not a human universal.* There is no universal notion of art. The Western conception of art arose only in eighteenth-century Europe.

Art as a behavior exists in all known human cultures.[2] Some philosophers of art claim that other times and cultures cannot have art because they lack "our" Western notion of art, the distinction drawn in eighteenth-century Europe between fine art for detached contemplation and mere craft. But the very concept that there is no non-Western art is a Western one, and there have in fact been many other traditions of fine art, in the Middle East, India, China, Indonesia, and Japan.[3] Neither the ancient Greeks nor Sepik carvers in Papua New Guinea have had a single word to match the modern Western word "art," but both peoples have practiced and have had concepts of art akin to some of the many notions of art currently available in the West.[4]

Most of us struggle to learn another language, but we can readily appreciate art across cultures. Dürer in the 1520s, encountering treasures from Mexico, commented that he had never in all his life "seen . . . anything that has moved my heart so much." Goethe, reading Chinese novels, observed: "These people think and feel much as we do." Chinese and Japanese audiences respond to Shakespeare and Beethoven with rapture.[5] And if audiences appreciate, artists appropriate. In the nineteenth and twentieth centuries Maori and Papua New Guinea carvers adopted Western tools and techniques as eagerly as Gauguin or Picasso borrowed from non-Western cultures.

The notion of art we need to work with, if we are to understand the evolutionary origins of art, has to run from lullabies to Led Zeppelin. If we accept art in this sense, as a behavior common across our species, how do we explain it?

We can attempt to do so without taking account of evolution.

> *Position 3: Art can be explained entirely by culture, by cultural traditions and cultural diversity.* Humans exist within and are

shaped by their own particular culture, rather than by some universal human nature. Behaviors that *we* may call art happen to be engaged in—but may not necessarily be considered art—in all known cultures, but within the culture the role of these activities can be radically different from case to case.

All human art indeed takes place within particular cultures or at the intersection of cultures; and no art can be explained without culture, without considering both particular art traditions and the society within which artists deploy them. Yet a purely cultural approach does not suffice. Culture cannot explain culture. Culture does not shield humans from biological explanation, since other species also have culture, behaviors transmitted by nongenetic means across populations. Birds and whales have traditions, dialects, fashions, and individual styles in their song. All known chimpanzee groups have their own distinct suites of cultural practices yet have nothing like human art.

Culture cannot explain why some species have culture, why some have it more than others, why it takes partly similar, partly different shapes across a given species, and, in the human case, why art forms such a central part of all cultures.[6] If art were entirely cultural, and culture not shaped in part by nature, art would occur in some groups and not in others—as I said at the outset, like chopsticks or forks. Instead it seems more like the universal human use of hands to prepare food and convey it to the mouth.

Most who accept art as needing more explanation than its being a product of culture alone have tried to explain it in functional terms.[7]

Position 4: Art can be explained in terms of its function or functions. The functions of art that have been proposed outside an evolutionary framework are as diverse and conflicting as providing *direct,* immediate experience—or offering an *indirect* reflection of the world; as a mode of *individual* display—or a means of *group* identification; as a reflection of the *familiar*[8]—or as preparation for the *unfamiliar*[9]—or as *defamiliarization;*[10] or as providing a sense of order *in the* world—or access to a supramundane world *beyond*. There could hardly be less agreement.

In order to cope with the vast range of options, philosophers often group the most common theories of art into the mimetic, expressive, formal, and communicative.[11]

Mimetic theories stress art's function as representing the world. But they cannot account easily for most music (what do Bach's Preludes and Fugues represent?). Nor can they account for abstract visual art, which has been part of art from the first (in scarification and tattooing, for instance), has persisted in decorative art throughout the ages, and became the dominant mode of high visual art in, for instance, Islamic calligraphy and design and twentieth-century abstract painting. Even in the case of representational arts, mimetic theories do not explain why one species has evolved to develop a compulsive delight in fictional representations.

Expressive theories see art's function in terms of artists' compulsion to express themselves in their art—although, as Karl Popper has noted, this claim is tautologous or trivial, since everything we do by definition expresses our selves.[12] The most eloquently expressive artist of all, Shakespeare, expresses himself so much through the mouths of others that we find it difficult to judge his stance on anything.

Communicative theories find art's function in the responses art engenders. But humans already communicate more effectively and richly in nonartistic ways, including nonverbally, than any other species. What in specifically artistic "communication" makes it useful enough or pleasurable enough for us to devote time to it that we could devote to other things, including nonartistic communication?

Theories of artistic *form* focus on the structure of a work of art as the product of the artist's design and the cause of the audience's response. But they find it difficult to explain why a letter by Keats, Pushkin, or Flaubert, or even Johnson's or Coleridge's table talk, may seem so much more artistic than a penny-dreadful novel.

In response to the challenge to definitions of art posed by modern art, from the urinal Duchamp called *Fountain* to Warhol's Campbell's Soup cans or more recent conceptual, installation, and perfor-

mance art, philosophers of art have for some time been as preoccupied as artists have with the boundaries of art, and have proposed
not functional but *institutional* or *historical* theories.[13] Art, the insti-
tutionalists claim, must exist within some institution of regarding
certain kinds of things as art. But, as many note, this reasoning is
circular and does not answer the key question: how did things begin
to be considered as art? A historical account of art attempts to evade
the circularity, a fatal weakness in definition but not in history. Yet it
also fails to account for the origins of art either in the human species
or in the individual infant attracted by and eager to participate in
pattern. An evolutionary account of art can clarify why the history of
art runs so deep that it has been ingrained in the psyche of the species and the individual.

There are good reasons to suspect that we may need biology as well
as culture to explain art: (1) it is universal in human societies;[14] (2)
it has persisted over several thousand generations;[15] (3) despite the
vast number of actual and possible combinations of behavior in all
known human societies, art has the same major forms (music and
dance; the manual creation of visual design; story and verse) in all;
(4) it often involves high costs in time, energy, and resources;[16] (5) it
stirs strong emotions, which are evolved indicators that something
matters to an organism; (6) it develops reliably in all normal humans without special training, unlike purely cultural products such
as reading, writing, or science. The fact that it emerges early in individual development—that young infants respond with special pleasure to lullabies and spontaneously play with colors, shapes, rhythms,
sounds, words, and stories—particularly supports evolutionary
against nonevolutionary explanations.[17]

Art's ubiquity and antiquity and its persistence across cultures in
distinct established modes suggest that its roots may be biological.
But maybe art's uselessness, not its usefulness, can explain art even in
biological terms?

> *Position 5: Art is a product not of* natural *selection for survival
> but of* sexual *selection for reproductive advantage.* Art's very

uselessness, its ornamental extravagance, shows that it has arisen not because of any survival advantage but because it appeals to members of the opposite sex.

Art can seem as showy and superfluous as a peacock's tail. That flamboyant fan costs its bearer energy to produce and maintain and makes the peacock both more conspicuous to enemies and less able to elude them. How could ornaments like that have evolved in a competitive world?

Darwin realized that such extravagant caprices appeared to challenge his theory of natural selection, but he explained them through his additional theory of *sexual selection*.[18] Males can compete for sex by repelling rival males or by attracting females. In the first case, if sexual selection persists, males evolve for fighting one another, developing in size (like a bull elephant seal) or armaments (like the horns of a stag or a stag beetle). In the second case the mere *sensory biases* of females can, over many generations, shape the appearance or actions of males as they compete to display to females through striking colors or forms (like the peacock's tail) or behaviors like song (in many songbirds), dance (in lekking birds), and bower-making (in bowerbirds). And indeed the history of art seems to reflect a greater male urge to display: it shows a preponderance of males among leading artists, and the evidence of musicians, classical, rock, and jazz, suggests that they are at their most productive in their most sexually fertile years.[19] But art is not confined to professional art for public display in highly specialized societies.

Sexual selection theory was extended and clarified in the twentieth century.[20] The theory of *parental investment* explains why females, not males, tend to be the choosier sex. Whichever sex invests more time and energy in producing offspring (usually the female, since by definition the female is the sex with the larger gamete) has more to lose in producing offspring with a partner with poor genes. Whichever sex has the lesser investment (usually the male) has more to gain by being chosen by as many partners as possible. Males chosen by many females can have huge reproductive success, since their investment in any partner can be brief and they can move on to others; but males chosen by none may fail to produce offspring at all.

Because of the great variance in male success, males face intense pressure for obtaining access to females, whether by repelling other males or by attracting females.[21]

Darwin had little to say about the origins of human art but thought that in humans as in other species "high cost, apparent uselessness, and manifest beauty usually indicated that a behavior had a hidden courtship function."[22] He ventured ("not too plausibly," comments Steven Pinker) that music developed "for the sake of charming the opposite sex" and that body adornments formed a beginning of human visual art.[23] Although others have suggested that art may owe something to sexual selection,[24] Geoffrey Miller was the first to propose in detail that sexual selection has been the driving force behind the expansion of the human mind and higher human behavior: intelligence, inventiveness, art, humor, kindness.[25]

Miller argues that the human brain is far more powerful than necessary for mere survival on the savannah. Brains consume energy out of all proportion to their size, and no other animal needs so large a brain merely to survive. Therefore, he argues, human brains must have been sexually selected. Within that overall hypothesis, he proposes that art was sexually selected as a social entertainment system for the savannah. But we do not need sexual selection as an explanation if a good functional explanation already exists. The human brain is not biologically superfluous. Many species lack wings, legs, arms, or eyes, but that does not prove these organs superfluous to the survival of the species that have them.[26] Large brains with particularly large neocortexes have enabled humans to fill many niches and become the dominant macrofauna around the world, in a way that has everything to do with survival.

And as parental investment theory explains, males can compete over anything, even over who can pee highest or belch loudest.[27] With so many potential means of display at hand, males' capacity for competitive display explains little about behavior as biologically improbable as carving likenesses or composing epics.

Miller notes that "male pigeons harass female pigeons with relentless cooing and strutting. If the females go away, the male displays stop. If the female comes back, the males start again."[28] The very difference between pigeon pouting and human art should give him pause. If art were sexually selected, this would predict that it is overwhelmingly male and directed to females, developing rapidly at puberty, peaking just before mate selection, and diminishing drastically afterward. Miller does adduce statistics to show that rock and jazz musicians produce most records in early maturity. But mothers of all cultures sing to infants; infants prefer their mother's singing to their father's; infants of both sexes engage in cooing and singing, clapping, and dancing as soon as they can; adolescent girls go wild over all-female bands like the Spice Girls or Destiny's Child; Hokusai in his seventies and eighties adopted the *nom de plume* Gakyō-rōjin, Old Man Mad with Painting, was still producing masterpieces in his ninetieth year, and pleaded on his deathbed for more time: "with even five more years . . . I could become a true artist."[29]

In insisting on fitness, Miller rejects questions of origins.[30] But in fact any complete evolutionary explanation needs to identify a pathway. The path suddenly becomes vividly visible in the cave paintings of Chauvet, dating to over 30,000 years ago, but these elaborate and accomplished images testify to a much longer track behind them, extending "probably hundreds of thousands of years."[31] The oldest of arts is likely to be music, since song occurs widely in many different lineages, in birds, in cetaceans (dolphins and whales), and in other primates (especially gibbons), and therefore stretches back deep into evolutionary time, and since song needs only the body's own resources.

Steven Brown shows that the most complex song outside humans, both in songbirds and in other primates, arises not through courtship, but through the maintenance of territory and relationships by several species of monogamous duetting tropical songbirds and by gibbons. He notes that duetting resembles human music in several ways that courtship cannot account for: (1) "responsorial, antiphonal, polyphonic and homophonic singing . . . [which] greatly increases the potential complexity of acoustic signals"; (2) both sexes sing to a more or less equivalent extent; (3) duetting is cooperative and coordinated, not competitive or disjoint: "Gibbon couples place a high premium on maintaining tight coordination and restart a duet if the appropriate level of coordination is not achieved . . . Duetting is not a contest but a display of cooperative strength"; (4)

duetting serves to defend year-round territories, as in many human tribes and bands, acting as "a highly ritualized 'keep out' signal accompanied by exaggerated physical displays"; (5) it plays "a significant role not only in defending territories but in maintaining social bonds." Brown adds that "none of the known primate calls is thought to be directly involved in courtship. Primates do not seem to exploit vocalization for courtship purposes, but instead rely on visual, olfactory and kinetic cues. Courtship calls are rare to nonexistent in hominoids, whereas territorial calls are ancestral to the entire group of species."[32]

Not only does Brown's ethological approach respect biological detail, and analogues to and precursors of human behaviors in other animals; it also respects the peculiarities of a human art.[33] He can show that qualities like pitch blending and isometric rhythms, central to music, can be explained by the need to coordinate sound between more than one participant, not by individual display.[34] Recent evidence even suggests that music *reduces* sexual inclination: singing *lowers* men's testosterone levels, a result compatible with a cooperative, not a competitive, account of music's origins.[35]

Both Miller's search for evidence in support of his hypothesis, and the search for counterevidence that his statistical work has inspired in Brown and others, are welcome ongoing programs. Sexual selection no doubt operates in *some* ways that have altered human forms. Wodaabe men in Nigeria and Niger are chosen by their women in the human equivalent of a lek dance, and are unusually tall, with strikingly big eyes, white teeth, and straight noses.[36] Such an example presents a stark difference from the human norm, but over thousands of generations sexual selection surely *has* played an important part in human life, especially, as Darwin and Richard Dawkins suggest, in the differentiation of superficial racial characteristics, like face and hair;[37] and it may also serve as one factor in human art, especially visual art, likely to be the next major form of art to have appeared after music and dance.

Ocher appears to have been sought for body decoration from as early as 120,000 years ago (indeed, evidence suggests that pigments have been ground for twice as long),[38] and other body modifications, such as hairstyling, tattooing, scarification, and body-piercing have been practiced around the world for tens of thousands of years.[39] As the recent fashions for body-piercing and tattooing in Western countries highlight, such activity peaks at the ages of maximum reproductive opportunity. It makes biological sense that the visual arts should have started with the kind of bodily display most likely to have a sexual payoff. But notice here the difference from sexual selection in other species. In prehistoric times, before mirrors, and even now in tattooing and other modifications, body and facial adornment often had to be not an individual practice but a social one. Songbirds do not chorus in support of their rivals, and bowerbirds do not assist other males to construct their showhomes. But from an early time, even in bodily adornment, the closest that human arts come to sexual selection, cooperation seems also to have been present in our highly social species. And elaborate body decoration in most societies serves primarily as a mark of affiliation and group identification.[40]

There may be an even earlier, and sexually selected, precursor of human visual art in the simple stone tools known as hand-axes. The first stone tools, of the so-called Oldowan industry, from over 2 million years ago until about 780,000 years ago, were roughly flaked on one side. About 1.4 million years ago, a new technology appeared: so-called Acheulean hand-axes, tools expertly knapped from flint or quartz into a shape like a flattened, two-faced teardrop, with a cutting edge right around. The sheer number of hand-axes found in some sites, the proportion that under microscopic examination show little or no sign of use, the high and perhaps excessive degree of symmetry and finish, the occasional selection and highlighting of incorporated fossils, and the existence of forms too large or small for apparent use—all suggest strongly that hand-axes may often have been refined to a degree far beyond need, in a way best explained in terms of sexual or social selection: as a display to others of prowess and judgment. Notice that this proposal, which Miller endorses, has a specificity absent in Miller's own arguments, and reveals an awareness of the slow increments by which the first impulse toward the visual arts may have developed.[41]

Art as sexual display does not explain *nothing* about art. But the very flexibility of human behavior suggests that sexual selection has been an extra gear for art, not the engine itself. Yet young men and women have one strong reason to look for what Miller calls a social "entertainment system" in each other:[42] the playful interaction between infants and mothers or others that arises from and ingrains more deeply in us the pleasures of patterned play and the unique importance of human shared attention (see Chapters 6–7). With that disposition taken into account, and admitted as an impetus for art, *then* sexual selection may explain an escalation of adaptations for sociality and for art.

Differential parental investment—higher male competitiveness, higher female choosiness—can then hint at part of the reason for the preponderance of males over females in art for public display, although women seem always to have participated in song, dance, weaving, and storytelling, especially near the home, as much as or more than men. In *The Tale of Genji*, Genji wins a painting contest against his friend Tō no Chūjō, and as a result the adulation of many women.[43] But this novel, the world's first, still considered the pearl of Japanese literature, was written by a woman—and a mother. Sexual selection can *help* explain the fact that most public art has been produced by men—although this weighting of course also reflects the protracted history of restricted opportunities for women, because of their greater average commitment to childrearing, and because of males' inclination to limit female freedoms.[44] But sexual selection cannot explain why men and women, boys and girls, in the home and outside, engage so compulsively in art as both artists and audiences.

6

ART AS COGNITIVE PLAY

To focus on High Art—on classic works of literature, music, and visual art, or on modernist challenges to classical modes—is to begin near the end of the story of art. Explanations need to start much further back: with chant or ocher, or with a timeless scene like a father playfully distracting his child by miming an ape and happening to amuse, and then choosing to play up to a few casual onlookers. Art stretches back far in time, spreads widely in space, recurs in cross-culturally consistent modes, and starts early enough in each child to suggest the need for some explanation beyond the purely cultural. How can we account for art in biocultural terms?

So far we have considered five positions: (1) art is not even a meaningful category; (2) art is not a human universal; (3) art is universal but entirely explicable by cultural traditions and cultural diversity; (4) art can be explained in terms of its function(s), without recourse to evolution; (5) art can be explained in evolutionary terms by its *lack* of function, through sexual selection.

Most nonevolutionary theories of art offer functional explanations but propose functions too vague and mutually contradictory to satisfy the stringent cost-benefit criteria required in biological explanation. Evolutionary accounts of art can try to explain art by their functional uselessness, as Geoffrey Miller does via sexual selection, *or* in terms of evolved functions, as an adaptive behavior.

An evolutionary *adaptation*, recall, is a feature of body, mind, or behavior that exists throughout a species and shows evidence of good *design* for a specific *function* or functions that will ultimately make a difference to the species' survival and reproductive success.[1]

If art is a human adaptation, it has been established throughout the species because it has been selected as a behavior for the advantages it offers in terms of survival and reproduction. It may offer one advantage, like echolocation in bats, or many, like an elephant's trunk, which evolved to sniff, dislodge, grasp, pull, deliver, push, twist, caress, trumpet, siphon, and squirt.

Steven Pinker, the most prominent proponent of the adapted mind —the mind as shaped in detail by evolution—has also been the foremost critic of claims that art is an adaptation. He throws down a stiff challenge to those who regard art as more than an evolutionary byproduct: "For the same reason that it is wrong to write off language, stereo vision, and the emotions as evolutionary accidents—namely their universal, complex, reliably-developing, well-engineered, reproduction-promoting design—it is wrong to invent functions for activities that lack design merely because we want to ennoble them with the imprimatur of biological adaptiveness."[2]

Here we come to *Position 6: Art is a byproduct of adaptive features of the human mind.*

Pinker accepts that narrative may be adaptive in enabling our minds to develop scenarios testing possible courses of action and their consequences. Otherwise, he considers art an evolutionary byproduct deploying our capacity for design to deliver high-energy treats to our cognitive tastes, concocting "cheesecake" for the mind, or developing "a useless technology for pressing our pleasure buttons" by "defeat[ing] the locks that safeguard" them.[3]

In explaining art as design skills working to gratify evolved human preferences, Pinker assumes that our ability to design developed only in purely instrumental modes. But why? Atop a Paleolithic spearthrower found at Mas d'Azil in France sits the carving of an ibex turning her head to look at two birds already perching on the turd she is extruding—an exquisitely intricate and playful carving that required far more design skill than the spearthrower itself. It seems at least arguable, and in fact highly likely, that art has helped ratchet up our interest in, our capacity for, and our confidence in design. A society whose members wove elaborate and superfluous designs because they were pleasing could more easily think up a woven eel trap

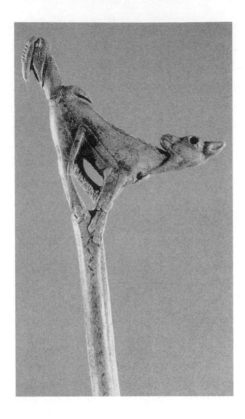

Spearthrower from Mas d'Azil,
France, ca. 14,000 B.C.E.

than a putative society focused exclusively on utilitarian technological solutions could think up decoration for clothing, containers, or coverings.

Pinker's metaphors (cheesecake, pleasure buttons, music as "a cocktail of recreational drugs that we ingest through the ear"),[4] like his conclusion that art is a byproduct, depend on seeing art as consumption. Gleefully offending the devotees of high art, he suggests that for an evolutionary account of art we need to think black velvet painting.[5] But before we respond to art we have to generate it. In modern society we can avail ourselves of ready-made art like black-velvet Elvises almost as easily as we buy ready-made cheesecake. But for most of human history and in most societies, art results from the efforts of all, as people weave and carve, sing and dance, tell and re-enact stories. The compulsion to engage in art needs to explain the compulsion to *make* art as well as to enjoy it.

Art has usually involved intense effort. It began when there were no art-supplies stores, when ocher had to be extracted or traded, when there were only stone carving tools that themselves had to be laboriously fashioned. Early visual art such as scarification, tattooing, and body-piercing often caused acute personal pain. Even with the comforts and conveniences of city life, creating art can involve prodigious investments of time and energy: Michelangelo's years on his back painting the Sistine Chapel ceiling, or the more than a century that has still not brought Gaudí's design for Barcelona's *Sagrada Familia* to completion. The cheesecake metaphor fails to explain why in every society the elaborate and often arduous efforts required to produce art have seemed worthwhile.

By stating so pungently the hypothesis of art as a byproduct, Pinker does the evolutionary explanation of art a great service. For the hypothesis fails—and therefore contributes to the case for art as not a byproduct but an adaptation: *If* art involved no benefit, if it only mimicked biological advantage, as drugs do, by delivering unearned pleasure, *yet* it had high costs in time, energy, and resources, *then* a predisposition to art would be a weakness that would long ago have been weeded out by the intensity of evolutionary competition. Nature selects against a cost without a benefit, as when it dispenses with sight in burrowing or cave-dwelling animals.

If the byproduct hypothesis were correct, then over thousands of generations and millions of births, individuals, and societies with a lesser disposition to art would have prospered because they did not incur the high costs of producing it and either simply had more opportunity to rest and harbor resources—like other top predators, such as big cats[6]—or had more to time and energy to devote to activities that *did* yield benefits, such as producing new resources or competing to acquire the resources of others. As Richard Dawkins notes: "natural selection is a predictive theory. The Darwinian can make the confident prediction that, if dams were a useless waste of time, rival beavers who refrained from building them would survive better and pass on genetic tendencies not to build."[7] Likewise with art: *if and only if* art were useless, more ruthlessly utilitarian and competitive realists with a lesser inclination to art would have survived and reproduced in greater numbers, and over evolutionary time their descendants would have supplanted those with a disposition to art.

Societies without any inclination to create their own dress, song, and story would have ousted those that did have these things. Individuals and groups without art—without shared songs and dance (including anthems and war-dances), without their own styles of dress and design (including face-paint, scarification, tattoos, uniforms, emblems, flags, or monumental architecture), without shared stories and sayings (including myths, heroic legends, proverbs)—would fare better than those with all these things, which art makes possible. But that seems never to have been the case. No human society lacks art, and the most successful societies have more art than ever before.

Social living offers advantages—it has been rightly called "our most powerful survival tool";[8] but it also produces tensions. Cities, social living at the extreme, offer the formidable benefits of specialization and concentration of labor and services, but they also have high costs in crowding, control, and coordination. If the byproduct hypothesis were true, if art offers only illusory benefits, people could live more successfully in cities, could cope better with the strains of urban life, *without* the pleasures of art: with only sterile serviceableness rather than aesthetically appealing design in dress, interior decoration, and architecture, without music, stories, parades, carnivals, concerts, shows. A bleak civic environment would outdo a vibrant one.

An obvious rejoinder would be that art has become possible because human technology has allowed us a surplus that provides a protective buffer against the selective pressures of evolution. But this argument cannot explain the persistence and the centrality of art (song, dance, story, design) among, for instance, the Aborigines, who have survived so long in the relentlessly harsh conditions of the Australian desert—and who, when given the opportunity to contribute to the world of modern specialized art, have produced, in the work of Clifford Possum Tjapaltjarri, Kathleen Petyarre, and others, some of the most sublime and haunting paintings of the last few decades.[9]

Or if art had no role to play in human survival, if it were useless in those terms, and its uselessness were the proof that art resulted solely from sexual selection, as Miller suggests, then we would engage in art overwhelmingly in our fertile years, and only so long as fertile individuals of the opposite sex were among their audience. An infant's

Clifford Possum Tjapalt-
jarri, *Yuelamu Honey
Ant Dreaming,* 1980.

delight in hearing nursery rhymes or lullabies, a mother's in croon-
ing them, a grandmother's pride in weaving designs in flax, wool, or
cotton, anyone's silent reading of fiction or keen interest in the work
of long-dead artists would be impossible to explain.

The implausible implications of both hypotheses suggest strongly
that art *has* been designed by evolution to serve some survival function. But what?

Before offering my own answer, let me first characterize art as a behavior. I suggest that we can view art as a kind of cognitive *play,* the
set of activities designed to engage human *attention* through their
appeal to our preference for inferentially rich and therefore *patterned*
information.

With the help of this notion I propose two principal functions for

Kathleen Petyarre, *Mountain Devil Lizard Dreaming (With Winter Sandstorm)*, 1996.

art. First, it serves as a stimulus and training for a flexible mind, as play does for the body and physical behavior. The high concentrations of pattern that art delivers repeatedly engage and activate individual brains and over time alter their wiring to modify key human perceptual, cognitive, and expressive systems, especially in terms of sight, hearing, movement, and social cognition. All of art's other functions lead *from* this. Second, art becomes a social and individual system for engendering creativity, for producing options not confined by the here and now or the immediate and given. All other functions lead up *to* this.

I would draw a distinction between the cognitive play of art, in which people engage in cooperative and open-ended ways, whether as creators, performers, or audiences, and closed games like chess or poker, or puzzles like crosswords or Sudoku. Play and art are non-zero-sum—that is, one side's gain is not another's loss, unlike in tennis, football, backgammon, or go—and that makes all the difference: the difference between competition, however playful, and the mutually amplifying effects of cooperation, ably summed up in the title of Robert Wright's *Nonzero: The Logic of Human Destiny* ("Indeed, in highly non-zero-sum games the players' interests overlap entirely").[10] Even in collective works of art, in which collaborators may have partially competing interests, like studio films, all involved aim to further the work of art, though perhaps according to different values; or jazz combos, in which all spontaneously react to one another's play, but for the sake of the work and the audience, not for the sake of a victory and the support of a partisan fraction of the audience. Not only is art non-zero-sum, but its effects are not closed, as in mathematical or verbal puzzles, but open ended. The cooperative amplification and the open-endedness make all the difference to art.

Before we reach play, what do we mean by "pattern"? The *Oxford English Dictionary* defines it as "an arrangement . . . order or form discernible in things, actions, ideas, situations, etc."[11] Pattern tends to signal regularities in the world rather than mere chance: the pattern that my head and my feet turn up in the same neighborhood is no quirk but part of the regularity that is me.

Because space teems with regularities from quarks to quasars and because life builds from the simple to the complex by endless recombination, we live in a world that swarms with patterns at every level, beside or within or across one another. Computers still fare dismally at pattern recognition, but because predicting what may come next can make life-or-death differences to living things, organisms—even unicellular animals, even bacteria and plants—have evolved to be pattern extractors, and at least the more intelligent animals, like higher primates and corvids, decidedly prefer regular, symmetrical, or rhythmic patterns.[12] In both space and time, in sight and sound, we therefore sense beauty in "the rule of order over randomness, of pattern over chaos."[13]

Information can be costly to obtain and analyze, but because it offers such an invaluable basis for action, nature evolves senses and minds to gather and process information patterns appropriate to particular modes of life, from sonar in bats or electroception in fish to touch in the star-nosed mole. Even plants that can detect the seasons or nearby growth can make informed "decisions" about how to react. Animals need much more rapid responses, and therefore have evolved minds to detect swiftly patterns meaningful to their kind of organism in their kind of environment—smells, for ants or dogs, ultrasound for bats, magnetic fields or nocturnal skies for migrating birds—and to coordinate the inferences the patterns allow.

Pattern occurs at multiple levels, from the stable information of spatial conditions and physical processes to highly volatile information about individuals and their moods, actions, and intentions. Pattern recognition lets us distinguish animate from inanimate, human from nonhuman, this individual from all others, this attitude or expression from another. The capacity to identify not only individuals but also higher-order tendencies in their behavior, personality, and powers allows for invaluably precise prediction.

Pattern recognition need not be explicit. Take "that all-important pattern, the face."[14] My computer's screensaver shuffles through thousands of photographs I have stored. Two of my young grandchildren, born two years apart, look uncannily alike, but even in the same baby clothes in photographs taken when each was the same age I can identify which is which the moment each comes on screen, although I could not say *how* I can identify them. Nor do I need to know. Pattern recognition does it for me, sifting data effortlessly and almost instantly into the forms that matter most to me: in this case, identifying a configuration of color patches on the flat screen not just as a person and as an infant, but as Emily or Ben.

Like other species, humans can assimilate information through the rapid processing that specialized pattern recognition allows, but we uniquely inhabit "the cognitive niche": we gain most of our advantages from intelligence. We therefore have what the Nobel physics laureate Edward Purcell has called an "avidity for pattern,"[15] for information that forms arrays from which we can make rich inferences. For that reason we like bright, distinct colors, crisp outlines, complex shape or surface design: think of our special fondness for butterflies and flowers. Zoologist Paul Weiss noted that our sense of beauty depends on our sense "of pattern over chaos . . . the beauty of forms rests on the lawfulness of their formation."[16] And we have an appetite for *open-ended* pattern, not only the forms we have evolved to detect automatically. Frogs react with an automatic flick of the tongue to small objects flying across their field of vision. That makes them swifter than you or I at catching insects, but they cannot respond to *new* kinds of patterns. Humans can. In addition to the configurations that evolution has programmed us to track, like the shapes or locations of objects, we search for patterns of many kinds, like, eventually, the chemical structures of insecticides that can make us more efficient killers of insects than even frogs have been designed to be.

We crave patterns because they can tell us so much: as chemist Peter Atkins noted, they are "the lifeblood of science and the seeds of theories."[17] Stephen Jay Gould has written of our penchant for pattern: "No other habit of thought lies so deeply within the soul of a small creature trying to make sense of a complex world not constructed for it."[18] Only humans have the curiosity to seek out pattern in the open-ended way that once led our ancestors to see constellations in the skies, then to infer first the revolution of the Earth from the motion of the stars and planets, then the expansion of the universe, then possibilities beyond our patch of our multiverse.

If information is chaotic it lacks meaningful pattern and we cannot understand. Extreme informational chaos, the absence of pattern, as in complete whiteout or dense fog, can even cause distress and loss of sensory function.[19] But valuable information can be concealed in the world's welter, like the causal configurations that medical researchers tease out to link particular chemicals or conditions to particular cancers. If on the other hand information is completely patterned, we need not continue to attend: if a stimulus remains unchanging, if the pattern can be predicted, the psychological process of habituation automatically switches off attention.[20] The most patterned novel possible would repeat one letter, say, *q*, over and over again—a queue no reader would want to wait in. But an *unpredictable combination* of *patterns* repays intense attention and can yield rich inferences, although finding how to ascertain all the patterns and all the meaning they imply may not be easy.

Art concentrates and plays with the world's profusion of interrelated or intersecting patterns. Take music:

> Whenever a noise exceeds our processing abilities—we can't decipher all the different sound waves hitting our hair cells—the mind . . . stops trying to understand the individual notes and seeks instead to understand the relationships *between* the notes. The human auditory cortex pulls off this feat by using its short-term memory for sound (in the left posterior hemisphere) to uncover patterns at the large level of the phrase, motif, and movement. This new approximation lets us extract *order* from all those notes haphazardly flying through space, and the brain is obsessed with order . . .
>
> It is this psychological instinct—this desperate neuronal search for a pattern, any pattern, that is the source of music . . . We continually abstract on our own inputs, inventing patterns in order to keep pace with the onrush of noise. And once the brain finds a pattern, it immediately starts to make predictions . . . It projects imaginary order into the future . . .
>
> The structure of music reflects the human brain's penchant for patterns . . . But before a pattern can be desired by the brain, that pattern must play hard to get.[21]

Or, to shift mental gears and artistic modes, take, for instance, Shakespeare's *Henry IV Part 1*. We respond almost immediately to patterns of character—Falstaff's shameless exuberant ebullience, Hal's controlled wildness, Hotspur's impetuousness, Glendower's impassioned boastfulness, and much more—and to patterns of action, like the plot and counterplot of the Gad's Hill robbery. Patterns set up expectations, which they may satisfy, overturn, or revise—especially, in this case, the mounting expectation that events will lead to a climactic battlefield confrontation between Hal and Hotspur.

By far the most salient patterns for us in story will involve agents and actions, character and plot, intentions and outcomes. Shakespeare offers such immediate patterns but incorporates others that emerge at different rates and levels, from the local and linguistic—like his attention-catching deviations from standard English phrasing, or the rhythms of his iambic pentameter, or the idiosyncratic stamp of his characters' speech, Falstaff's largesse, Hal's nimble-minded quickness, Hotspur's fiery gallop, Pistol's fustian ranting, Mistress Quickly's endearingly unruffled confusions, and much more—to larger-scale patterns of character or scenic contrast, of image, idea, or structure, like the motifs of rebellion or of the robber robbed: patterns sufficiently compacted and overlaid that some have taken centuries to tease out explicitly. Some patterns we can easily understand and describe; others, like patterns of meter and structure, may remain unconscious yet still shape our response.

Our perception of pattern and of deviation from pattern produces strong emotional reactions.[22] Art appeals to our appetite for pattern at multiple levels, in producing or perceiving bodily movement, shapes, surfaces, sounds, words, or miniature worlds. Like play it therefore provokes us to continue the activities it offers long enough and to resume them often enough to modify our neural circuitry over time.

As searching human intelligence evolved out of curiosity in other animals, so human art, I propose, has evolved from animal play.[23] Play occurs not only in every mammal species in which it has been looked for, and in many birds, but even in some fish and reptiles, and perhaps even in invertebrates (octopi).[24] Especially in animals with protracted parental care, like birds and mammals, offspring can be comparatively secure in the presence of parents.[25] Nature has made the most of juvenile dependency by having species with flexible behavior play repeatedly and intensely when young.

Although much remains to be discovered about the biology of play, almost all researchers concur that a behavior involving so much energy expenditure and risk of predation and injury, yet so highly motivated and so widely conserved across species, must serve powerful functions. Pinker rightly stresses that "to demonstrate that X is an adaptation, one can't simply show that people like doing X." Rather, one has to specify functions—survival and reproductive benefits—in advance, as if in engineering terms, rather than in terms of how they matter to us.[26] Let me describe play in these terms.

Evolution can install general guidelines for action—nature's factory settings—but for some behaviors fine-tuned choices and wider ranges of options that can be deployed at short and context-sensitive notice make a decisive difference. This applies particularly to the volatile sphere of social relations, and especially to the most urgent situations, flight and fight. Such behaviors can be fine-tuned by experience and the range of options extended by exploratory action. Creatures with stronger motivations to practice such behaviors and to explore new options in advance, in situations of low danger and adequate resources, will fare better than those without. The more pleasure that creatures have in play in safe contexts, the more they will happily expend energy in mastering skills needed in urgent or volatile situations, in attack, defense, and social competition and cooperation. This explains why in the human case we particularly enjoy play that develops skills needed in flight (chase, tag, running) and fight (rough-and-tumble, throwing as a form of attack at a distance), in recovery of balance (skiing, surfing, skateboarding), and in individual and team games.

The more often and the more exuberantly animals play, the more they hone skills, widen repertoires, and sharpen sensitivities. Play therefore has evolved to be highly self-rewarding. Through the compulsiveness of play, animals incrementally alter muscle tone and neural wiring, strengthen and increase the processing speed of synaptic pathways, and improve their capacity and potential for performance in later, less forgiving circumstances.[27] Since scientists cannot yet probe inside the skulls of children at play, their research focuses on other species. All animals they have so far studied grow more neural tissue in enriched than in impoverished environments, with more friends, toys, ladders, and wheels to play with.[28] Monkeys trained on a particular task develop in the relevant neurons finer discrimination, more rapid response, faster processing, swifter recovery times, better coordination, and clearer signaling.[29] Play is more difficult to monitor than directed training, but scientists have recently discovered that in young rats play generates a specific compound that drives genetic transcription in the amygdala and frontal cortex, rapidly growing areas of the social brain—and we can expect further such discoveries.[30]

In order to fine-tune actions for rapid activation in emergencies, play often involves exaggerated or extreme behavior, like losing balance and then recovering, or risking defeat in play-fighting until attacker and defender swap roles.[31] (Aptly, the name of the Japanese kabuki theater derives from the obsolete verb *kabuku,* "to lose one's balance," "to be playful.") The amount and intensity of play in different species correlate closely with the species' intelligence and flexibility of behavior.[32]

Animals eagerly initiate, solicit, and persist in play. Chimpanzees that learn sign language use it mostly to ask for what they want: especially food—and play.[33] Play stimulates the release of the neurotransmitter dopamine (central to the brain's reward system and a key motivator of evolutionarily positive actions like eating and sex), which encourages further play.[34] But like curiosity, play has its built-in stopping routines, including fatigue and loss of interest.[35] Animals, especially adults, with less to learn from the mental and bodily training that play offers, can therefore avoid expending too much energy in play.

By the time of the first massive expansion of hominid brains, about two million years ago, our ancestors had already evolved uniquely long childhoods and therefore the long stretch of security under parental care that we still enjoy. Within such security, play affords us the opportunities to acquire routines that we can make effortless or almost automatic by practice, repetition, and experiment, freeing our minds to cope with more novelty at a higher level.

As for play, so for the cognitive play of art we can specify the design conditions in advance. If there are cognitive capacities in which flexible fine-tunings and widening the range of options deployed at short and context-sensitive notice can make decisive differences—and our aural, visual, vocal, manual, and social skills all qualify—

then individuals with stronger motivations to practice such behaviors in situations of low danger and adequate resources will fare better. A predisposition toward the patterned cognitive play of art will establish itself in a species in which cognitive skills are paramount.

Art as a kind of cognitive play stimulates our brains more than does routine processing of the environment.[36] It offers what biologists call a supernormal stimulus, an incentive more intense than usual, in this case a rush of the kinds of patterned information that our minds particularly crave. Because neural connections establish themselves piecemeal through experience, and because we find art self-rewarding, because we engage in it eagerly and repeatedly, art can over time fine-tune our minds for rapid response in the information modes that matter most to us.

Michael Merzenich, a neuroscientist famous for his studies of brain plasticity and for developing successful brain-training software, remarks that "the Internet is just one of those things that contemporary humans can spend millions of 'practice' events at, that the average human a thousand years ago had absolutely no exposure to. Our brains are massively remodeled by this exposure—but so, too, by reading, by television, by video games, by modern electronics, by contemporary music, by contemporary 'tools,' etc."[37] But our predisposition for art has long ensured that we have grown up in and constructed our own especially enriched environments. For tens of thousands of years our play with the intense patterns of intricate arts like weaving and carving, song, dance, verse, and story has enticed us to repeat "practice" events and so remodeled human minds.

Only focused attention and incremental remodeling can make the most of the brain's plasticity,[38] and art can reconfigure minds only so long as it rewards us enough, like play, to hold our attention again and again.

To see the close relationship between art and play, and the self-rewarding nature of both, consider chimpanzee painting. In the 1950s, when Desmond Morris supplied chimpanzees in his care with paint, brushes, and paper, they threw themselves into painting provided they received no external reward. Those who *were* offered food would make a few perfunctory strokes and break off quickly to seek another tasty morsel. But those whose motivation remained uncorrupted by "payment" developed a fierce commitment to painting. They painted intensely, persisting, while the session lasted, until they thought a sheet finished, though they would never glance at their work later.[39] But whereas chimpanzees engage in art only if the materials are supplied, even infant humans will seek out ways to stimulate themselves through art.

Human art often remains or returns close to the play from which it emerged. Comedy verges on and even aims at play. Music, too, can shade into play. The Kalinga of the northern Philippines participate in a repetitive modulatory game with bamboo sticks in the shape of tuning forks that generates music of a kind—but participants seem to regard the activity as more game than performance. Few filmmakers have produced such searing explorations of the human condition as Ingmar Bergman, but he described filmmaking as like returning to childhood, a game, a kind of play.[40]

Art becomes compulsive because it arouses pleasure, and it arouses pleasure because, like play, it fine-tunes our systems. In small-scale societies—including most of the human past—all participate in the arts of the community, in song and dance, weaving and carving, makeup and costume. In large-scale societies with considerable specialization of labor, professional artists can achieve new levels of performance by dint of highly focused practice and effort. After childhood, most adults in such societies may become consumers rather than producers of most kinds of art, except in singing and storytelling for their children. For both those who produce and those who consume it, art staves off boredom—itself an emotion that evolved to reactivate curiosity—and counters habituation.[41]

Art as cognitive play augments our capacities so that we can, at least in the domain on which each art focuses, efficiently produce ideas or actions: sounds, movements, visualizations, or representations, and, in the case of story, scenarios for reasoning about our own and others' plans and actions.

Art's origin in self-rewarding cognitive play explains its therapeutic power. John Carey answers the title question of his *What Good*

Are The Arts? mostly in the negative. High art, he claims, serves no particular good, certainly not the exalted functions claimed for it. But art practiced by prisoners offers them a sense of accomplishment that they have rarely had.[42] And art, as practiced by all in the days before it could become a status badge or a consumer product, rewarded all, *because* it sharpened minds and skills.

Just where does art start? We enjoy the patterned cognitive play we call art because our unique predilection for pattern combines with the widespread mammalian impulse to play. As children and throughout life we are each internally driven to engage in art. But we are also reliably prompted in this direction, in a uniquely human manner, by the behavior of others, especially our parents.

Parental care extends throughout the mammalian and avian lines. Parental responsiveness to infants is an elaborate and necessary evolved behavior in mammals, where usually only the female engages in sustained parenting, as well as in birds, where often both parents play an active role in rearing their young. In the human case, too, parenthood involves sustained contributions from both parents. It is also uniquely intimate in a way that bootstraps infant brains and fast-tracks their predisposition to shared and patterned play.

From infancy humans seek to command the attention of others, to shape it more finely, and to share it more fully, than do any other species. Unlike other primates, humans have eyes whose elongated shape and whose color contrast between iris and sclera make eye direction easy to see: in short, humans have evolved eyes that reveal rather than conceal their direction.[43] Monkey babies lack the stimulus tools to capture and hold their mothers' attention.[44] Chimpanzee mothers rarely gaze into their babies' eyes or communicate with them, though they will respond when babies initiate play by biting, and will tickle and laugh in tender reply.[45] But human mothers and infants attend to one another from the first. Infants' eyes after birth can focus only about eight inches away, the distance between the mother's breast and her face, and unlike infants in any other species they maintain eye contact while suckling. Newborns prefer to attend to faces more than to any other visual stimulus, and under laboratory conditions they have been shown to be capable of imitating humans, but not animated models, within an hour of birth.[46] At only a day old they will wriggle their limbs and orient their faces in time with taped female speech, but not with disconnected vowel or tapping sounds.[47]

So it continues. For the first six months, infants have a love affair with human faces, voices, and touch, a factor that allows adults to bootstrap their children's intelligence by focusing their attention.[48] Across the world older humans instinctively address infants in "motherese," whose higher-pitch appeals and whose simple but exaggerated emotional contours, coupled with highly contrastive visual signing, can be decoded without language and can therefore recruit and maintain attention.[49] Minds find more attractive what they can process more easily,[50] and the highly patterned and stylized language of motherese simplifies understanding and invites attention.

By about eight months, parent-infant "protoconversations"— "more like a song than a sentence"[51]—set the scene for the special nature of human sociality and for art. Aptly described as multimedia performances, since they use eyes and faces, hands and feet, voice and movement, these protoconversations consist of rhythmic, finely attuned turn taking and mutual imitation, involving elaboration, exaggeration, repetition, and surprise, with each partner anticipating the other's response so as to coordinate their emotions in patterned sequence.[52] Though taking different forms in different cultures, protoconversation in all human cultures resembles a duet in which child and adult "seek harmony and counterpoint on one beat to create a melody."[53] If art offers the mental stimulation of patterned cognitive play, it begins reliably for individuals in normal environments through the social scaffolding provided by playful, patterned parental prompting.

From about nine to twelve months, infants tune into the attention and behavior of adults in new ways, and try to have adults tune into theirs. Although all apes note where others look, on the basis of head direction, only humans, from a year old, track eye gaze as well as head movement.[54] Human one-year-olds engage in joint attention—following others' hands or eyes and checking to see that the others follow theirs—and in protodeclarative pointing—indicating objects or events simply for the sake of sharing attention to them, something that apes never do.[55] They expect others to share interest, attention, and response: "This by itself is rewarding for infants—apparently in a way it is not for any other species on the planet."[56]

The presence of an infant releases special tenderness in human mothers. Infants prefer recordings of lullabies sung by mothers while a child was present to those recorded without children present, and adults also judge these versions more loving.[57] Although all female primates find babies fascinating, and the younger the better, they are not automatically nurturing, so that infants will fare better the more they are attuned to mothers or others.[58] A caregiver in turn will be able to stimulate an infant most aptly, in ways neither too simple nor too complex, neither too familiar nor too new, when simply having fun with the child.[59]

The unprecedented attunement of human infants' attention to that of adults, inside and outside play, and even before language, reflects familiar quirks of human evolution. The human infant is unique in being born so undeveloped, partly as a result of the narrowing of the female pelvic canal because of permanent bipedalism in early hominids, partly as a result of the first rapid enlargement of the brain in archaic *Homo* about two million years ago, and partly as a result of the general change in developmental rates, leaving humans with by far the longest infancy and childhood of any animal, presumably as a consequence of selection for increasing sociality, flexibility, and intelligence.[60] Mothers and others provide a social entertainment system for infants, apparently because evolution has selected for both adults and children who can turn childhood dependency into mutual delight.

The behavior of older humans that reliably prompts infants' disposition toward patterned play, like motherese, protoconversation, and lullabies, owes something to culture but is too similar *across* cultures, too interculturally recognizable, and especially too spontaneously engaged in, without any need for training, to be understood only in terms of culturally acquired behaviors. Just as evolution has designed parents to respond in species-appropriate ways to the physical needs of infants, it has designed human parents (and grandparents and siblings) to respond in the ways that help ensure that infants develop their predilection for the patterned cognitive play of art.

7

ART AND ATTENTION

SOCIAL COHESION AND INDIVIDUAL STATUS

IN THE 1950S PSYCHIATRIST Cathy Hayes raised a young chimp in her own home. In late infancy Viki, the chimp, began to trail an arm behind her as if pulling a toy on a string, and would even pretend to catch the string on obstructions and then release it again. After several weeks of this behavior, Viki one day appeared to entangle the imaginary toy around the knob of the toilet, and cried for help. Hayes pantomimed untangling the rope and returning it to her, to be rewarded with what could have been either "a look of sheer devotion" or "just a good hard stare." A few days later, when Hayes decided to invent a make-believe pull-toy of her own that clacked on the floor and swooshed on the carpet, "Viki stared at the point on the floor where the imaginary rope would have met the imaginary toy, uttered a terrified 'oo-oo-oo,'" leaped into Cathy's arms, and never played that game again.[1] For all Cathy's quasi-maternal care, Viki's playfulness, and Cathy's playful engagement with her, Viki was totally unprepared to share attention with Cathy.

In Viki the sharing of attention so natural in humans, even in an infant's protoconversation with its parents, seems still far away. Protoconversation in turn seems a long way from the *Mahabharata* or Mozart. Yet sharing attention, if not quite the start of art, proves indispensable.

To explain art we need to attend to attention. Art dies without attention, as people since Aristotle have noted, both within and outside evolutionary explanation.[2] Art alters our minds because it en-

Samburu girls, Kenya:
art as pattern.

gages and reengages our attention from nursery rhymes to rest-home singalongs. But art has rarely been considered in terms of the special role that engaging attention has *evolved* to play in human lives.[3]

Art begins in infants' predisposition to engross themselves in patterned cognitive play. But the unique pull of human *shared* attention also compels parents to engage infants in protoconversation, which directs their children's predilection for pattern toward the rudiments of song, music, and dance. Without shared attention, the sustained self-stimulation of cognitive play could lead into hermetic private worlds like Viki's that would be disastrous for creatures whose strength lies in their sociality. But *with* the force of human shared attention, the benefits of cognitive play ramp up in powerful ways, social and individual, to produce a plethora of functions.

Sharing Attention: Art and Attunement

Art offers us social benefits by encouraging us to share attention in coordinated ways that improve our attunement with one another.

Evolution is competition, at multiple levels: one gene or organism or group against another. Nevertheless nature has also repeatedly evolved cooperation, since genes, cells, and organisms working together can often achieve much more than they can acting individually. But competition remains, including competition for cooperation (within groups) and cooperation (at the group level) for competition (against other groups).

All social species prosper more together than alone, or they would not remain social, but humans take this to another level, ultrasociality,[4] the most intense cooperativeness of all individualized animal societies. Not endowed by nature with formidable strength or speed, we have been able for hundreds of thousands of years to coordinate our activity sufficiently to kill large prey—and, for thousands of years, to construct pyramids or cathedrals and settlements of thousands or even millions.

Social life offers rewards but engenders strains. Conflicts arise from purposes not quite shared or even radically opposed, and the advantages of cooperation can always be exploited by those who accept the benefits without paying their share of the costs. The larger the group, the harder the problem of maintaining cooperation. Yet the larger the group, the more chance it has of overcoming smaller groups, provided it can ensure its own cohesion. Substantial rewards await groups that can somehow solve better the fundamental problem of social life, preventing cooperation from being infiltrated and undermined by the short-term advantages of selfishness.[5]

Partly through language, humans have learned to coordinate their activities in precise and flexible ways. But they have also been able to *motivate* continued cooperation on larger scales and in greater detail than other species. Art has played a key role in training and motivating us to share our attention in ever more finely-tuned forms.

Attention has been central to the rise of sociality. An initial reason for adopting a social existence lies in the vigilance of others, in attending to others' sudden attention to threats.[6] Survival rates plummet for social creatures that stray from the shoal, flock, or herd, since these loners become much more vulnerable. In tighter kinds of sociality, from honeybees to apes, attention to the attention of others becomes progressively more important. First, animals read others' attention to resources as well as risks: a patch of pollen-rich flowers or ripe fruit, a route to reaching it, a technique to access or process it, a safe territory. Then, as the advantages of social living multiply, creatures attend to others in the group and *their* mutual attention in order to monitor affiliation and alliance and the rank needed to avoid constant conflict over access to resources.[7]

As social relationships become still more flexible, as in corvids, dolphins, or chimpanzees, shared attention makes possible open-ended coordination within groups and against rivals or prey.[8] Groups that coordinate swiftly and freely can outdo other groups. To do so, they need to share and check attention and intention as they pursue flexible goals: as Frans de Waal notes, "The selection pressure on paying attention to others must have been enormous."[9] And in order to *motivate* complex cooperation, animals need to derive interest and pleasure from attending to each other, as dolphins do in their synchronized play. Yet even near the top of this cline, in chimpanzees, attention to others remains far coarser than in humans, who can follow and direct others' attention with unmatched precision in order to learn, teach, plan, cooperate, or stimulate.

How do we understand the benefits of the closest social attunement? Biologists often use *analogy* to clarify functions by comparing them across multiple species or lineages: flight, for instance, among insects, birds, and bats, or vision, intelligence, or sociality in many animal lines. They also use *homology,* similarities in structure, to establish lines of descent, such as the four-limbed pattern of amphibians, reptiles, birds, and mammals. Analogies and homologies help illuminate the role of fine social attunement.

Groups become more cooperative through selection for the cooperative dispositions of their members. Coyote pups disinclined to play bond less tightly to others in the group, strike out more often on their own, and prove three times more likely to die young.[10] In the human case, those who do not follow local traditions can become outcasts: acceptance matters for social animals as much as do oxygen and food.[11] Conversely, creatures that wish to cooperate attune themselves to one another. Parrots, duetting songbirds, gibbons, and humans tend to act and sound like those they wish to ally with.[12] Dolphins have individual signature whistles, but male dolphins in close cooperative bonds, an advantage both in hunting prey and in pursuing females, converge toward each other's signature whistles as their bond strengthens.[13]

Social attunement reaches its closest analogy to human art in other highly social species that also benefit from close, but open-ended, coordination. Dolphins work together in stable alliances in synchronized behaviors like releasing air bubbles to herd fish toward the ocean surface, where they consume them in mass. They appear to take pleasure in coordination, in swimming, diving, and leaping in tight unison, or even, as we have seen, in toying with air bubbles in ways that serve only the psychological and social purpose of interesting themselves and others, in a kind of protoart. Chimpanzees, too, can work in concert, whether hunting monkeys or charging together against other chimpanzee groups.[14] They can also act in playful or artlike ways that appear to generate intense and even frenetic excitement from the sheer pleasure of coordination, as when they brachiate, dance, pant-hoot, stamp, or drum in rowdy carnivalesque unison.[15]

Since humans benefit from collaboration far more complex and on a far larger scale than that of dolphins or chimpanzees, we have a much greater need to motivate cooperation and master coordination. We achieve much of our advanced cooperation through language, but even language emerges from the increasingly flexible and finely-tuned attention to others evolved among highly social mammals.[16]

Although language permits us to share intentions, we can also use it for competitive rather than cooperative ends. But a system of unconscious emotional contagion that works before language and below conscious awareness and has evolved along with sociality allows us to coordinate genuinely cooperative intentions. Mirror neurons, whose function was discovered only in the early 1990s, fire when we see others act or express emotion as if we were making the same action, and allow us through a kind of automatic inner imitation to understand their intentions and attune ourselves to their feelings.[17] Summarizing much recent work on social intelligence, psychologist Daniel Goleman notes:

> Empathy—sensing another's emotions—seems to be as physiological as it is mental . . . the more similar the physiological state of two people at a given moment, the more easily they can sense each other's feelings . . . When people are in rapport, they can be more creative together and more efficient in making decisions . . . Shared attention is the first essential ingredient. As two people attend to what the other says and does, they generate a sense of mutual interest, a joint focus that amounts to perceptual glue. [The second ingredient needed is good feeling, cooperative intent; the third] coordination, or synchrony . . . We coordinate most strongly via subtle non-verbal channels like the pace and timing of a conversation and our body movements.[18]

Animals learn best to attune themselves to one another through the pleasures of play, taking turns, checking to see that their partners still seek the "high" of intense mutual engagement.[19] Humans, more flexible in behavior and dependent for much longer in childhood, broaden and lengthen social play, not only in childhood but right throughout adult life. And they extend it, crucially, into the cognitive play of art.

Art allows us another dimension for minds and emotions to meet. Although children play with words, movements, and pretense on their own, art from the first, from the protoart of protoconversation, can be social.

Our unique predisposition to social learning makes art a matter of attunement as well as of self-stimulation. Learning from others benefits many species with flexible behavior, but humans take matters much further. From an early age we imitate and practice more than any other species.[20] Even in a schoolless hunter-gatherer society, individuals learn more than 99 percent of fifty core skills with help from others.[21]

Social learning can greatly reduce the time it takes to find successful strategies. In the case of art, social learning starts with protoconversation and continues as children swiftly recognize and replay rhymes, songs, dance, designs, and stories as they become initiated into the art modes of their group. They saturate themselves in supernormal stimuli vivid enough to engage and reengage their attention, and compulsively request or reenact the same stories or songs. They enjoy the freedom of cognitive play without needing to begin from scratch, as infectious cultural models like dragons and monsters or fairies and genies lower the costs of invention and help them master fears or understand the gap between wishes and actualities.

We play with sound and image and story on our own; we learn efficiently from one another; and we enjoy sharing the pleasures of art with others. Engagement in art, as participant or as spectator, has the same self-rewarding nature as play, and since its very goal is to capture and reward attention, it can succeed to the point of compulsiveness. In the days before professionalized art and mechanized reproduction, we served as one another's social entertainment systems, and if we can now pick up an iPod or a novel, we still do so to enjoy engaging with other minds. And despite video and sound recording we still respond more intensely if we form part of a large audience that listens, claps, sings, sways, dances, laughs, or cries together.

Each of the main modes of art builds social coordination in its own way from the foundation laid by protoconversation. Just as the impulses within a single nervous system must mesh with millisecond accuracy if a creature is to move smoothly toward its goal, so creatures working in close cooperation need split-second timing and flawless synchrony to pursue joint aims efficiently.[22] But organisms with open minds can attend at any moment to a vast number of features of the external world or their own internal states. In order to prepare for action toward common goals, individuals need to synchronize actions and check intentions. Close cooperators like monogamous gibbons or grebes signal back and forth in the patterns of arboreal song or aquatic dance to confirm their joint purpose. In the same way in the time-based arts of human music and dance, we synchronize feeling and movement, learn how to coordinate in time and tone, and draw comfort and strength from our physical and emotional attunement.[23] In visual art the architecture, costume, textiles, pottery, sculpture, and iconographic traditions of a group—not to mention body adornments like hairstyles, scarification, and tattoos—serve as an omnipresent reinforcement of shared norms.[24] And through pretend play and fiction we learn to try out the positions of others, we attune ourselves spontaneously to the shifting emotions of an unfolding story, and we encounter story after story that embodies prosocial values in memorable, emotionally compelling images, actions, and outcomes (for more detail, see Part 3).

Art, even if it diverts energy from immediate survival or reproductive needs, can improve cooperation within a group enough for the group to compete successfully against others with less inclination to art. We should think in the first place not of art galleries or concert halls (though these too raise community confidence and lower alienation), but of chants, drums, dance, body-markings, costumes, banners, and the like.[25]* But art also simply fosters more intense sociality within the group, before it enters into competition or even active conflict with other groups. Individuals more motivated to catch the attention and stir the responses of others through carefully designed appeals to shared preferences, and individuals more motivated to respond to these appeals, are more likely to *want* and to *be able* to form more tightly coordinated and therefore more successful groups.

In the case of chimpanzees celebrating community through excited cries or matching movements, we recognize both analogy and homology. Like us, our closest evolutionary relatives can also engage in flexible cooperation against others and derive a rich emotional response from harmonizing attention among themselves through pattern and rhythm, chant and dance. Human chant and dance still fire up group spirit. Historian William McNeill, who himself recalls the "sense of pervasive well-being" that he experienced in the army drillyard in 1941—"a strange sense of personal enlargement; a sort of

---

* David La Chapelle's 2005 documentary, *Rize,* on the South Central Los Angeles dance forms known as clowning and krumping, invented in the 1990s, offers a striking recent example of the power of art to bind groups, especially in competition with other groups.

Surma girls, Ethiopia: art as *play* with pattern. From an early age Surma
children imitate their elders' fanciful body-painting in chalk and earth. Best
friends often show their close bonds by painting their faces with identical
designs: art as social cohesion.

swelling out, becoming bigger than life, thanks to participation in
collective ritual"—has shown how human societies "since the beginning of recorded history have used synchronized movement to create harmony and cohesion within groups."[26] The day after September
11, 2001, the U.S. Congress assembled on the steps of the Capitol to
sing "God Bless America." Song and dance "rouse the emotions and
stimulate like nothing else the production of opiates to bring about
states of elation and euphoria" and a swelling sense of kinship with
those around us.[27]

Humans can fine-tune the emotional impact of shared attention
on cooperation. We can draw on explicit shared models, exemplars,
and admonitions, on myth, stories, and ritual reenactments. As
Kwame Anthony Appiah writes, "The *Iliad* and the *Odyssey*, the *Epic
of Gilgamesh*, the *Tale of Genji*, the Ananse stories I grew up with in
Asante, weren't just read or recited; they were discussed, appraised,
referred to in everyday life."[28] Stories not only make norms explicit,

but also invite us to attend, rouse emotions amplified by our social attunement, and solve what economists call the problem of common knowledge (I will do something only if you will, and vice versa: but how do I know you will, and how do I know that you know that I will?) by making us feel that we share these values and react in much the same way.[29] Humans who play together stay together, like the Didinga women of Uganda singing as they work—"We mold a pot as our mothers did"[30]—and affirming and reinforcing, in the pleasure of song, their commitment to their tribe and tradition.

Stephen Jay Gould reported feeling overwhelmed when he sang in a full performance of Berlioz's *Requiem.* Conceding that his reaction could conceivably be explained by neurobiology or sociobiology, he insisted that neither could ever capture "anything of importance about the meaning of that experience."[31] An objective explanation necessarily differs from a subjective sensation, but surely it can only *add* to rather than detract from the emotional impact of the experience to view it in the light of the ecstatic carnival choruses of chimpanzees, frenziedly hooting, drumming, and running or swinging about together—or in the steady glow of our own need for shared attention, from the moment we are born, and the emotional charge that nature has therefore built, on top of social sympathy in other mammals, into human shared attention.

SHAPING ATTENTION: ART AND STATUS

Sharing attention makes all the difference to the *social* consequences of art, but it also opens new routes for *individual* consequences. Individuals can benefit from social life both by belonging to successful groups and by maintaining or raising their status or prestige within their groups.[32] The social advantages of art involve both: getting along (improved cooperation, and therefore participation in more successful groups) *and* getting ahead (improved status within one's own group).[33]

Our human dependence on cooperation does not eliminate competition. Cooperation allows many benefits that we cannot obtain individually, but we will then compete for those benefits: indeed, as Darwin pointed out, "competition should be most severe between allied forms, which fill nearly the same place in the economy of nature."[34]

Social hierarchy relaxes the tension between cooperation and competition by reducing conflict over precedence—expensive in terms of time, energy, and injury. Since after a hierarchy has been established those of higher status have better access to resources and hence usually enjoy greater reproductive success, the desire for status often intensifies over evolutionary time.[35] Humans naturally pursue status with ferocity: we all relentlessly, if unconsciously, try to raise our own standing by impressing peers, and naturally, if unconsciously, evaluate others in terms of their standing.[36]

Yet because no one prefers inferior positions, we often cooperate to resist hierarchy. Hunter-gatherer societies manage this well, since without agriculture no one can easily accumulate resources or concentrate power. All known hunter-gatherer societies remain more or less egalitarian, not because their members lack the desire for higher status, but because none of them wants lower status and because they can act together to ensure that no one else establishes ascendancy. Although they notice differences in strength and skills, they use ridicule, ostracism, and even expulsion to thwart any individual's attempt to earn special treatment for special qualities, in a strategy that anthropologists call reverse dominance.[37]

But in societies with agriculture, surpluses can be hoarded and disparities grow. Status differences can deepen to produce chiefdoms, kingdoms, and empires, until people slowly discover new routes to reverse dominance—riots, revolutions, representative government— in order to resist the steepest hierarchies. These routes tend to prove serpentine, for where resources can be accumulated, even attempts at ruthlessly redistributing them still result in status differences. Status continues to matter more to humans than to many other species.

In most species, status correlates simply with size, but in primates it can be attained in other ways, especially through the capacity to earn social capital, through grooming, for instance, or through the ability to interpret and relate to others and to cultivate alliances, especially with those of higher status or prestige in socially valued qualities.[38] De Waal reports that the chimpanzee Dandy, when he was the youngest male in the Arnhem chimpanzee colony, did not always gain access to food if housed at night with the other adult males. The others would threaten him away.

After a few months the keeper reported that in the twenty minutes or so between entering the cage and feeding time, Dandy was always unusually playful and often engaged the whole male band in play. Arriving with food, the keeper would find them romping around, piling straw on each other and "laughing" [hoarse guttural sounds associated with play]. In this relaxed atmosphere Dandy would eat undisturbed, side by side with the others.[39]

This example, incidentally, aptly answers Miller's contention that social selection has negligible force compared with sexual selection.[40]

The more dominant a primate, the more attention others direct toward him or her. Primatologist Michael Chance recognized that subordinates pay disproportionate attention to dominants, glancing at them far more than the dominants at the subordinates. He defined dominance in terms of being the *focus of attention* of subordinates, and proposed that the social organization of attention has been a crucial factor in human evolution.[41] He observed that hierarchy establishes itself rapidly among children, whose status can be ranked accurately according to the frequency with which they are looked at by three other children simultaneously.[42] The technology has changed, and the degree of political power, but the special attention earned by Elizabeth II or Bill Clinton or Bill Gates is little different from that earned by Elizabeth I or Ramses I or II.

We all seek attention as a mark of the acceptance, respect, and even status it betokens. But attention is at a premium. Each of us needs to attend to what matters for us now. The emotions sensitively track changes in our environment, awarding a positive or negative valence to events (the smell of food or smoke, say), overriding whatever we might be doing and ordering us to "pay attention now!"[43] We do not want to have our attention diverted by others unless the distraction proves its worth.

So not only do we seek to win others' attention; we also resist our own being commandeered by others without good reason. A chimpanzee uncertain that it can gain rank through force or threat "can often improve its status by other attention-getting devices—doing tricks and 'showing off.' Children do the same thing, often accompanied by cries of 'Look at me.'"[44] But children also soon learn about others' emotional resistance to the undeserved usurpation of attention, and our consequent dislike of showoffs or bores.

In conversation we do not typically offer information to others in the hope that they will eventually offer us some in return. Instead, we often seek status through what we say, according to its relevance and value to others. We may impart information in conversation less as a service than in order to be accorded status by listeners; we "reward one another in the currency of status."[45]

In spontaneous conversation, we earn acceptance, respect, and status for the relevance to the situation of what we say. Often we may do little more than establish or maintain polite sociality or relaxed camaraderie. In art, by contrast, we work at forms specially elaborated to hold attention in ways that appeal enough to override the criterion of relevance. No musician or writer can know just who will hear his potential hit tune or read her blockbuster, or where. Art's special capacity to earn and hold attention *despite* the diverse situations of audiences makes it a potent means of earning the currency of status. Michelangelo had commissions from popes and princes, and died with as much gold under his bed as in a duke's stronghold.[46] Shakespeare was able to buy one of the largest houses in Stratford and a family coat of arms. Both attracted attention in their day and, four or five centuries later, earn the attention of millions every day. Only religious leaders like Christ, Buddha, and Mohammed keep a stronger and longer hold on more minds. And in more mundane currency, J. K. Rowling has captured the imagination of enough people in ten years of writing fiction to earn more than Queen Elizabeth II.

We crave acceptance and, if possible, respect, prestige, and status because of the difference they can make. Since the attention art can command offers both a first payout of status and a base for later dividends, artists can be strongly driven by the desire for wide, high, or long-lasting attention. As H. G. Wells observed, "A mad millionaire who commissioned masterpieces to burn would find it impossible to buy them. Scarcely any artist will hesitate in the choice between money and attention."[47] The tradition from Horace's "Exegi monu-mentum aere perennius" (I have erected a monument more durable than brass) through Shakespeare's "Not marble nor the gilded mon-uments / Shall outlive my powerful rhyme" to Pushkin's "Ya pamyat-nik sebe vozdvig nerukotvorniy" (I raised a monument to myself not made by hand) testifies to artists' craving for attention and status not merely during their lifetime, but for long after, a kind of earthly im-mortality. If seeking posthumous canonization seems unrelated to any possible biological advantage, it is "no more surprising that we crave respect when there is no reward than that we should have sex when it won't lead to babies."[48] In fact artists may earn a very con-crete biological reward: if they continue to reap respect even after death, their descendants, and hence their own genes, can benefit di-rectly from the enduring prestige of their art.

If art can engage the attention of the artist, its appeal to our com-mon appetite for pattern may also attract the attention of others. But one person's ploy for attention can be rapidly *re*deployed by others, because we all seek the status that attention confers, and because we all have an innate capacity for imitation. One of our main cues for social learning is "Imitate the successful." We also have a strong dis-position to conformism, which enables us to learn from what others have discovered and to operate within a cohesive group: our other key heuristic is "Imitate the most common."[49] A new individual ini-tiative therefore can become first a model, then a fashion, then a tra-dition, and eventually even a jealously enforced norm, especially if it finds itself co-opted by the powerful social cohesive of religion. To this we turn next: to the perplexing nexus of art and religion.

8

FROM TRADITION TO INNOVATION

ADAPTATIONS CAN HAVE multiple functions. Human hands evolved from primate forelimbs that began to be used not only for locomotion, for scurrying up trees or along branches, but also for reaching for fruit and for grooming. Our common ancestors with chimpanzees also used their hands to knuckle-walk, brachiate, and fight. As our immediate ancestors became bipedal, they could no longer easily walk on all fours or swing through trees, but their hands evolved to manipulate objects finely, to wield or throw them as weapons, to create and use complex tools, and to point and signal. Like hands, art has multiple functions, some reflecting its origins in long-established animal behavior like play or social coordination, others leading toward functions hitherto unknown.

Some of the older functions reflect art's once intimate alliance with religion. In Western societies we now tend to think of art in secular terms. Even in the West, that was not the case when cathedrals towered over towns, when stained glass enshrined society's most hallowed stories and illuminated manuscripts augmented the preciousness of sacred scripture. For the distinguished film scholar David Bordwell, preoccupied with a recent art form, the close association of art and religion may at first seem questionable.[1] But for thousands of generations art has not only continually served religion but has also deepened its impact on human minds and societies through religion's power to intensify intragroup cooperation and intergroup competition.[2] We cannot understand the force of art in the past, and in many small-scale societies even its force in the present, without understanding its close relationship to religion and the resultant pressure to conformity. We cannot understand the changing functions of art in more modern times, its increasing association with creativity and innovation rather than with conformity and tradition, without taking note of its past. And without recognizing art's relation to religion we cannot understand why it remains linked with an emotion like elevation—which modern psychology has only begun to take seriously,[3] despite its importance in religion and art—or why, as the West became increasingly secular, religions of art arose in Romanticism at the end of the eighteenth century, in Symbolism at the end of the nineteenth, and in Modernism at the beginning of the twentieth.

ART AND RELIGION

Art's power to strengthen social cohesion, discussed in Chapter 7, accounts for the close connection between art and religion in cultures throughout history and in explanations of art. Many object to applying the Western term "art" to objects and practices in other cultures that see them in a religious rather than an artistic context. And many simply maintain that art grew out of religion.[4]

This claim seems highly unlikely. Although archeological evidence is inevitably patchy, no one has found signs of grave ritual, the earliest plausible hints of religion so far discovered, much older than 90,000 years ago.[5] But ocher seems to have been used for body decoration for 120,000 years or perhaps twice as long, Acheulean hand-axes have been refined beyond practical needs and in impractical display sizes for hundreds of thousands of years, and chanting, drumming, dance, and song, which have analogues among chimpanzees or gibbons, may have formed part of the human repertoire for even longer. Art appears to have long preceded religion.

Indeed those who claim that art derives from religion do not explain how it could do so. If a group of early humans had begun to believe in supernatural forces, why *would* and how *could* they then have invented art to serve the purposes of religion? Why would they have thought of art as a next step, if there had not *already* been ways of embellishing surfaces and altering shapes and producing sounds and movements that elicited a deep response in human eyes and ears? Religion, on the other hand, needs art as a precursor. Without the existence of stories that diverge from the true, without the first fictions, religion could not have arisen. Religion depends on the power of story.

We crave one another's attention, but no one savors stories confined to the banal and expected: we know we would be better off attending to the real world. To merit attention, stories select the striking, unusual characters or events or both. Recent research shows we remember best those stories with characters that violate our categoric expectations, crossing one animal kind with another, or combining human and animal, or separating the psychological from its usual physical constraints.[6] Even now we earn attention in stories by confounding categories, in aliens, mutants, robots, and vampires. And creatures with psychological powers but not limited to normal physical laws—spirits or gods—have been central to story from as far back as we can tell.

We see our own agency as our prototype of cause: we want to move something, and we do.[7] As we will find in more detail in chapters to come, we make an early and lasting distinction between agents and nonagents, between the animate and the inanimate, and we are prone to overattribute agency: it is safer to suppose a bush a bear than the other way around.[8] And because we pay such close attention to one kind of agent, our own species, we have evolved an understanding of false belief—a capacity to see that others, or we ourselves, may conceive a situation differently from the way it really is.[9] Because we understand false belief, because we can appreciate that we might not know the full situation, we seek the whole story; we want an explanation that sees behind the visible. Spiritual agents as unseen causes not only make stories memorable but also appear to promise the extra explanation we have accidentally evolved to seek.

To retrace art and religion, we need to keep in mind a distinction long central to human understanding: the distinction between the physical and the nonphysical—the psychological, in modern terms, or the spiritual, in older ones. In one sense, this distinction is crumbling as science investigates the intricate connection between mind and brain. In another sense, cognitive anthropologists and developmental psychologists now explore it in new ways as a key to human apprehension of the world, even before language. New techniques make it possible to study how infant minds distinguish the ontological domains of inanimate and animate. They bring different expectations to and draw different inferences from the two different domains, called "folk physics" and "folk psychology" in one set of terms, "theory of things" and "theory of mind" in another.[10]

From this and other recent findings, significant implications follow for seeing art in relation to religion and ritual. Art has no immediate physical function, but only an immediate psychical one. A decoration on a bowl does not change the bowl's physical capacity but does change its psychological appeal; a harvest song does not by itself gather crops but alters the attitude of the harvesters; a story does not bring about its own outcome but causes an audience to feel and respond in some ways as if they had witnessed the events.

In the initial and default case, across the world, art affects human beings. But many peoples believe that it will also affect other kinds of beings, unseen spirits or gods, presumed to respond in ways similar enough to *human* spirits that they, too, will be moved—and moved, perhaps, to intervene for, or not intervene against, those who have made the artwork or who accord it respect. Two points need to be stressed here. First, the impact on human beings is there from the first: songs, shapes, and stories are after all *designed* for human attention. Second, the impact on imagined other beings also depends on art, on the prior existence and power both of story and of the spiritual forces it conjures up.

Because *we* are moved by song, images, and stories, we may suppose that these unseen forces will also be moved. Trobriand Islanders, for example, decorate their canoe prows to attract and hold the attention of humans *and* spirits—who they assume perceive as humans do.[11] And because we can envisage the future vividly enough to become anxious about uncertainties, we may be ready to move the unseen spirits to act more in our favor, or less in our disfavor, with the help of the art that so effectively catches *our* attention and stirs *our* response.[12]

Even if we can't see spirits, they may we watching us at any moment. A societywide belief in such supernatural surveillance can help solve the problems of cooperation inherent in any individualized society.[13] According to the anthropologist Malidoma Patrice Somé, elders in his own Dagara culture (West Africa) say that the "real police in the village is Spirit that sees everybody. To do wrong is to insult the spirit realm. Whoever does this is immediately punished by Spirit."[14] As recent research has shown, large-scale cooperation can be fragile in the absence of punishment yet relatively easily established and maintained *with* punishment.[15] A human society unified by religion, especially if it believes in supernatural vigilance and punishment, can therefore usually solve problems of cooperation more readily than another without.[16] Unreal beliefs in unseen forces acting for or against one's group, and for or against oneself insofar as one supports or contravenes the collective good, are much more likely to motivate action than are modestly real beliefs.[17] Since religion has much stronger effects on social cohesion than art alone can produce, and since cohesion amplifies the advantages of sociality, religion tends to appropriate to itself much of the power of earthly art.

If religion strengthens social cooperation, a society will seek to ensure that its members genuinely commit to shared beliefs. One way of doing this can be through what biologists have studied in animals as "costly signaling," behavioral signals that incur high costs. Costly signals can be used in sexual competition but can also have a powerful impact in many species in reinforcing group cooperation.[18] If a signal has low cost (as in the case of a mere display, promise, or claim), it can be easy to fake. High cost in terms of time, effort, or resources can serve as a guarantee of allegiance (only those genuinely committed to the group will be prepared to undertake what the ritual requires), and biological and historical human case studies show that groups that cement their cohesion by costly ritual can outcompete groups without such ritual.[19]

Costly signaling theory alone does not explain why such a costly activity as religious ritual should take an *artistic* form in humans. After all, ritualization of practices with high cost but little sensory appeal—prostration, prayer, recitation, offerings, tithes, fasting, sacrifice, mutilation, pilgrimage—can also serve as cohesive social signals. But ritual with art has several advantages over ritual without. Art may increase the time and energy costs in ritual preparation, and therefore the signal value of the commitment: cathedrals can take a century or more to build. Art promises pleasure, commands attention, stirs the senses and emotions, and arouses pride and awe at the effects produced and the mastery exhibited. The very improbability of artistic practices—the fact that they could hardly be generated by chance—makes art a distinct marker, a contrast to the natural and to other rival groups, and hence in both respects a source of pride.

Art begins in engaging *human* attention but can readily be deployed both to engage the attention of putative spirits and to ensure social cooperation at the human level, whether in the service of the gods or not.[20] Traditional art can be far from nonutilitarian: it can have what seems the highest practical purpose possible, securing both the goodwill of the spiritual world and the group's focus on these powerful unseen agents. And as we have just seen, this "practicality" is not an illusion: case studies confirm that the advantages of social cohesion can easily repay the effort invested in ritual practice and outweigh the disadvantages of belief in nonexistent spirits.[21]

Religion and ritual cannot explain the *origin* of art, but once art began, traditions of artistic elaboration, including ritual, could be co-opted for social cohesion. This could become a powerful sustainer of art and indeed perhaps its main *function,* even in strict evolutionary terms, in many small-scale societies without specialist artists. Indeed, the very power of art to move the spirit—a Dogon sculptor reported that he occasionally created a work "that made everyone who saw it 'stop breathing' for a moment"[22]—makes art natural to associate with religion and ritual. Art has played a central function in human lives not only in itself but also in giving rise to religion and *then* reinforcing religion's power to cement group cohesion by augmenting the impact of ritual.

Yet even in traditional societies art also persists in modes closer to play or trade than ritual. Tribal societies distinguish between inviolable core myths and stories that they call, like the Kuranko of Sierra Leone, "play" *(tan)* stories, made-up stories, which offer room for individual initiative and invention.[23] Especially as societies expand and diversify, as the division of labor spreads, and as other cultural forms develop to solve problems of cooperation in other ways (improved communication, trade, policing, and justice, for instance) and it becomes possible to tolerate a greater range of religious options, art can become professionalized and secularized as well as communalized and spiritualized. At its highest, even secular art may retain religious art's sense of offering not just intense interest but also deep explanation and exaltation, and of drawing upon a spiritual power somehow linking us through our artistic heritage. Or art may remain closer to a less exalted, less spiritualized, more playful form of catching attention, in popular and folk arts and crafts. Or in a highly specialized world it may lead even to avant-garde art, to questioning and debunking traditional values and traditional ways of earning attention.

ART AND CREATIVITY

I have proposed that (1) art begins as solitary and shared patterned cognitive play whose self-rewarding nature reshapes human minds, and that it intensifies its impact by raising (2) the status of individual artists and (3) our general inclination to cooperate closely with one another, with or without religion. Out of these three functions, I further propose, there gradually emerges another major function of art, at odds with the social-cohesive role of traditional art: (4) creativity. That art leads to creativity will seem obvious, almost axiomatic or tautological, to modern sensibilities—though much less so in traditional societies, in which art appears to embody time-honored, god-given, or ancestor-approved forms.[24] But that nature has *evolved* art to *create creativity* will seem anything but obvious.

Creativity may seem to us like a purpose of life and especially of art, but it cannot be a purpose of evolution, merely a surprising consequence of the way evolution's blind logic has worked on Earth. Evolution has no foresight or purpose, it does not aim at creativity or even diversity, and in natural selection no future usefulness counts.[25] *We* may appreciate the scores of thousands of orchid species, but nature could get by without them. *We* may value the wide diversity of styles of European painting in the 1910s—Cubism, Futurism, Vorticism, Constructivism, residual academicism, Impressionism, Fauvism, and so on—but humanity could have survived had they never been invented. Yet evolution spawns not only diversity but also complexity, intelligence, and, in art, at last, creativity.

Steven Pinker observes that despite its competitive basis natural selection does not forbid cooperation, but only makes it a difficult engineering problem.[26] Creativity poses a still more challenging engineering task. Creativity for what? By what criteria?

Most novelty matters little. Every move we make is in some sense new, yet unlikely to be of lasting moment.[27] But the steady logic of evolution allows for the *accumulation* of *successful* novelty. A system that generates blind variations on existing forms, tests them against their environment, generates further variations from those that have survived so far, and continues the process indefinitely, gradually accumulates lineages of progressively more successful design. Note that designs need not *remain* successful, since even apart from unpredictable catastrophe the new variations in each round, in art as in nature, themselves alter the environment in which all forms contend.

Evolution itself has evolved in numerous ways.[28] Not least, it has given rise to subsidiary processes of "blind variation and selective retention" or subcycles of "generate-test-regenerate," especially valuable when the environment proves predictably unpredictable, changing in ways too fast to be tracked by selection on genes.[29] One such second-order process is the human immune system, built to cope with the swift evolution of pathogens that can reproduce through many generations in a day. Another is the overproduction and selective retention of synaptic connections within the growing human brain, a process that allows it to be fine-tuned by the unforeseeable particulars of individual experience.[30] Such systems of evolutionary processes built by evolution have been called "Darwin machines."[31] They cannot find *the* right answer beforehand: there is no single right form of life, no single right antibody, no single right synaptic link, no single right move.[32] But they can generate possibilities that the environment tests. Survivors generate new variations and face new rounds of tests, so that even without preplanning, success accumulates.

Art constitutes another Darwin machine, an evolutionary subsystem effectively designed, in this case, for creativity. Art shows evidence of good design to generate and accumulate successful novelty:

1. "Darwin machines" or evolutionary processes depend on the blind generation of variation. Randomness, nature's way of exploring new possibilities, seems an intrinsic part of brain function.[33]

2. But without selective retention, randomness alone could not generate creativity that accumulates in force: as in dreams, a cascade of new ideas would take and lose shape almost without trace. Art involves not just private ideas but patterned external forms, sound, surface, shape, story, durable or at least replicable, like the patterns of melodies and rhymes that make music and verse memorable and transmissible. Pure imagination, on the other hand, alters unstably and irretrievably as brain activation spreads.

3. Because art appeals to our cognitive preferences for pattern, it is self-motivating: we carry innate incentives to engage in artistic activity.

4. A testing as well as a generating mechanism operates within art-makers' minds. The low cost of testing increases our opportunities to refine what we do through online feedback. And unlike all or almost all other species, we have a capacity for self-monitoring and rehearsal in order to achieve desired results.[34]

5. Since we pay close attention to others, art can earn interest and esteem and hence offers an incentive to override the stopping routines in other goal-directed activities. You do not continue to kill your prey once it is dead, but if you know that your work can impress an audience you may keep honing your Acheulean hand-ax or your sonnet and seeing new possibilities as you do so.

6. Because art involves external forms, the testing mechanism operates also in the minds of *other* humans, in terms of *their* interest. Attention provides the selective mechanism of art. If a work of art fails to earn attention, it dies. If it succeeds, it can last even for millennia. Criticism can smart, but art is selected against much less brutally than attempts at creativity in the wider, nonhuman world, from making new potions to launching new projectiles (more people were killed designing the V-2 than in the rocket's attacks). Nevertheless external audiences

provide art with an efficient testing mechanism undistorted by the self-deceiving biases of the artist.

7. Unlike other species, we can imitate closely and therefore follow established forms. Crucially, we *need* to imitate in order to innovate. Building on what came before underlies all creativity, in biology and culture. Starting again from scratch wastes too much accumulated effort: far better to recombine existing design successes. Even adopting a high mutation rate, changing many features at once, would rapidly dismantle successful design.[35]

8. Established artistic forms reduce invention costs by posing well-defined problems and offering partial solutions.

9. The existence of established forms also reduces audience attention and comprehension costs, since familiar forms carry both an open invitation ("Come and join the play of art") and a formal or generic set of instructions, a prefocusing of expectations ("Listen to this song/sonnet/sonata/story").[36]

10. In a system designed to secure attention, habituation (the loss of attention through the persistence or repetition of a stimulus) encourages innovation.[37] Since repeating exactly the same thing over and over again guarantees it will lose its impact, art faces a consistent pressure for novelty. Over generations traditions hone their attention-grabbing power (the symphony, the murder mystery, the computer game), but new successes raise the bar for still newer entrants and reward still further novelty.

11. We appreciate even minor variations within established forms as worthy of attention and response. With our senses highly tuned to basic patterns, we enjoy repetitions and variations on a theme in art as in play.[38]

For all these reasons, art efficiently generates cumulative novelty. A Lamarckian or instructionist model of evolution envisions the environment as somehow instructing organisms to fit it, as if there were some foreknown optimum. A Darwinian system, by contrast, remains open, unpredictable, and free. It cannot presuppose a best option, an ideal fit, a goal that can be precisely determined beforehand. It gropes in the unknown in the vastness of possibility space. Organisms blindly produce—from within themselves, and not from the environment[39]—new possibilities from which they select and on which they build. In art understood as a Darwin machine, works are not somehow created to fit the cultural environment. Instead they are generated, unpredictably, in the minds or actions of artists, and selected first by them in accordance with their intuitions about their social world, and then by this world itself.[40]

Art is superbly designed for creativity, but not for generating materially *useful* new moves. If cognitive stimulation, social cohesion, and individual status were not already in place as the start of art, then a "useless" creativity in itself would be insufficient to establish art as a behavior. But with these individual and group functions already operative, then art's additional function of creativity can also develop gradually, especially as traditions compound and lineages lengthen.

Individual artistic moves have slight practical consequence in comparison with discoveries in technology or science. But they exist robustly enough to extend the adaptive functions of art. And art's creativity can have additional consequences that over time become significant even in solely material terms.

Art has long provided incentives for social exchange. Ocher employed in the first body decoration was traded over 120,000 years ago. Early long-distance trade, when transport was still difficult, as in the Natufian culture of the Near East shortly before 10,000 years ago, was in small prestige items with a high ratio of value to mass, like jewelry, tapestry, and other "splendid and trifling" goods, in Edward Gibbon's phrase about later antiquity.[41] The modern book, music, film, and recreational software industries drive vast economies. Art has repeatedly provided incentives for discovery, in materials, processes, products: in weaving and pottery, for instance, in the invention of new techniques, looms, furnaces, dyes, glazes, and textiles. Art can encourage creative habits of mind like drawing or model-building that help technological discovery, and itself benefits from technology (fiction from print, drama from film, music from recording and electronic amplification, cartooning from computer animation) in a positive feedback loop.[42]

For us, artistic creativity offers a good in itself. The diversity of forms of music, design, and story across cultures and times, and the achievement of the most outstanding works—those that take the breath away, in the Dogon sculptor's terms, or make the hairs on the spine stand on end, in A. E. Housman's—amply justify themselves. And art intensifies creativity throughout human culture.

Art develops in us habits of imaginative exploration, so that we take the world as not closed and given, but open and to be shaped on our own terms. In and through art, we readily turn the actual around within the much larger space of the possible, the conditional, and the impossible. Art opens up new dimensions of possibility space and populates it with imaginative particulars. To revise Lewis Mumford: "If man had not encountered dragons and hippogriffs in stories, he might not have conceived of the atom."[43]

By fostering our inclination to think about possible worlds, art allows us to see the actual world from new vantage points. It therefore enables science, both in general, as Robert Root-Bernstein has persuasively argued ("artistic thinking produces possibilities that scientists can evaluate for efficacy in the here and now"),[44] and in many specific instances. Among the instances Root-Bernstein cites in recent science are anamorphosis in art—distorting a shape in painting so that it can be readily seen from only one point, most famously in the skull in Hans Holbein the Younger's *The Ambassadors*—which has helped biologists D'Arcy Thompson in *On Growth and Form* and Julian Huxley in *Problems of Relative Growth* understand evolutionary and embryological processes as anamorphic distortions; pointillism, in Georges Seurat and others, which sowed the seeds for computer pixillation; and the technique of false-coloring objects, which began with Émile Bernard, Paul Gauguin, and the Fauvists at the end of the nineteenth and the start of the twentieth centuries, and which scientists now use "to emphasize inobvious elements of data."[45]

Art builds our confidence, at the individual and the group levels, in shaping our own destinies. In art we can choose simply what moves us. Slumdwellers or impoverished sharecroppers can affirm with a picture on the wall that they can shape at least a part of their lives to please.[46] In the dark days of Hitler, psychologist Wolfgang Köhler wrote a newspaper article against the purges of the universities. He and friends spent the night after its publication playing chamber music together, in an avowal of ultimate value and order and control, as they waited for the fatal knock on the door—which, luckily, did not come.[47] The sense of individual and collective mastery that art provides—in composition, performance, and the mere comprehension of its complex patterns—stands at the opposite pole of experience from the will-sapping dysfunctionality of learned helplessness, the belief that we have no control over our situation.[48]

By refining and strengthening our sociality, by making us readier to use the resources of the imagination, and by raising our confidence in shaping life on our own terms, art fundamentally alters our relation to our world. The survival consequences may be difficult to tabulate, but they are profound. We have long felt that art matters to us. It does, objectively as well as subjectively. By focusing our attention away from the given to a world of shared, humanly created possibility, art makes all the difference.

PART 3

EVOLUTION AND FICTION

"Suppose we get the Professor to tell us a story."

Bruno adopted the idea with enthusiasm. "*Please* do," he cried eagerly. "Sumfin about tigers—and bumble-bees—and robin red-breasts, oo knows!"

"Why should you always have *live* things in stories?" said the Professor. "Why don't you have events, or circumstances?"

"Oh, *please* invent a story like that!" cried Bruno.

The Professor began fluently enough. "Once a coincidence was taking a walk with a little accident, and they met an explanation—a *very* old explanation—so old that it was quite doubled up, and looked more like a conundrum—" He broke off suddenly.

"*Please* go on!" both children exclaimed.

The Professor made a candid confession. "It's a very difficult sort to invent, I find. Suppose Bruno tells one first."

Bruno was only too happy to adopt the suggestion.

"Once there were a Pig, and a Accordion, and two jars of Orange-marmalade—"

Lewis Carroll, *Sylvia and Bruno Concluded* (1893)

9

ART, NARRATIVE, FICTION

WHY DO WE SPEND SO much of our time telling one another stories that neither side believes?

To explain fiction fully we cannot merely explain narrative. First, because narrative depends heavily on understanding events, and event sequences, before we cast them into narrative. As we will see, we share much of that understanding with other species, but it takes important new forms in humans.

Second, because narrative requires our unique capacity for metarepresentation: not only to make and understand representations, but also to understand them *as* representations.[1] This develops in children, without training, between their second and fifth years. Our representations are not confined to language: they may involve action and objects or images and music rather than, or as well as, language.

Third, because our predisposition for fiction seems at first glance counterproductive in biological terms, unlike our capacity for true narration. In any natural data-gathering system like the human mind we might expect an "appetite for the true." Narrative extends the kinds, depth, and value of true information we can have at our disposal, and we reap obvious advantages from informing one another about the activities of others in the group, about recent challenges or opportunities, or about old solutions to newly recurring conditions. Gossip and history make immediate biological sense, but why do we find fiction just as compelling? The "appetite for the true" model "spectacularly fails to predict large components of the human appetite for information."[2] In modern societies, most people prefer novels to textbooks, fiction films to documentaries. Not that we do not care to distinguish true information from false: in communication professing truthfulness, we remain vigilant against being misled. But we often simply prefer fiction to fact.

Fiction, like art in general, can be explained in terms of cognitive play with pattern—in this case, with patterns of social information—and in terms of the unique importance of human shared attention. Music and visual art each appeal directly to one of our main senses, and therefore have antecedents in other animals, in the song of humpback whales or nightingales, in the bodily or architectural displays of birds of paradise or bowerbirds. But since fiction appeals to our craving for higher-order information, it has slighter precursors in other species. Even more than other social species, we depend on information about others' capacities, dispositions, intentions, actions, and reactions. Such "strategic information"[3] catches our attention so forcefully that fiction can hold our interest, unlike almost anything else, for hours at a stretch.

A biocultural approach to fiction will focus especially on the shared understandings that make us able and eager to tell and listen to stories. It therefore operates upon premises different from other approaches current in literary studies. It suggests that the most fruitful research program will consist not in looking for codes in the language of narrative, in the structuralist mode, or in analyzing how ideology or the contestation of ideologies determines narrative, in the poststructuralist mode, or in erecting a taxonomy of possibilities (of communicative positions, temporal relations, or character roles), in narratology. It will focus instead on the deep, specieswide competences that enable us, *first,* to understand events in ways similar to but richer than those of other animals (Chapter 10); *second,* to form and follow representations of events, or narratives (Chapter 11); *third,* to invent stories as a kind of cognitive play, in this case with social information (Chapter 12). It will show how these capacities and inclinations build up slowly from simpler predecessors but alter markedly as they pass through the major evolutionary transition of human ultrasociality.

Despite the "linguistic turn" of thought late in the twentieth century, narrative does not depend on language. It can be expressed through mime, dance, wordless picture books, or movies. And although such narratives are often predicated on or elaborated through language, they need not be. Languageless adults in Mexico—deaf, never taught sign language, living together on the fringes of society —mime narratives for one another.[4] They can do so only because even without language humans share an ability to understand and represent events in complex ways.

Narrative becomes easier with language, and we can certainly test people's comprehension of and response to narratives most easily in verbal form. But over recent decades comparative and developmental psychologists have found ways to test event comprehension in animals and infants.

We can understand events through competences difficult to detect because they usually work so efficiently and rapidly that we remain unaware of them. Our responses seem automatic and immediate, as if we read events straight off the evidence. In fact we understand so much only through the specialized efforts of elaborate but largely unconscious inference systems. It has taken the difficulty of programming computers to construe natural-language narratives and of analyzing how human infants and other animals understand events *before* language to clarify just how complicated the task is and yet how efficiently evolved minds manage.[5]

Throughout Part 3 I stress that neither event nor narrative comprehension can be primarily a matter of convention.[6] Such conventions as we find in narrative tend to reflect the regularities most important to human lives and minds. We are not *taught* narrative. Rather, narrative reflects our mode of understanding events, which appears largely—but with crucial exceptions—to be a generally mammalian mode of understanding. The many culturally local conventions of human behavior and explanation tend to be adjustable parameters within common cognitive systems.

Only after establishing the biological roots of narrative and fiction can we take the next step and consider the functions of fiction, to which we turn in Chapter 13.

10

UNDERSTANDING AND RECALLING EVENTS

CAN WE UNDERSTAND EVENTS from birth? Do we learn to do so along pathways prepared for by evolution? Or do we learn only from scratch, piecemeal, through individual experience? Can we understand events only once we acquire language and local narrative conventions?

Under controlled conditions, as rated by independent observers, experimenters have been able to induce babies as early as forty-five minutes after birth to imitate simple actions like opening their mouths or eyes wide and closing them; sticking out their tongues; or making *aah* sounds. Human infants can see and act on such simple actions performed by a human model (but not a robot) almost from birth. Within two days, they can imitate surprised, sad, and happy faces.[1] Such responses require perceptual, motor, emotional, and motivational areas in newborns' brains somehow to coordinate, even within bodies that can barely act, in a world they barely know. We come preequipped to attend to, relate to, take our lead from, and understand other humans and, along with other animals, to understand a world new to the individual but long familiar to the species.

UNDERSTANDING OBSERVED EVENTS: INTUITIVE ONTOLOGIES

Beginning with Locke in the seventeenth century and Hume in the eighteenth, associationist psychology has considered minds to be equally open to any potential associations between whatever the senses perceive. These early investigators argued that repeated conjunctions of phenomena create and strengthen mental connections that suffice for us to learn to understand our world. Much later, behaviorist psychologists in the first two-thirds of the twentieth century assumed that associating any two percepts, like a bell and food, could "condition" a response like the salivation famously demonstrated in dogs by Pavlov.

But a series of experiments by John Garcia and colleagues in the mid-1960s showed that rats could develop an aversion to drinking harmless colored water in a *single* exposure, if the experimenters induced nausea afterward, but would *not* associate drinking colored water and electric shock after *any* number of tries.[2] Behaviorists sneered: "No more likely than birdshit in a cuckoo clock."[3] But despite the scornful response, it makes sound evolutionary sense that opportunistic omnivores like rats should associate new foods and negative bodily reactions on a first encounter, even if the ill effects begin hours after eating: such a dedicated learning channel lets them link a food with danger without needing to run the perhaps fatal risk of "repeated associations." But since it makes no natural sense to pair drinking water with electric shock, rats cannot associate *these* factors even after endless repetitions. Animals cannot build up associations of any kind in the course of their exposure to their world. Instead minds within a given lineage evolve to learn regularities relevant to *their* mode of life.

Evolution builds many specific learning tracks into minds, preparing them to expect certain kinds of situations and to understand them by making rich inferences from particular information patterns. Objects will be important to many species, so senses and minds recognize the edges of objects (many mammalian visual systems have specialized edge detectors) or deduce the whole of an object from seeing it at an angle or in part. Minds are "bundles of expectations, preferences, ways of inference," "geared to the greatest possible cognitive result for the smallest processing effort."[4] And because evolution has shaped minds to leap to complex conclusions from slight or ambiguous information, we can understand even others' unpredictable behavior.

Event comprehension builds up through a nested hierarchy of cognitive processes, as information that satisfies one pattern meets information satisfying another at a neural convergence zone, and information satisfying patterns at this level meets other higher-level information at convergence regions.[5] The evidence seems strongly against explaining this system merely through repeated individual associations. Even with massive mental computing power, unconstrained mental associations would not suffice to construct a system of event comprehension in the course of an individual's development or to process event comprehension in real time. A mind equally open to all possible links would have no way of deciding what information would prove useful: the color of people's eyes or their socks? the shape of their noses or their shoes? the number of syllables in their name?[6]

Minds exist to predict what will happen next. They mine the present for clues they can refine with help from the past—the evolutionary past of the species, the cultural past of the population, and the experiential past of the individual—to anticipate the immediate future and guide action.[7] To understand events as they happen, with limited time, knowledge, and computational power, minds have evolved to register the regularities pertinent to particular species and to infer according to rough-and-ready heuristics.[8]

We cannot study such inference systems by introspection, because like many mental processes they are so rapid, automatic, and reliable that we think we simply read ready-made facts and do not realize how much we infer. Wittgenstein wrote: "'We *see* emotion' . . . We do not see facial contortions and *make the inference* that he is feeling joy, grief, boredom. We describe a face immediately as sad, radiant, bored, even when we are unable to give any other description of the features."[9] He was correct that we need not make *conscious* inferences; but our evolved pattern-matching neural processing does the inferring for us, rapidly and automatically linking perceptual, motor, and emotional networks in the brain to provide immediate recognition and evaluation. Even in higher-level processing, when tested on our reactions as readers, we find it hard to distinguish between what we have actually read and what we have only inferred.[10]

We understand events especially through our *intuitive ontologies,* our implicit theories of different classes of things. Evidence for intuitive ontologies has converged from several different fields: from ascertaining what nonhuman animals, infants, and children can understand, from experiments on human adults, neuroimaging of brain activity in humans and other animals, and clinical studies of brain dysfunctions.[11] Studies of infant cognition, for instance, adopt new procedures that habituate children to particular kinds of stimuli. Habituation upon repetition occurs in any organism with a nervous system: *any* stimulus gradually ceases to arouse.[12] But when something strikes an infant as new or anomalous, its interest revives: it looks longer at a scene or a screen or sucks harder on a teat as it listens to sounds. This revived interest indicates the infant's wordless categories or expectations: only something it considers sufficiently different to belong to a new category will cause it to respond more intently. Such procedures allow developmental psychologists to discover how infants understand events.[13]

We share with other animals a rudimentary sense of space, of up and down versus across, built into our sense of balance and kinesthesia (our monitoring of our own positions and movement). Our understanding becomes finer-grained as we encounter not just space but objects in space. Infants apprehend objects very early, although their understanding continues to develop in something akin to theorylike stages.[14]

Experiments with infants indicate that they have more or less reliably developing "theories" of objects, which begin, long before language, with expectations about object trajectories, then object coherence and cohesion, then the causal force of object support and contact, then object persistence, concealment, containment, occlusion, protuberance, and so on.[15] This early intuitive understanding of physical objects depends not on manual skills, as the pioneering developmental psychologist Jean Piaget thought, but on more basic cognitive input, at a succession of processing levels: from, for instance, lateral inhibition in the retina, to edge detection in visual processing, to the comprehension of objects *as* objects.[16]

Clearly, we share with many other creatures this much of folk physics, not vastly different, even if unformulated, from the intuitions that shaped Aristotelian physics: rules of thumb accurate enough to guide the actions of mid-sized creatures in a world of mid-sized and middle-distance objects.[17] Developmental psychologists estimate that infants achieve an adult level of understanding of all basic physics by eight months.[18]

Unless they already know its name, both children and language-trained chimpanzees will assign a new word like "apple" to an indicated object rather than to its color or other features.[19] By about eighteen months children have become fascinated with naming objects.[20] But even before they can name them they can classify them according to their developing *theory of kinds*. Indeed all warmblooded vertebrates apparently categorize objects at a variety of levels.[21]

Localized neurological deficits in adults indicate that humans make major ontological distinctions among conspecifics (other humans), animals, plants, and artifacts.[22] A certain kind of neurological patient may, for instance, have severe difficulties in naming living things, but a relatively well-preserved ability to name all other objects.[23] Even children classify living things as such on the basis that they derive from other living things,[24] and artifacts on the basis that they are designed for a particular function. We distinguish animate objects from inanimate when we see that they move with nonrigid self-propelled motion and can interact without contact.[25] We automatically draw different inferences from each category. Color matters first for a child or a monkey looking at food, but shape counts first in a tool.[26] We will seek the purpose of an artifact, but we will expect a living thing to have the essence of its kind; we will look for the head or face of an animal but not of a plant; and we will process face-recognition cues in humans but not in other animals.

We make the first crucial distinction between animate and inanimate far below the level of conscious awareness: neurons have been found that respond to biological but not to nonbiological motion.[27] And we understand the distinction consciously long before we can express it linguistically. Infants recognize it, through cues of self-propelled motion, by at least three months old. By at least twelve months, they have different expectations of, and make different causal inferences about, animate or inanimate things.[28]

Long before anything like narrative became possible, animals have had to be aware of other agents as volatile and potentially urgent threats or opportunities. Agency itself therefore catches attention.[29] Although most of us now see far more cars than we do animals, and face more danger from them, and have been warned since childhood about the risks they pose, experiments show that we detect change of position far more quickly and accurately in animals than in cars.[30] Long before narrative, too, animals have needed to distinguish one organism from another at first sign: by smell, sound, or especially, in the human case, by sight. Hence human children have an innate fascination with identifying animals of all kinds, out of all proportion to the likelihood of their encountering alligators or zebras.[31]

We humans overdetect agency,[32] since it is safer to mistake a twig for a snake than vice versa. And we will interpret something as story if we can. Babies and adults alike cannot help seeing a sequence of moving dots in terms of animate causality.[33] In a famous 1944 experiment students were shown a short silent black-and-white film and asked to write down afterward what they saw. The film shows three shapes, two dark triangles and one dark circle, moving about a screen featureless except for a stationary rectangle with one side partially opened out. Without being cued by the experimenters, all the students except one reported in terms not of a series of spatial shifts but of individuals in conflict, usually two men competing for the attentions of a woman, who found one of them too aggressive and favored the other. Only one student described the film in terms of the displacements of different shapes in a plane.[34] More recent studies have produced similar results.[35] Our minds naturally mature to interpret events in the most powerful way we can, automatically sending information that satisfies the right patterns—in this case, the complicated patterns of motion, relative to one another, of shapes with nothing except their movement to suggest agency—to higher data-processing levels and the richer inferences they allow.

Beyond rough-and-ready heuristics for agency and common patterns of movement like flight, pursuit, fight, or mutual approach, animals make much subtler discriminations among their own species.[36] Our human distinctions are marvelously subtle—as are those of dogs, dolphins, or chimpanzees within *their* species, in ways *we* cannot easily detect, despite the various links between these species and ours. Many kinds of behavior become *ritualized*, standardized, exaggerated, and in intensified contrast to other kinds of behaviors. Children have an early interest in mastering human "rituals" as sim-

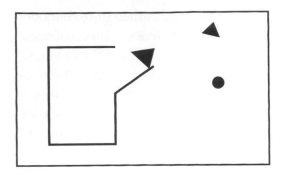

A frame from Fritz Heider and Marianne Simmel film, 1944, showing a geometric shape "trying to escape" from a space.

ple as greetings and farewells, comforting, or play. In the famously closely-studied case of Emily, "the first and primary function of the narrative form" in her crib monologues when they began at twenty-one months "was to establish the canonical events of the child's world from which other narrative types emerged . . . [and] search for stability in the dynamic of experience."[37] More elaborate standardized behaviors by which we interpret recurrent situations have been studied in artificial intelligence in terms of *scripts*, standard expectations of standard situations, like going to a restaurant.[38]

Many animals understand others in terms of goals and desires, the beginnings of a theory of mind. David and Ann Premack's chimpanzee Sarah shows that she can understand goals: she correctly selects photographs that provide the solutions to problems a videotaped actor tries to solve.[39] Human infants at just under a year undergo a first cognitive revolution when they understand the goal-directedness of others' actions,[40] though they cannot yet understand desire. By two, children have a "surprisingly sophisticated" grasp of others' desires, though not yet of their beliefs.[41]

Empathy arises from recognizing others' goals and desires, a capacity necessary in an individualized social species. Darwin noted that "many animals certainly sympathize with each other's distress or danger,"[42] as modern laboratory research confirms. Even rats and monkeys have a default response to others of their kind, not just their kin, whom they see in distress.[43] As species become more flexibly social, such sympathy deepens: although monkeys appear never to offer consolation to victims of aggression, chimpanzees readily do.[44] Human children "by one year of age . . . spontaneously comfort people in distress."[45] Distress at the sight of another's pain is "an impulse over which we exert no control: it grabs us instantaneously, like a reflex."[46]

Social animals readily categorize other conspecifics according to age, sex, and class or status, but in individualized societies they must also be able to identify *individuals.* Biologists had assumed that invertebrates could not recognize individuals, but even mantis shrimps have now been discovered to do so.[47]

The more flexibly social a species, the more individuals need to discriminate, to know others' position in relation to themselves or even to third parties. Species that cooperate through reciprocal altruism also need a good memory for who owes them and for those whom they owe favors. In a species as dependent on sight as ours, we recognize other individuals through a dedicated face-recognition subsystem in visual processing. Not only can we recognize and remember faces for many years, but emotion saturates recognition. As an automatic part of sensory processing the amygdala, the brain's emotional router, attaches an affective weighting to faces or voices important enough for us to recall later. We keep mental "files" on individuals without confusing them,[48] in ways that make it easy for us—and extremely hard for computers—to track individuals in life or in literature. Except in the most careless language, we have no problem following the antecedent of a pronoun; even in the most careful writing, computers programmed to analyze discourse easily lose track.

Animals not only recognize other individuals but discriminate among them. They evaluate personalities and prowess from past behavior and use their own past emotional response to individuals as a guide to ongoing associations and interactions. Those individuals who could more accurately assess others as associates or opponents would on average leave more descendants.[49] Primates such as baboons and chimpanzees discriminate among individuals on the basis of character and adjust their behavior accordingly—even to the point of female baboons' accepting males as partners according to their demonstrated social sensitivity.[50]

Humans are tireless social evaluators.[51] A markedly positive or negative weighting enhances our memory for others and can produce a palpable warmth or chill in our reactions, in life as in fiction.[52] Comparison clarifies judgments: even in animals as simple as guppies the cognitive principle of the contrast effect aids discrimination.[53] Fiction naturally makes the most of this capacity, deliberately pointing up character contrasts, often along thematically relevant axes of personality, like Penelope as the constant wife, Helen as the inconstant.

A final aspect of *theory of kind* we could call a *theory of bonds* or social relations, an intuitive sociology. Less studied in developmental psychology than more basic forms of cognition, except in the attachment between infants and caregivers, it has been closely investigated in social animals.

One aspect is *affiliation and association* (kinship, friendship, alliance, group identity). Even barnacles and tadpoles distinguish kin from non-kin.[54] All humans can and want to understand not only kinship but all these relations from an early age, however variable the terms of wider kinship in different societies.

A second aspect is *hierarchy, status,* or *rank.* Though undesirable from an egalitarian human point of view, hierarchy seems to some degree unavoidable in larger societies to avoid incessant squabbling over resources. It persists in democratic societies and soon establishes itself in new forms in societies that try to impose revolutionary egalitarianism. All social animals effortlessly track the unique terms of hierarchy in their own species.

A third aspect of the theory of bonds is *social exchange.* In the human case this extends far further than in any other individualized society and has led to the evolution of a complex suite of moral emotions to solve problems of trust and commitment.[55] We can observe the basis for human moral emotions in other animals, especially primates: empathy, which as Darwin noted makes individuals much more able to live in groups;[56] a sense of fairness and self-righteous indignation, recently found experimentally even in capuchin monkeys,[57] and demonstrated cross-culturally in humans;[58] forgiveness and reconciliation, needed to repair relations, observed over the last twenty years in many species;[59] emotions like generosity and gratitude;[60] and a capacity for detecting cheating in social relations, which sharpens our attention to and understanding of social exchanges.[61] Children as young as a year and a half spontaneously give toys, proffer help, and try to comfort the visibly distressed.[62] The evidence for the social and moral emotions in other primates, not even our nearest relatives, suggests that "human behavior has very old evolutionary roots."[63] Nature has endowed us with a moral capacity "much like a gyroscope at rest," and culture's role is "to spin it and establish its orientation."[64]

THEORY OF MIND

Many animals understand other conspecifics' behavior and relationships. Many species of birds and mammals also have the rudiments of a so-called theory of mind, an understanding of others in terms of goals, intentions, and perhaps desires. But humans appear to have a uniquely elaborate theory of mind that enables us to read one another, and therefore social events, in a far finer-grained way than any other species. Research on theory of mind in animals and humans has been intense, especially in children and those with autistic spectrum disorders. Since the term was introduced in 1978, thousands of papers have been published on the subject.[65] Yet all this astute theoretical, observational, and experimental work on how we understand other minds has until very recently been almost entirely ignored in literary studies, despite the fact that trying to understand why others do what they do matters so much in both human life and literature.[66]

Over the last forty years many have proposed that the pressure to understand others has been a major driving force in the growth of higher, especially primate, intelligence.[67] We need to infer others' predispositions and their likely intentions and actions, which may in turn be based in part on *their* attempts to guess *ours*. If this hypothesis is correct, higher intelligence emerged primarily as social intelligence, through a cognitive arms race to understand conspecifics and to reveal to or conceal from others our own beliefs, desires, and intentions. Because theory of mind was first opened as a field to explore within primatology, the social intelligence hypothesis was first called the Machiavellian intelligence hypothesis. Chimpanzees seek to understand one another especially in order to *compete*, sometimes by deception, for food, sex, or status. But researchers have begun to realize that we have probably reached our unique *human* level of theory of mind through pressures to *cooperate* more closely against other groups: we understand one another so much better because unlike chimpanzees we have crossed a cooperation divide and from infancy have a greater motivation to social engagement and shared attention than any other species.[68]

Moreover, since theory of mind was first explored, a lucky accident in the early 1990s led neuroscientists to discover the existence of mirror neurons, first in macaque monkeys and then in chimpanzees and humans. Mirror neurons in the brain's premotor cortex fire not only when an animal performs an action but when it sees another (especially a conspecific) perform the same action, or even when it sees less than the whole action but can see enough to infer the other animal's intention. Some mirror neurons need exact triggers, like a precision grip but not a whole-hand grip; others may fire under less stringent conditions; and some can be cross-modal. Certain mirror neurons in a monkey who merely hears a peanut being broken will fire in the parts of the motor cortex that would be activated to *make* the movement needed to break a peanut. Mirror neurons in monkeys, chimpanzees, and human preschoolers and adults allow an effortless, automatic understanding of the intentions of others through an almost reflex inner imitation. As one mirror-neuron specialist remarks, in a phrase that could not be more relevant to storytelling, "when we (and apes) look at others, we find both them *and ourselves*."[69]

Nevertheless a fully human theory of mind requires a capacity for interpreting others not simply through outer actions and expressions, and even through inner states like goals, intentions, and desires, but uniquely also through *beliefs*. This last layer of the rich human theory of mind may require specific genetically designed mechanisms or be simply a robustly channeled consequence of normal human social development.[70] Even in the latter case, human development depends on a coordinated suite of biological adaptations for our ultrasociality: our primate legacy of mirror neurons; our helplessness and underdevelopment at birth, yet our preparedness from the first to engage with human faces and voices, even while suckling; our eyes, whose color and shape have evolved to reveal rather than conceal direction and therefore intention; our long childhoods, and our motivation as children and parents to engage intensely with one another, through stages like protoconversation, pointing, and sharing attention for the mere pleasure of it, late in the first year, joint attention (my checking back and forth to see that others know where I am directing their attention and to confirm the target and the mutual engagement) at about twelve to fourteen months; and language.[71] Perhaps our capacity to attend to and recognize the different perspectives of others that begins to appear in the second year of life and deepens with language provides a sufficient base for continued development toward a fully human theory of mind.[72]

The uniquely human level of theory of mind builds on a platform of simpler capacities, from the earliest evolving and most basic, detecting animacy, to the more subtle, reading head and eye orientation and the changing expressions of faces and attitudes of bodies, to the still more complex, intuiting or inferring goals, intentions, and desires from these outer signs. Even at this level, it allows a much finer-grained understanding of events, since "the exact same physical movement may be seen as giving an object, sharing it, loaning it, moving it, getting rid of it, returning it, trading it, selling it, and on and on—depending on the goals and intentions of the actor."[73] At the fully human level of theory of mind, we pass one crucial step further: we also infer what others *know* in order to explain their desires and intentions with real precision.

Because we normally acquire theory of mind so effortlessly, we cease to notice anything to master. We assume that we simply see situations as they are. We need to appreciate the difficulties that those with autistic spectrum disorders have with theory of mind and the difficulties normal children have with what seem like "no-brainers" to adults, to recognize how theory of mind transforms our comprehension of events.

Theory of mind is no mere byproduct of higher general intelligence, but a psychological specialization that may be either relatively spared despite serious mental retardation, as in Williams syndrome, or relatively damaged, perhaps with no or mild retardation, in autistic spectrum disorders. Children with full autism appear to lack theory of mind; unlike other children with mental handicaps at the same mental age, they have severe problems in inferring others' attention from reading the direction of their eyes, or their emotions from their expressions, or their knowledge from what they can perceive. They therefore have great difficulty understanding social events and the sources of people's knowledge—and in understanding or telling stories, or interpreting films in social ways.[74] Temple Grandin, autistic but a high-performing agricultural engineer, as a child felt that there seemed to be something "going on between other kids, swift, subtle, constantly changing, a swiftness of understanding so remarkable that sometimes she wondered if they were all telepathic."[75]

Simon Baron-Cohen, a leading researcher into autism, shows how different our natural interpretation of events is from the way an autistic child might construe a simple situation such as he happened to observe in his local park:

> Joe and Tim watched the children in the playground. Without saying a word, Joe nudged Tim and looked across at the little girl playing in the sandpit. Then he looked back at Tim and smiled. Tim nodded, and the two of them started off toward the girl in the sandpit.

He offers two possible readings of the situation, one sinister, one innocent:

> Maybe Joe and Tim had a *plan* to do something nasty to one of the children. Joe *wanted* Tim to *know* that their victim was to be the little girl in the sandpit, and he *indicated* this by the direction of his gaze. Tim *recognized* Joe's *intention* and nodded to tell Joe he had *understood* the *plan*. Then they went over to tell the little girl, who was *unaware* of what was about to happen.

Or:

> Maybe Joe *wanted* to point out to Tim who it would be fun
> to play with. Tim *agreed* with Joe's *idea,* so they went over to
> ask the little girl in the sandpit if she *wanted* to play.[76]

All the terms he highlights, he notes, would be mysterious to an autistic child, and neither interpretation available.

Although it is more sophisticated than the other prototheories we
have considered, theory of mind is not a "theory" in the sense of an
explicit and culturally produced structure of propositions. It begins
in the evolution of social species and emerges naturally over the first
five years in normal human minds, not only in Western societies but
in children around the world.[77] We can trace three stages of the early
individual development of fully human theory of mind: (1) First,
infants up to about eighteen months have a single updating model
of reality, of the world they can see *now.* (2) From about eighteen
months to about four years, they can hold multiple models of reality in their minds, but do not understand the process fully. They can
call up memories of the past, anticipate the future, have some notion
of the wishes and intentions of others, understand and respond to
pictures or stories, enjoy pretend play, and distinguish these other
representations—memories, wishes, anticipations, fancies, pictures,
stories—from the world of the here and now. Other animals, especially chimpanzees and perhaps other great apes, dolphins, and other
cetaceans seem to share some of these capacities. (3) During their
fifth year, children grow into a distinctively human theory of mind,
capable of metarepresentation—of understanding the process of
representation—and involving *beliefs* as well as desires, goals, and
intentions, a state that no other animal appears to reach. At this
higher, exclusively human level, theory of mind becomes not only an
intuitive psychology but an intuitive epistemology. No wonder it is
hard to attain, even though as adults we employ it effortlessly.

A key test for this specifically human level of theory of mind is the
false-belief test, often called the "Sally-Anne" test. The experimenters
enact a scenario before each child, perhaps with puppet dolls called

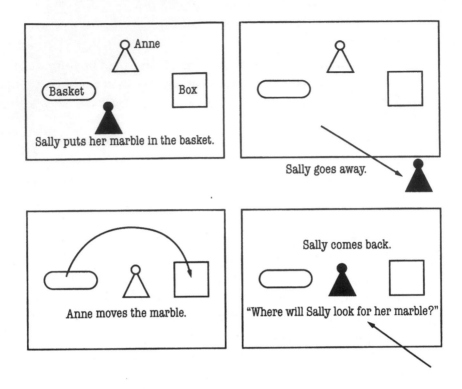

Diagram of the Sally-Anne test for false belief (theory of mind).

Sally and Anne and a toy room as visual aids. Anne is in the room as Sally places a marble in a basket. Anne leaves, and the door closes behind her. Sally then takes the marble out and puts it in a box. Each child subject is asked: "Where do you think Anne will look for the marble when she comes back into the room?" Three-year-olds regularly fail this test: they think that Anne will look in the box, where *they know* the marble now is. Beginning at some time between four and five, children regularly pass the test. They understand first-order false belief, the capacity to realize that someone can have a different idea about something than what is actually the case.[78]

Children have difficulty understanding their *own* knowledge. In another test they are asked what they think is in a Smarties tube, and naturally answer "Smarties" (a British candy like M & Ms). If they are then shown pencils rather than candy inside, and asked what they *now* think is in the Smarties box, they answer "Pencils." But when they are asked what they thought was in the box when they were *first* asked, three-year-olds will also say "Pencils," since they now know what's there; five-year-olds will say "Smarties"; four-year-olds show mixed results. They are not merely embarrassed to admit to having been wrong: though normally happy to point out other children's mistakes, three-year-olds are just as sure that *other* children first said "Pencils" when in fact they had said "Smarties." Children seem to come to understand their *own* understanding of things as they come to understand others'.[79]

Before the age of five, children cannot understand or remember clearly either their own or others' mistaken beliefs. Nor do they appreciate the bases for knowledge. One experiment involves a tunnel constructed so that children can either lift its lid to see an object inside or put their hands inside to feel the object. The children are shown two piggy banks, one full and one empty. When one is placed in the tunnel, they are asked how they can tell which it is, the full or the empty. If they cannot answer, they are prompted: "Do you have to *see* the piggy bank or *feel* the piggy bank?" Most three- and four-year-olds fail, and, when challenged to prove their wrong answer, "just gave a blank stare, seemingly unable to comprehend why they could not tell whether it was the full or empty piggy bank even though they were *looking* at it."[80]

Because they cannot understand the bases of knowledge, children also perform poorly at games of hiding or deception. When three-year-olds are asked to play a game in which the adult experimenter has to guess in which hand they have a marble, many either hold the marble in one open palm (with the other hand perhaps formed into a closed fist) or, as soon as the adult prepares to guess, show where the marble is.[81]

Young children cannot clearly anticipate what others know and therefore how they might behave on the basis of that knowledge. But in their fourth year they become curious about others' knowledge. Heinz Wimmer, one of the first to study the child's theory of mind, recorded his son's first inquiry, at three years and eight and a half months, about how people know something: "Heinz: 'Today will be nice.' Theo: 'Where do you know that from?' [German: *Woher weißt Du?*]." By the age of five, this concern "can become a veritable obsession":

> Sybille at 5 years and 1 month was visited for the first time by Judith (a friend from kindergarten) and her mother, who live in a distant village. After the visit Sybille asked her father:
>
> Sybille: "How did they know where I live?"
>
> Father: "Maybe they asked in the kindergarten."
>
> Sybille: "But Sigrid [the kindergarten teacher] doesn't know where I live."
>
> Father: "Maybe Judith's mother looked up our address in the phone book."
>
> Sybille: "But how did they know the way?"
>
> . . . .
>
> Sybille: "From where did they know which door we live at?" (in a house with three families and three separate entrances)
>
> Father: "They could have looked at the nameplate, where it says 'Spangler.'"
>
> Sybille: "Does it just say 'Spangler' there? Then how did they know that Sybille also lives there?"
>
> and so on.[82]

Theory of mind becomes uniquely human between ages three and five and continues to deepen over the next few years. Although from an early age children appreciate happiness in terms of what people want and what they get, they remain confused about another basic emotion, surprise, until five or six. Since surprise involves what people believe and what is really the case, five-year-olds usually fail to anticipate when a character in a story will be surprised.[83] At around this age children begin to master second-order false belief, understanding one person's thoughts about another's. Imagine that after Anne leaves, she peeks back through the keyhole as Sally moves the marble. Each child is asked: "When Anne returns, where do *you* think *Sally* will think *Anne* will look?" Five-year-olds usually fail this question; six-year-olds pass.[84] Over the next few years children also refine their capacity to detect sarcasm, bluff, irony, double-bluff, and to read emotional information from around the eyes.[85] By early adolescence they can cope with third- or fourth-order false belief, close to the human limit. Even adult error rates suddenly rise to about 60 percent on fifth-order tasks.[86]

Theory of mind has many implications for human brains and behavior. The major computational cost of evolving language may have been not syntax but theory of mind, the capacity to handle multiple perspectives, multiple orders of belief or representation. Theory of mind and not language may have been the driving force behind the unprecedented expansion of the human brain. Indeed the parts of the brain especially associated with language, like Broca's and Wernicke's areas, are significantly smaller than those associated with theory of mind abilities, in the prefrontal cortex, the brain region most disproportionately enlarged in humans.[87]

With our fully developed theory of mind, humans can engage in much more finely-tuned social behavior than other species. We can predict the behavior of others far more accurately. We can inform others of what we can infer they do not know or cannot themselves infer, when we wish to cooperate, and we can conceal information if we wish them not to know, when we wish to compete. With advanced theory of mind, deliberate teaching also becomes possible.

Theory of mind allows a much more precise and multiperspectival understanding of social events. Because we understand beliefs as the basis for forming desires, goals, and intentions, and because we understand the sources of belief, we automatically and effortlessly track what others might know about a situation, and can therefore understand their behavior much more finely, even if it involves X's reactions to Y's reactions to Z's thoughts about A. Almost automatically we track what others can know, and that makes all the difference to our capacity to cooperate and compete.[88] No wonder, then, that point of view and dramatic irony play such central roles in fiction, or that the gap between appearance and reality is such a widespread theme. But as we will see, the implications of theory of mind for fiction extend much further still.

UNDERSTANDING EVENT SEQUENCES

How we understand episodes and still longer *sequences* of events has been studied less than our understanding of objects or physical microevents. Much of the detailed information to date derives from human developmental studies.

So far we have focused on increasingly sophisticated abilities to understand events, especially social events, in terms of understanding the *entities* involved: first as objects, then as inanimate or animate, then as allo- or conspecifics, and finally as individual conspecifics apparently desiring *p* and believing *q* and therefore intending *r*.

We can follow a partially similar cline of development in the capacity of children to understand *events* themselves. As they mature from infancy to adolescence, children advance from perception to deeper inference, from description to explanation, from identification to temporality to causality to goals, from isolated events to whole episodes and then links between episodes.[89]

Infants first register objects that are stationary or move along a continuous trajectory. Children's first words mostly name objects. Reacting to stories, three-year-olds identify and describe objects, announcing the presence of one item or another, or reporting elements in a listlike manner. Responding to a wordless picture story about a boy searching for his lost frog, used to test children cross-linguistically and cross-culturally, a typical three-year-old indicates the stable features of the story world: "There's a frog here . . . Here's a moon. Those are boots."[90]

Children can detect temporal connections, causes, and goals in events before they can follow them in narrative, and they can follow such connections in narrative before they can discuss them there or produce them in their own purely verbal narrative. But in enacting events, or in pretend play, young children perform better than they can explicitly articulate.[91]

Evidence now suggests that at less than a year infants can understand some of the temporal and causal structure of events.[92] One- and two-year-olds can imitate action sequences in the order in which they were presented.[93] Reasoning about cause, even in novel circumstances, not just through repeated associations, has been demonstrated even in rhesus monkeys, who show no surprise when an apple and a hand with a knife disappear behind a screen and two apple halves emerge, but stare in apparent disbelief when an apple and a glass of water pass behind the screen and the apple reappears in two halves.[94]

As we mature, we pay progressively more attention to cause, recalling proportionately more events attached to, rather than detached from, causal chains at the age of four, still more at six, and more still as adults.[95] In redescribing events, however, although four-year-olds will have progressed beyond three-year-olds' naming of objects to a focus on the temporal, on action ("The boy and the dog slept while the frog quietly go out of his jar"),[96] they do not yet refer spontaneously to causes, internal states, or goals, as five-year-olds do ("and after went out calling for the frog").[97]

From early infancy children can appraise and revise their own behavior in terms of goals.[98] From thirteen months, if not earlier, they can interpret the behavior of others in terms of goals, and by twenty months they recognize causes and goals connected with concrete physical actions.[99] As we mature, we continue to interpret events progressively more in terms of goals.[100] Again, *explicit* spontaneous reference to goals lags behind *implicit* skills: it does not develop until five years,[101] when theory of mind has also begun to establish itself firmly.

As we mature, the right hemispheres of our brains seek explanation at a deeper level and coherence on both a local and a larger scale.[102] By age nine, children seek explanation in terms of cause and goal.[103] By late adolescence, they look for coherence beyond the local episode. With that search for explanation and an increasing awareness of the dangers of false belief comes the likelihood of realizing we may know too little, and the reassurance of reaching for the most readily available cultural understanding that seems to promise in-depth explanation. It is only at this point, past early childhood, that children begin to acquire the local, culturally inflected, religious, or scientific explanations for phenomena as different as dreams and disease.[104] But such local and readily revisable cultural explanations could not grow except on the rich substratum of universal, naturally developing event comprehension. This event comprehension reflects the regularities of the world—things, kinds, minds—and is already well established in many kinds of minds quite without human culture and in human minds largely independent of any deliberate teaching.

CONNECTING AND RECALLING EVENTS AND CHARACTERS

To understand longer sequences of events, not just within but between episodes, we need considerable memory resources. Psychologists distinguish semantic from episodic memory.[105] *Semantic* memory stores general knowledge, like my knowledge of trees or words like *tree, root, oak, deciduous,* memories often overlearned, automatized, and rapidly accessible through parallel searches.[106] *Episodic* memory records particular events that I remember as *experienced,* as *mine,* and can more or less locate to a specific place and time in my past, like my memory of climbing a particular macrocarpa in childhood, or breaking off a poplar branch to use as a knight's lance. Although such memories can be triggered, in Proustian fashion, by sensory similarities (a whiff of eucalyptus brings back memories of Stanford or Canberra), deliberate search for a particular earlier memory tends to be serial and time-consuming in comparison with semantic memory.

Psychologists also distinguish in another way between two stages of memory: the constricted space of short-term or working memory and the capaciousness of long-term memory. To understand sequences of events longer than episodes, if a new episode has no apparent connection to its predecessor, we need to search long-term memory, first for another episode somehow connected with the event in focus, then, if that yields nothing, for general knowledge in semantic memory.[107]

Many have thought that infants cannot form memories and that animals have very limited memories and cannot form episodic memories. But testing for cognitive capacities in ways that do not depend on their verbal expression has revealed much that animals and infants can do. Australian freshwater rainbowfish live for only two or three years in the wild but can form procedural (how-to) memories that last for at least eleven months.[108] Sheep have "semantic" memories that can recall individual faces of up to fifty other sheep and ten humans for at least two years.[109] Scrub jays may have episodic memories.[110]

Human infants of thirteen months can store temporally ordered memories that last at least eight months.[111] Toddlers form semantic memories, expectations about a class of event, from only a single occurrence.[112] Young children can "verbalize their memories, even if the events on which they are reporting occurred before the advent of productive language capability" and occurred only once.[113] Infants retain memories for several months without language, and can later superimpose language on previously encoded preverbal memories.[114]

From the first, storing and accessing episodic memories or envisaging immediately future options must have needed the capacity to distinguish memories from present perceptions or future imaginings, or the consequences would have been immediately catastrophic. Since we now know that animals can plan strategic futures, whether in chimpanzee politics or in crow tool-making, we have reason to think that these capacities developed before the human mind, that minds could already distinguish memory and future projection from perception—as of course our own fluid memories or fancies differ from the vivid and steady detail of our perception.

Frederic Bartlett first showed in 1932 that our memories are not eidetically exact but partially reconstructed, reshaped by the mind at every stage: in initial perception, in encoding, during storage, and in retrieval.[115] We remember only selectively, according to what we attend to and its recency, salience, and emotional impact. Neurologically, total recall would be prohibitively expensive: it would almost always be useless for us to remember trivial details years later, and the existence of such a vast data bank would catastrophically slow memory search for urgently needed information. Memory, it turns out, decays according to what mathematicians call a power-law function, matching the fact that events also recur in a power-law way: the more recent the event, the more likely it is to recur, and the more recent the memory, the more likely it is to be relevant and therefore worth accessing.[116] In optimal information-retrieval systems, items should be recovered only when their relevance outweighs the cost of retrieving. And that is exactly how memory works: we remember common and recent events better than rare and long-past ones.[117]

We tend to remember deep rather than surface factors, the "gist" rather than the detail, just as in stories we remember not words but our inferences about sequences, causes, and goals.[118] We recall not surface impressions but implications for action. We remember information *across* rather than *within* sentence boundaries.[119] We sort events so rapidly into sequence and causal sense that both children and adults recall events in chronological order even if they have been told them *out of* order and instructed to recall the information *as presented.*[120] At three years, children cannot mentally reassemble an out-of-order sequence of pictures. At four, they can, but inflexibly. By six, they can construct the events forward and backward and sort them into hierarchical categories.[121]

We form general expectations of individuals ("traits") or situations ("scripts"), and do not need to retain what conforms to those expectations, since we can simply access the general pattern in semantic memory. But we retain episodic memories partly so that we can reevaluate past incidents if we encounter new information that challenges our evaluations, and perhaps revise our understanding of this part of the past.[122] We search in memory for explanations beyond the immediate context according to the salience of the event we wish to understand, its causal connections with other events, and the time we have to search.

Because we need to remember particularly well the characters of those we encounter, we have evolved a sophisticated and efficient system for assessing and recalling others. Experiment shows that we form a swift, detailed, nuanced sense of a new acquaintance, usually highly consistent with the impression that others take from the same evidence. We do not need to wait for multiple encounters before forming a response, and do not have to dredge past memories to compute another's character at the moment we realize we need an assessment. We file a summary judgment effortlessly even on a first encounter, and we retain and recall this trait summary from semantic memory remarkably accurately, much longer than episodic memory seems to hold the details that led to the judgment. But our trait summary of another may prove wrong in the light of new evidence. Evidence that contradicts a trait summary primes us to recall from episodic memory details we would have otherwise found hard to remember, so that we can mentally revisit the grounds for the initial impression, reconstrue the situation if need be, and recompute our trait summary accordingly.[123]

Once we have developed full theory of mind, we will want to keep track of what others know, especially about strategic social information, in order to predict how they might behave. We can take an almost automatic running tally of who could see what when,[124] and we use this to interpret real and fictional events. Although this process is not infallible, its automaticity and efficiency are telling signs of an advanced adaptation. When we combine our inferences about what others know with our sense of their character traits, we have at our disposal a powerful social calculus.

REMEMBERING AND IMAGINING EVENTS

A "virtual explosion" of very recent work on memory, thought, and imagination has made it possible to see the close affinity between these three aspects of mind and our experience—including our experience of stories.[125] The literary analysis of narrative has tended to stress words and conventions. Equipped with the results of recent cognitive psychology, a naturalistic account of narrative can now probe deeper into what happens in our minds, and why, as we encounter the world of fact and the worlds of fiction.

The emerging notion of *grounded cognition* shows that we think via our experience, which in turn depends on evolved mental processing.[126] We encounter the world multimodally, through our multiple senses, our emotions, our actions, and our reflections. Language, however, is amodal: the word "chair" retains its senses regardless of whether we meet it in spoken or written mode, printed on paper or screen or spelled out in neon lights or alphabet blocks, in Morse code or Braille. Cognition, recent neurocognitive research shows, begins not with the amodal symbols of language but with multimodal simulations of multimodal memories of multimodal experiences. In simulation, explains Lawrence Barsalou, we reactivate "perceptual, motor, and introspective states acquired during experience with the world, body, and mind . . . the brain captures states across the modalities and integrates them with a multimodal representation stored in memory (e.g., how a chair looks and feels, the action of sitting, introspection of comfort and relaxation). Later, when knowledge is needed to represent a category (e.g., chair), multimodal representations captured during experiences . . . are reactivated to simulate how the brain represented perception, action, and introspection associated with it."[127]

In other words we think, remember, and imagine by mentally simulating or reactivating elements of what we have previously perceived, understood, enacted, and experienced. To take some simple examples from fine-grained neurocognitive research: my seeing a cup handle "activates a grasping simulation that inadvertently affects motor responses on an unrelated task";[128] my accurately judging the weight of what another person lifts requires simulating the lift in my own motor and sensory systems; my hearing a word activates motor areas associated with my saying it;[129] my perception of space is not neutrally three-dimensional but is shaped by my body, its relation to its environment, and its potential to act there.[130] Brain damage to a particular region makes it more likely that I will lose mental categories that rely on it for processing. For instance, visual areas are especially active when we think about animals and motor areas when we think about artifacts (hammers, cups, chairs). Damage to visual areas of my brain will make it more likely that I will lose animals as a category, and damage to motor areas more likely that I will lose the category of tools.[131] Mirror neurons, which fire in the appropriate part of our own motor areas when we see another perform an action, form the basis for the simulations that underlie our rich social cognition so central to narrative.

Not only cognition but memory and imagination also depend on simulation. Since Frederic Bartlett, psychology has realized how much memory constructs rather than simply records and retrieves, but over the last few years new evidence arriving from neuroimaging and from differences between normal and amnesiac individuals has proven highly pertinent to understanding narrative. Episodic memory's failure to provide exact replicas of experience appears to be not a limitation of memory but an adaptive design that helps us to retrieve and recombine memories in order to run vivid simulations of future experience. Where the exact memories of savants allow remarkable recall of the past, but at the expense of their coping with the future, our normal constructive episodic memory system "can draw on elements of the past and retain the general sense or gist of what has happened. Critically, it can flexibly extract, recombine and reassemble these elements in a way that allows us to simulate, imagine, or 'pre-experience' . . . events that have never occurred previously in the exact form in which we imagine them."[132]

Tellingly for this *constructive episodic simulation hypothesis,* imagining the future recruits most of the same brain areas as recalling the past, especially the hippocampus and prefrontal, medial temporal, and parietal regions[133] to provide a form of "life simulator"[134] that allows us to test options without the risk of trying them in real life. The vividness allowed by reconfiguring multimodal memories allows us to evaluate advance simulations rapidly and almost automatically through the emotional weightings our bodies assigned at the time to the initial experiences.[135]

Stories employ words and conventions, but long before most narrative conventions emerged, we evolved a capacity not only for reexperiencing the past in memory but also for flexibly reconfiguring it to offer concrete simulations of future situations. And when we engage with stories, our response does not restrict itself to the medium of presentation or to a kind of inner verbal retranscription. Rather we create on the fly a mental world that we keep track of by experiencing it through semisimulation: "As people comprehend a text, they construct simulations to represent its perceptual, motor, and affective content. Simulations appear central to the representation of meaning."[136]

Psychologists have used simple stories to understand how we follow events in ordinary life—a cognitive process much more complex than psychology labs can usually probe, and much closer to fiction as well as to the facts of experience. Experiments show that in understanding stories our minds keep extraordinarily close track of agents, especially of a principal agent, and especially of his or her active goals. We simulate the focal agent's situation, taking what psychologists call either an observer (outside) or field (inside) position, as we can in our own memories and dreams.[137] As we track focal characters, simulation allows us to make swift inferences about their situation from goal-relevant information that we amplify by keeping it active in working memory. Once a goal has been achieved, our activation of information related to it quickly decays in working memory; if a goal is postponed, we automatically inhibit any information relevant to it to enable better tracking of the new, currently active goal.[138] Like simulation in general, this process is neither learned nor culture-specific and would seem to have mental roots—like mirror neurons, for instance, which respond even more strongly to goals than to actions[139]—much deeper than narrative. The capacity to track other agents effortlessly surely derives from the need of any flexible agent facing potential predators, prey, partners, rivals, or allies to infer the maximum information about the likely next behavior of those who could make a decisive difference to its fate.

Our capacities to comprehend events and to recall and reconfigure them in memory develop in us naturally, and to a considerable extent without language. But that we can handle events so well individually does not prevent us from trying to find ever more interesting ways to *relate* events, if we have good reason to—as we do.

11

NARRATIVE: REPRESENTING EVENTS

ARISTOTLE, THE FIRST GREAT analyst of narrative, called story-telling the imitation of an action.[1] He famously had a biological bent, but how can a post-Darwinian biology deepen our understanding of narrative? How can we explain narrative from the ground up, by considering the interplay of cooperation and competition in social life? Why and how do creatures share information? Why and how would they share *narrative* information? What benefits make narrative so compulsive for human narrators and narratees, and what are its costs?

Narrative need not involve language. It can operate through modes like mime, still pictures, shadow-puppets, or silent movies. It need not be *restricted* to language, and often gains impact through enactment or the emotional focusing that music offers in dance, theater, opera, or film, or the visual focus in stage lighting, comics, or film.[2] But language of course makes narrative more precise, efficient, and flexible.

Narrative need not involve language, but it does need external representation, not merely internal representations such as our diffuse, distractible mental "representations" of events as we witness, recall, anticipate, imagine, or dream them. Lately it has become almost a truism to speak of the self or of experience as fundamentally narrative. Despite the near-consensus, we have little reason to think that this is true in either case.[3] Event comprehension lies at the core of understanding experience, but we do not represent all of our experience—even all the day's experience—in narrative form, even to ourselves, and we do not *need* to represent it in narrative form to construe it as experience or to be conscious of our selves. A neurological patient with a particularly severe form of memory loss, who has retained almost no detail of any event or person in his life, nevertheless seems to have a normal sense of self within the here and now.[4]

It would be burdensome to tell ourselves continuously the story of ourselves. But why should we tell *any* stories to others?

Sharing more information than we can glean from our own efforts has been a major incentive for social life. Animals gain more from passive observation the more they incline to remain near others, and that mere preference for company can then build more active forms of cooperation.

Active communication, especially via voice, allows the rapid transmission of detailed, complex, contingent information. Although such signals remain comparatively cheap, they cost senders in time, energy, and risk. If a vervet monkey utters an alarm call, it makes itself conspicuous to the leopard it senses, and therefore increases its own danger; but if each group member emits a call as soon as it detects a threat, all improve their survival chances. While warning or food signals offer a group benefit, individuals would earn maximum short-term gain by *not* sending signals themselves but heeding those of others. Yet if all adopted that strategy, all would become more vulnerable to predation or hunger. How, then, does cooperative communication establish itself? And how can we explain the much more complex and costly communication of narrative?

Richard Dawkins and J. R. Krebs argued in an influential 1978 paper that communication should arise more for competitive than for cooperative reasons: we should expect the manipulation rather than the accurate transmission of information.[5] But competition thrives best on *concealing* information: a predator silently stalking its prey, an ambush catching enemies unaware. Cooperation, by contrast, usually stands to gain from communication.

Manipulative and deceptive signaling does occur widely in nature, especially between species, but proves not to be the basis for *efficient* communication. Signals that evolve through competition tend to be costly, as arms races develop between insistent senders and resistant receivers. Messages become louder, longer, more repetitive, massively redundant, like the roars of red deer stags or Superbowl advertisers.

Signals used for cooperative purposes, by contrast—"conspiratorial whispers"—will be energetically cheap and informationally rich.[6]

This is what we find in highly social individualized species with elaborate communication systems. Dolphin social signals allow tightly coordinated behavior even in unpredictable circumstances. Primates emit food, alarm, and social calls to maintain, protect, and regulate their bands. Humans, the most cooperative of vertebrates, have the most efficient and flexibly conditional forms of communication: pointing, gesture, and especially language. Computer modeling provides compelling evidence that low-cost, efficient communication like human language could have evolved only from exchanges in which "both parties stand to gain a high payoff from effective communication."[7]

A predisposition to share information is not enough. How *can* animals share information about events?

The social pooling of information begins in present threats or opportunities. The waggle-dance of honeybees and the alarm cries of vervet monkeys are unspecific and prenarrative. The bee's dance appears indeterminate between identification, report, and instruction, between "there is . . . ," "I have come from . . . ," and "you should go to . . ." "a flower patch 100 meters away at 30° to the right of the sun."[8] Like a child's first responses to narrative ("There's a frog here . . . Here's a moon"), the vervet's call indicates a particular kind (leopard, eagle, snake), its presence now, and, in this case, like the bee's dance, an immediate action hearers should take.[9]

Bee dances and vervet calls reflect only stereotypical behaviors occurring *now*. But other animals can partially represent events that are neither immediately present nor typical. Imitation occurs widely in social species, since individuals can gain by learning from the experience, perceptions, and behavior of others. Dolphins, for instance, derive competitive advantage—against prey, predators, or potential mates or rivals—from being able to coordinate and synchronize. This presumably explains why they can also readily imitate other species such as seals, turtles, skates, penguins, and humans. One dolphin has been observed spontaneously imitating the routine of the diver who would come to clean its pool, by taking up a feather in its beak to scrape algae off the observation window and emitting sounds to match the diver's regulator and bubbles to mimic its exhaust air.[10]

If event representation begins in imitation, humans are "the consummate imitative generalist[s]."[11] They can imitate, as we have seen, within an hour of birth, and do so more readily than any other species.[12] Two psychologists who hand-reared a chimpanzee with their son to see if it might become enculturated by example found that instead their son began to emit chimpanzee grunts—at which point they abruptly called off the experiment.[13]

Unlike other species, humans rehearse and refine a broad range of actions in order to imitate others more exactly.[14] Humans also begin to understand imitation *as* imitation or representation through play, like the exaggerations of mock chase and peek-a-boo. In the next two chapters we will consider the importance of advanced pretend play to the origins of fiction.

Humans not only imitate others more compulsively and more closely than animals of other species—and admire those who do it well—but have discovered other ways of representing events. Since imitation occurs in species as remote as cetaceans and primates, imitatory enactment appears to have been the first move toward representing actions. But with human language there emerged much more flexible and efficient ways to represent events. As we saw in the last chapter, our species also has a richer understanding of actions within our kind than any other species within theirs. Our precise understanding of our actions makes possible our precise representation of human behaviors *and* our precise comprehension of those representations. And we not only are more *capable* of representing events to one another, but we also have uniquely strong *motives* for doing so, with our unique predilection for social engagement.[15]

Brains evolved not to give humans rich mental lives—though we are delighted they do—but to permit creatures that have them to make better decisions. Our minds must mine the present for clues, refine these clues via information saved from the past, and turn them into intimations of the future on which we can act.[16] But surely the best guide to what's next is to know what's happening now? That is evolution's own conclusion: we still pay attention primarily to the here and now, and our senses simply cannot see into the past, only at most into present traces of the past like a footprint, a photograph, or a fossil. But how do we benefit, how do we know better what to do in the future, when our minds divert us from the present to images of the past? What is the function of narrative?

Narrative can offer us either particular social information to guide immediate decisions or general principles we can apply in future circumstances. It focuses overwhelmingly on "strategic information": on whether Jack is sleeping with Jill, not merely how soundly Jack is sleeping.[17] In flexibly social species, such strategic information counts toward survival and reproductive success as much as non-social information does. It also changes more unpredictably: that bunch of fruit might become ripe enough to eat tomorrow, but fights or reconciliations or other slips in the social landscape could happen even today.

Chimpanzees not only watch one another closely but can also draw the attention of others to the current actions of still others. Among the examples de Waal notes at the Arnhem chimpanzee colony: Dandy sees Luit paying court to Spin while the current alpha male, Yeroen, sits a long way off. Barking excitedly, Dandy attracts Yeroen's attention and leads him to the other two mating.[18]

In humans, behavior becomes still more flexible and social, and social monitoring correspondingly more intense and more collaborative. We scrutinize one another from the crib to the deathbed. We observe one another not with cool detachment but with warm engagement: "Being affected by others is a design feature of human beings."[19] Through mirror neurons and other systems we are wired for emotional contagion. We half imitate what we see others doing, although an inhibiting mechanism stops us from actually moving while we simulate. We automatically have empathy for others. We know how they feel because we literally feel what they are feeling.[20]

Color and motion have a popout effect within the visual system. So, too, for humans does any pattern that touches the tripwire "human": a form, a face, an expression, an action, an interaction—even the relative motion of two triangles and a circle on a screen. In experiments showing films of reflective patches attached to people moving under low illumination, so that only moving dots are visible

In an experiment by Gunnar Johannson, viewers see light sources attached to moving bodies not as isolated dots but as people.

on screen, "you hardly see dots at all. What you see *through* the dots is a person. There is no doubt about it—you are watching a person doing things." Within one-fifth of a second "the displays home in on a brain mechanism that detects people."[21] Autistic children find it harder than others to match photographs of different people with the same mood, but they can match upside-down photographs of the faces faster than other children. They can calmly match patterns with no emotional meaning to them, whereas normal children feel confused by the emotional impact of the patterns we are programmed to feel, like the angle of lips or brows, yet cannot make sense of the patterns in faces upside-down.[22] Outside the autism or upside-down experiment, we swiftly observe and interpret our world in terms of patterns of agency, humanity, individuality, personality, action, and interaction.

We attend to one another compulsively, with lightning-fast emotional radar, but since we live in groups that disperse and reassemble fluidly, we also miss much of what goes on. Since we also have advanced theory of mind and understand false belief, we know that we often *don't* know about particular items of strategic information; that *others* may know when *we* don't know; and that we *and* they know what a difference it can make to know or not. We therefore listen eagerly to those with strategic information they think we will value.

In our ancestral environment strategic social information would

Understanding Minds

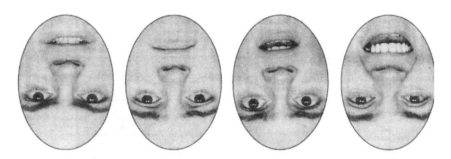

Upside-down faces used to test children with and without autism.

almost always have been about people we had already met and would often meet again.[23] We therefore have an endless fascination with character information, since it helps us to predict the behavior of those we interact with and remains relatively stable over time.[24] Nowadays screen close-ups make many care about the lives of media stars, however unlikely we are to meet them. Just as our continued craving for sweet and fat reflects old circumstances, so our sometimes indiscriminate appetite for social information reflects a time when we were likely to encounter repeatedly everyone we heard about. And we especially ingest information about the powerful, because their decisions and actions could influence our lives, and about those who command attention, since those who could do so were usually social leaders.

Our thirst for news about others may not always serve a purpose in our supersaturated modern environment, but there is every reason to believe that our craving for strategic social information is adaptive: it is universal, present from birth, can be selectively impaired in autism, and served and often still fulfills a clear function. Nature could not possibly arrange for each of us to be interested in precisely those who eventually matter to us, so it made our appetite intense and broad. Although nature could not foresee megacities, globally mobile populations, or mass media, most of the time the fierce curiosity about others it developed in us still serves us well.

So far we have considered narrative as an extension of our compulsive, emotionally-attuned social monitoring. But narrative may help us to make better decisions even without supplying immediate information. We talk about the past not only to disclose currently relevant social particulars but also to provide tools for reasoning about action. Narratives may record practices previously successful or unsuccessful. Among the Himba in Namibia, the indoctrination of obedience and heroic virtues forms a normal part of everyday life. When men visit, they sit together and start praise-singing, hailing the heroic deeds of their ancestors.[25] Narratives may offer explanations—outcome *O* occurred because we performed action *A*—or analogues, parallels, or partial models for reasoning, as when indirect memories of Easter Island, the Holocaust, or Hiroshima guide our decisions today.[26] Or they may merely introduce us to more of the range of human behavior, so that we have a richer context for understanding when we encounter something new.

Individual memory first evolved because sufficient regularities exist in the world for roughly similar situations to recur frequently. But since situations never repeat exactly, minds need to match similarities loosely. Through something like massively parallel distributed computer programming, advanced brains became able to pattern-match memories and present predicaments, to allow them to bring to bear on present choices even distantly remembered events and the emotional weightings attached to them.[27]

As our brains expanded, we could apply the past to the present and future still more flexibly. But we were still trapped within what we had witnessed and remembered ourselves. With narrative we could, for the first time, share experience with others who could then pass on to still others what they had found most helpful for their own reasoning about future actions. We still have to act within our own time, but with narrative we can be partially freed from the limits of the present and the self. And our ability to see connections between accounts of the past and present or future action prepared us for some of the core fascinations of fiction (see Chapter 24).

NARRATIVE: SOCIAL AND INDIVIDUAL BENEFITS

Narrative can provide listeners with clues to the present, hints from the past, examples or analogues for reasoning about future decisions. But how does narrative benefit tellers—without whom listeners learn nothing?

Should we see sharing social information as a form of reciprocal altruism? I give you strategic information that may benefit you, either because it allows you to choose better courses of action or because you can pass it on to others for whom it will be directly relevant. I do not extract an immediate payment, but when you know you have strategic information worth my attention or my passing on to others, I trust you will inform me in return.

In most forms of reciprocal altruism, the main temptation is not to repay the full value received. But in passing on gossip or news, we rarely feel tempted not to offer information for information. We do not view gossip as a chore or cost to avoid. Rather we *compete* to report the latest news. Why?

Gossip can offer both individual and group benefits. Within-group cooperation gains from the punishment of noncooperators, and "gossip is overwhelmingly critical, and primarily about the moral and social violations of others. Only about 10% of gossip is about the good deeds of others."[28] Gossip can therefore serve as "a policeman and a teacher" by drawing the attention of others to breaches of cooperation and other social norms.[29] While direct punishment can be costly, comparatively low-cost gossip may suffice in small-scale societies to change the behavior of noncooperators or those who interact with them. In hunter-gatherer societies, gossip and ridicule serve to keep in check those who violate their strong egalitarian norms. Insofar as those who gossip are seen to offer a social service—and not to use gossip for manipulative personal gain—they will be valued by other members of their groups.

Offering news in general benefits the giver as much as the receiver. I know I can earn a future credit on the social information exchange, often at little cost, for reporting something I happen to have witnessed or heard. I should "sell" this information while I still have something to sell, before another informant or a change in the situation makes my news worthless. And I want my news to be worth something to you, because then you will listen to me. I want your attention.

Normally, and for good reason, we attend to the here and now. We need to remain alert within the present, because danger may lurk even in the safety of our homes (what's that burning smell? that creak on the stairs?). Nevertheless much ambient information is of low value. When we are not on the move, the world around us probably looks much as it did five seconds or five minutes ago. But if I report an event, you absorb it within the present but focus your attention on the past. You know that I think it worth diverting your focus from the here and now, so you expect me to offer information of value, with the compression that narrative makes possible. Unless I have misjudged, or I muff the delivery, I have your attention and can direct it where I want as long as the story lasts.

As we have seen in Part 2, the capacity to command attention in social animals correlates highly with status—and the incapacity to gain attention marks low status, the inability to avert negative attention augurs danger (mockery, reproach, attack, or punishment), and the withdrawal of attention (ostracism, isolation) constitutes a severe punishment in itself. Status in social animals also correlates with survival and reproductive success. We therefore seek attention as a good in itself and compete to tell stories. Studies show that we spend more than half our casual conversation time in gossip.[30] If narrative were primarily altruistic, we would hold back and wait for others to produce their gossipy gift for us. Instead we normally compete not to *hide* social information but to *divulge* it.[31] And when others recount a story, we often compete to offer additional details, alternative explanations, equivalents ("You think *you've* had a bad day! . . .") or parallels ("But when *we* were in Bolivia . . ."). We can support our friends, challenge enemies, or make our *own* claim on the attention of others.[32]

Apart from earning immediate credit on the information exchange, social narrative can earn a general *status* for those who consistently provide information of high value and low cost: information not already known, relevant to listeners' situations, unexpected (not "Jill woke up this morning," unless Jill has been in a coma), accurate, and vividly, lucidly, and economically expressed. We will judge an individual with repeated access to valuable new social information as occupying a significant node in the social network.

Those who offer information about the powerful can earn status from those with less information. (As with all these factors, we also adopt counterstrategies: we welcome insider knowledge, but we dislike name-droppers trying to gain undeserved status.) Those who can offer information *to* the powerful can benefit directly from their patronage, as the chimpanzee Dandy did with the alpha male Yeroen by drawing his attention to a threat to his sexual rights, or as Polonius tries to do by offering Claudius news of and reasons for Hamlet's seeming insanity.

Social information need not be news to earn status. Narrative may offer not just immediately relevant social information but also general instances of human behavior to guide our reflections and decisions. Information not directly related to the short term must earn interest in other ways. In preliterate societies elders accrue status for their depth of historical knowledge and the pertinence of the examples they can adduce for reasoning about the present and future. Even in modern institutions those with a rich store of apt precedents will be prized as repositories of institutional memory. Those who can recall and recount the past in ways that shape audience responses can have considerable influence over decision-making.[33]

NARRATIVE AND STRATEGY

Although humans are uniquely cooperative, we also compete within as well as between groups, even in narrative. As both tellers and listeners, we use narrative strategically.

Much animal communication depends on environmental triggers, and for many species food and warning calls can be hard to suppress. Narrative, however, since it can refer to any *other* place or time, remains highly independent of the here and now. We can normally choose what stories to tell, and when, and to whom. And we can often predict what others know and might do with any additional information we could disclose. We have reason to pass on honest information if we want honest information in return, but we may disclose, withhold, or distort information strategically, even to the point of inventing purportedly true stories.

Narrative always bears at least a trace of strategy. We have to judge whose attention to catch, and when, and with what. We will seek to maximize audience interest and impact and to minimize audience effort or resistance. Take the case of a professional: the bard-priest *(dumsa)* of the Kachin people in Burma traditionally adapts his stories to suit the audience that hires him.[34] Or an amateur: an immigrant woman who arrived in Chicago late at night spontaneously recounted the evening's events to a group of women as a near-rape experience, appealing to female solidarity in the face of vulnerability, but reported the same events to a mixed-sex audience as an encounter with a colorful group of weirdos, to avoid making part of her audience defensive and to appeal to their appreciation of diversity and novelty.[35] We may wish to maximize the time of telling (and thereby of holding the audience "in the palm of our hand") *insofar as* doing so squares with not losing the audience's attention: most of us sense how compelling we are as raconteurs, and how long we can hold the floor. We may wish to tell a story in order to move our listeners to a particular conclusion—and we might recall that the art of rhetoric was considered the core of formal education in the West for two millennia.

And we may wish to reshape or even *invent* a story to stir exactly the kind of response we seek.

Most of us do not lie habitually, but we may do so in response to circumstance. As children we discover the possibilities for primitive exculpation and inculpation: "*I* didn't start it, Daddy! *He* did!" "No, *he* did! We were just . . ." The urgency of the situation and our knowledge that no one else saw can embolden us. Even later, in tight situations or with little risk of detection, we may lie or "embellish" the truth. But as in other forms of cooperation, most of us come to take a longer view of things.

Advanced cooperation depends in part on the evolution of emotions encouraging it. To progress still further we need also the cultural invention of values and practices that arise from and reinforce these social emotions: punishment for transgressions, for instance, and judicial systems to remove the need for personal revenge. The

evolution of narrative depends in part on the evolution of our inclination to detect and reject abuses of communicative cooperativeness: our anger at dishonesty, our discomfort at flattery, our outrage
at unfounded denigration, our indignation at the betrayal of con-
fidences, our annoyance at boasters and self-promoters, our irrita-
tion at bores who claim our attention but offer little of interest in
return. Each society forms its own cultural values to reinforce these
responses. Homer has Odysseus strike Thersites for backbiting, to
roars of approval; Dante assigns flatterers and slanderers to the lower
depths of his Inferno; schoolchildren chant "Tell-tale tit! Your tongue
will split!" And each society creates its more or less successful institutions for testing and punishing violations: corroboration or challenge; oaths and perjury laws; divination, confessions, judicial and
forensic investigations; witnesses, seals, double-entry bookkeeping;
and the like.

We reach for a first lie *in extremis*. We experiment with lying:
"I didn't go near that chocolate cake!"—despite the incriminating
smears. But most of us realize that while we wish to maximize the
benefits for ourselves in what we tell, others also seek to maximize
their interests. If we lie, manipulate, boast, flatter, or blab, they will
resist and resent us. In the long run we serve our own interests best
by earning reputations as tactful tellers of reliable, relevant, and, we
hope, sometimes riveting reports.

If we seek to manipulate our audience, we will try to escape detection so as to circumvent resistance. We can appeal to their pride
without seeming to flatter openly. We can appeal to cooperative instincts, to prosocial values we all wish to encourage in others, even
though we might not be able to follow them perfectly ourselves. Indeed since everybody loves an altruist,[36] appealing to cooperative behavior has a good chance of appealing to all. Our good strategic reasons to voice prosocial values in narrative make it all the likelier that
we *will* air such values and then feel them to be shared. The publicity
that prosociality can earn through narrative can in turn deepen the
base of trust on which cooperation depends.

If we view narrative as a strategic social process—as we all do intuitively—then one of the chief interests in studying narrative will be
to study the strategies involved. And the most interesting strategists

of all not only recount events but invent and shape them to appeal to an open-ended audience in their own time and place—and perhaps all times and places.

As audiences, too, we actively strategize. We assess a story first of all for interest. When we observe the world for ourselves, we do not expect every sensory detail to be relevant, only to have the potential for relevance. We will suddenly focus attention on a looming danger or an emerging opportunity, but otherwise much of what we perceive will have no bearing on our current purposes. But if someone proposes to tell us about another time and place, we want something worth the redirection from our here and now. Is it relevant enough? Does it involve people or events important, close, new, or unexpected enough to affect us? Does the story need to take so long or sound so convoluted? We resent, interrupt, walk away from, or avoid altogether bores who relaunch common knowledge or old gossip or information never worth imparting. Boys, with their full testosterone tanks, deal particularly roughly with one another as listeners: "Storytelling, joke telling and other narrative performance events are common features of the social interaction of boys . . . The storyteller is frequently faced with mockery, challenges and side comments on his story. A major sociolinguistic skill which a boy must apparently learn in interacting with his peers is to ride out this series of challenges, maintain his audience, and successfully get to the end of his story."[37]

If we feel that stories continue to merit attention, we weigh them as they unfold. Do they seem reliable? Do the tellers have reputations for honesty? Are they telling the truth, boasting, slandering, flattering, tattling? If so, can we correct for bias or exaggeration? Can we sift direct observation from inference or hearsay? Some languages even build in evidentiary markers for past events so that audiences can know immediately whether the speaker claims to have directly witnessed, merely heard of, or only inferred this or that detail.

We will also seek to make stories count for *ourselves* as much as possible. We may apply their information to our own decision-making, seeking direct links to situations or people we know. If we see no immediate nexus, we might still check whether reports offer bases for reasoning about action: direct models for behavior or the oblique inferences of parable.[38] We will face selection pressure to extract clearer or more pertinent information or implications.

We are not passive receptacles but highly active reconstructers. Our common understanding of events and language allows us to follow stories from slight cues. As we make sense of stories, we also *respond* to what we understand of the events reported. The social emotions first evolved in direct encounters with others, as a guide to whom we should deal with and whom we should avoid, and in what circumstances. As animals increased in social intelligence, these emotions could also become activated through observing others interacting between themselves, without the observer's direct involvement. The social emotions could also be *re*activated, with little loss of force, by *memories* of others or their actions.

And in the human case, the social emotions could also be activated by default in response to *accounts* of the behavior of others that we have not witnessed directly. Because emotional responses to events evolved long before representations of events, and therefore before the representation of untrue or unreal events, our emotional systems did not evolve to be activated only on condition of belief. We have an "interested party" response to narrative.[39] We need not *identify* with individuals we hear about (after all, a number of them may be of interest to us, and we cannot see ourselves as several at once, especially if they are more or less in conflict),[40] but we respond to them almost as if we were witnessing the scene, and had a share in the outcome of their actions—especially since in the early human environment most reports would have been about people with whom we were likely to interact for the rest of our lives.

We refine much of our advanced social understanding through narrative. As young children we naïvely seize on the opportunities that lies offer, only to discover how they can be found out and how others regard them. As we mature, the conflict between others' narratives and our own helps us understand in emotionally charged ways the differences among people's perspectives on events. We learn such lessons partly through real-world narrative, through being caught up in the conflict of real-world story and counterstory; through the negotiations of play, which like real-world events involve active participation rather than passive observation, but allow more leeway for harmless trial and error, and therefore learning; and in fiction, which does not involve active participation but can be powerfully and pointedly shaped to dramatize the force of narrative and counternarrative.

Indeed one account clashes with another in fiction from the *Ramayana* to *Rashomon* or the latest soap opera. Competing narratives about events that seem real to the characters matter to fiction for good reason. Characters cannot be aware of every action and need to be informed of what has happened elsewhere. Who was in a position to know, who informs whom, and how the information is interpreted can powerfully guide actions and reactions.

We are all highly sensitive to the roles that report and counterreport play in our lives and to the way we assess people through the stories they tell. But like many things we handle with ease and take for granted, these prominent features of human life have received comparatively little formal investigation. Fiction writers, however, have noticed and made the most of them, from boast and counterboast in Homer to con and countercon in David Mamet. Much subtle evidence for human psychology has accumulated in the world's literature,[41] and perhaps nowhere more than in the way story and counterstory weave together in the web of human life.

Although it seems highly unlikely that language evolved *in order to* pass on gossip,[42] the advantages of reporting events to others would have been manifest as soon as semantics and syntax became advanced enough to support it. And the flexibility of modern human language was surely enhanced by the pressure to communicate complicated social information involving multiple points of view.

Human language, as the recent field of conceptual semantics shows, appears uniquely designed to represent what we understand of events. Language's universal subject-verb-object combinations, however variable in sequence and detail, divide the world into stable entities and relatively transient actions.[43] Even in nonhuman affairs, human actions shape the way we refer to matter (mass nouns like *sand* or *clay* as raw substance, versus countables like *glasses* or *bowls*, which have been molded from substance into objects), space (prepositions across languages reflect "an intuitive physics of fitting, supporting, containing, covering, and other ways that humans put objects to use"), time (the future tense, for instance, across languages often incorporates the notion of volition or movement toward), and cause (prototypically an impetus transferred by a potent agent, especially a person, to a weaker entity that would rather stay put).[44] Language can report events so well because our overwhelming interest in human actions has itself shaped language.

To understand *social* events with the precision we need, our observations also need to incorporate theory of mind. We need to master not just event representation, but also metarepresentation, the ways in which particular perspectives on events represent them *as* understood, anticipated, recollected, or imagined by someone, or portrayed by something, in a particular way in a particular situation. Many aspects of language seem especially well designed not merely to represent events, but to move an audience to all sorts of possible, rapidly shifting, vantage points on events or circumstances.

Even something as basic to language as verbal tense does this.[45] We effortlessly compress multiple viewpoints in time into a single tense marker and then easily unzip these successive or overlapping perspectives:

> (1) *Trog is twenty summers.* (2) *He has lived with various women.* (3) *In the year of the big dry, he lived with Grota.* (4) *In the year of the big wet, he would set up with Trunga.* (5) *He would then have lived a year with Grota.*

In sentence 4, "he would" looks forward from the vantage point of "the year of the big dry," which became our temporary base in sentence 3, to an event *then* in the future but now in the past. In sentence 5 we adopt this new event, setting up with Trunga, as a base from which to look back at the duration of Trog's relationship with Grota.

Even in ordinary language, even in something as basic to syntax as tense or negation,[46] we nimbly shift our own and our audience's perspectives from our shared starting point to other provisional vantage points and foci with great efficiency, speed, and accuracy. Language reflects the shifts our minds use to make sense of events and their sequence, once we have theory of mind and a capacity for metarepresentation. And it can impart to others that sense, and those changing perspectives of time, place, and person, with minimal confusion.

Narrative arises from the advantages of communication in social species. It benefits audiences, who can choose better what course of action to take on the basis of strategic information, and it benefits tellers, who earn credit in the social information exchange and gain in terms of attention and status. That combination of benefits, for the teller and the told, and the intensity of social monitoring in our species, explain why narrative has become so central to human life.

The events that narrative reports may be directly related to present or future choices of action, to situations or people that listeners may become involved with. Or they may offer ways of reasoning about action: analogues or "parables" to guide our social planning; models to emulate or spurn; or merely images of the range of human character, situations, and behavior, and, in ancestral environments, perhaps also of the behavior of predators and prey.[47]

Narrative is always strategic, for both teller and listener, in ways that can range from the callously selfish to the generously prosocial. Because natural selection occurs at multiple levels, it can assist individuals or groups at different levels in their competition with other individuals or groups. But narrative especially helps coordinate groups, by informing their members of one another's actions. It spreads prosocial values, the likeliest to appeal to both tellers and listeners. It develops our capacity to see from different perspectives, and this capacity in turn both arises from and aids the evolution of cooperation and the growth of human mental flexibility.

But maximum flexibility, in humans as in others, depends on play.

12

FICTION: INVENTING EVENTS

ANIMALS LOVE TO PLAY, even with other species. Rats rarely tire of being tickled by humans, dogs of chasing and fetching. But only humans would play simultaneously with dinosaurs and ducks or dragons and skeletons. Only we immerse ourselves as children in pretend play and emerge through and beyond childhood into a world surrounded by fiction, a world of actuality surrounded by possibility.

Young children do not often spontaneously produce stand-alone stories, and cannot easily articulate in words alone what they understand of stories they hear. But in pretend play they readily make up stories and often act out what they understand of a story better than they can frame it in words.

Children's social pretend play lacks form as story or drama. It proceeds in fits and starts; it lacks consistency and direction. Children easily and naturally "break frame," stepping outside the action—an utter no-no in modern adult improvisatory theater—to narrate the story, or to act as codirectors or scriptwriters, as they negotiate the next development. (This phenomenon, incidentally, refutes the narratological tradition that insists that only narrated and not enacted stories count as narratives. For children, direction, narration, and enactment flow readily and naturally into one another. So long as the play-story continues, consistency of medium or mode does not matter.) Developmental psychologist Keith Sawyer has undertaken a longitudinal study of preschool children at play. Here he records Jennifer and Muhammed playing with toy animals at a sand table. Jennifer has a little duck, and Muhammed a dinosaur.

JENNIFER

 1 Oh big dinosaur   *high-pitched voicing, as duck*

 2 I cannot

 3 (screams)

MUHAMMED

 4 No, you can get on me   *in deep, gruff voice*

 5 I just won't care

 6   *J puts her duck on the dinosaur's back*

JENNIFER

 7 He said, he said

 8 You bad dinosaur   *in deep voice*

 9 Quickly, she hided in the sand *pushing the duck into the sand*

10 so the dinosaur

MUHAMMED   *have a story by*

11 No, pretend he killed her   *they can even verbalize it*

12 Ow!

13 He's already killed

Though inept as drama or story, this exchange makes perfect sense as play. The continuity of location, props, and pretend identities provides enough consistency for the "story" to lurch to its next point of interest or contrast. It engages attention. The children make the most of their theory of kind, their appreciation of the differences between species, as their story and the pitch of their play-voices highlight the contrast between the two animals. Like children everywhere, Jennifer and Muhammed are fascinated by the power and violence of the dinosaur. Like children and adults—not only in our own species—they have a special fondness for childlike features in humans and other animals.[1] Ducks with their rounded beaks therefore particularly appeal among birds, and toy ducks with their yellow plumage appeal as the youngest of ducklings. Even if the two toys are not too dissimilar in size, Jennifer and Muhammed make the most of the gap between the booming, imagination-haunting dinosaur and the squeaking, vulnerable little duck. They cannot manage gradual development, but they can pick dramatic moments: challenge *(You bad dinosaur),* flight *(she hided in the sand),* and violent death.

In later childhood and beyond we can understand and represent both actual and counterfactual events flexibly and almost effortlessly. Although language helps, our suppleness in social supposition does not derive *from* language. It begins to develop in us very early, before language, theory of mind, and metarepresentation are far advanced, in the form of pretend play. We cannot retrace far into our species past the origin of our capacity to comprehend and represent events, but we can, fortunately, follow the origin of stories in our individual past. And since our protracted childhood makes human life-history unique, it is appropriate that childhood offers our clearest window—through the one-way glass of the developmental psychology labora-tory—on the origins of our interest in story.

Animals play, and, despite difficulties in defining it, human ex-perts and nonexperts alike easily recognize play.[2] Most biologists agree that it must be adaptive, since it is so widespread within and across species, since it consumes valuable energy, since it puts players at increased risk of predation or injury, yet remains eagerly antici-pated, solicited, and maintained. But they do not often agree on its adaptive role.

Let me return by another route to the account of play I proposed in Chapter 6. Pleasure is nature's way of motivating creatures to per-form an activity *now*—and, in the case of play, to expend energy ea-gerly in mastering skills and acquiring strengths they might need later in urgent or volatile situations, in attack, defense, and social competition and cooperation. Animals and humans not only look as though they enjoy play, but their brains release dopamine—the main neurotransmitter for the anticipation of rewards—before and dur-ing play.[3]

Juvenile play deprivation among both rats and humans correlates with serious social malfunction in later life. Young rats experimen-tally deprived of play grow up unable to judge how and when to de-fend themselves and veer between being far too aggressive and far too passive.[4] In humans such experiments have yet to be tried, but in a large-scale study of sociopathic murderers in Texas, researchers were surprised to find *no* common background factor other than an absence or an extremely reduced amount of play in childhood in 90 percent of the perpetrators.[5]

Concerned to reject false romanticization, researchers complicate our notion of play as essentially inventive by stressing its frequent repetitiveness.[6] But this feature not only should be expected; it lies at the heart of play. Play's compulsiveness ensures the repetition that allows time to reconfigure minds and bodies. If play incorporates routines not required in the ordinary course of events, but needed in escape, pursuit, or attack, like (in the human case) accurate throwing of projectiles, then the skills it develops can become automatic and refined. Through overlearning actions, we can take them to a new level of control and flexibility, as now in playing tennis or the piano, or as once in wrestling and throwing at targets. (According to an eighteenth-century explorer, the Hottentots of southwestern Africa "know how to throw very accurately with stones . . . It is also not rare for them to hit a target the size of a coin with a stone at 100 paces.")[7] The pleasure of mastery and the particularly strong male pleasure in competition further hone skills learned in play.

All participants must understand behaviors like chasing and rough-and-tumble *as* play and not real attack.[8] To initiate play, canids have a ritualized play bow, particularly stereotyped in the young, like the "Once upon a time" that signals to a human child a partial suspension of the rules of the real.[9] All movements and postures become loose in play.[10] Baboons have a gamboling gait and a relaxed open-mouthed play-face—closely related to the human smile—to indicate "This is not in earnest." Play constitutes a first decoupling of the real, detaching aggression or any other "serious" behavior from its painful consequences so as to explore and master the possibilities of attack and defense.[11] In play we act as if within quotation marks, as if these were hooks to lift the behavior from its context to let us turn it around for inspection. Within the frame of play, animals make a first step toward the representation or re-presentation of the real that thought and language provide and that allow us to rotate things freely in the mind, exploring them from new angles. Play permits detachment, yet does so by engaging players intensely. Inviting, engrossing, energetic, this self-rewarding concentrate of ordinary action makes it possible to develop rapidly skills that it would be dangerous to learn, and impossible to overlearn, within the urgencies of the real.

Play is widespread among animals, but *pretend* play appears to be an almost exclusively human activity. Infants engage compulsively in pretend play, alone or with others. At twelve months they can manipulate objects as if they were something else.[12] Developmental psychologist Peter Hobson recalls his year-and-a-half-old son in his high chair, playing with a spoon, catching his father's eye mischievously, and "driving" and vocalizing the spoon as if it were a car: "He seemed aware that he and I were both looking at a spoon and not a car—yet, either separately or together, we could choose to relate to the spoon *as* a car for the purposes of pretend."[13]

Children's capacity to see a part of the world as if other than it is begins before they can use language or clearly attribute mental states to others. By the age of two, often after picking up cues from adults or older children, they find pretend play easy and fun, and they understand and enjoy the pretend play of others, like a mother talking into a banana as if it were a telephone.[14] Although this marks an advance over the infant's single updating model of reality, children at this age can entertain multiple models of reality, like past and present, or the real world and a pretend part of it, but cannot yet understand the nature of representation. A child laying its head down in play-sleep on a cloth is not *representing* the cloth as a pillow: acting *as-if* is not the same as using one thing to represent another.[15]

Play, like story, develops by stages as a child's discriminations extend from objects to kinds to minds. At first infants see toys as mere objects to be pushed and pulled, to be tested for their fit and feel. Even dolls are just shapes in human form. But gradually they become persons, first the passive objects of the child's care, then, with the child's help, active agents, until eventually the child imagines the toys as seeing and feeling and at last attributes to them knowledge and belief.[16]

Developmental psychologist Janet Astington traces a similar trajectory in her daughter's interest in stories, from things and their qualities to individuals and their actions, intentions, and beliefs. At two, her daughter's favorite "story" began: "Here is the farm. Here is the stable. The horse lives here with one little foal. Here is the cow-shed. The cow lives here with two little calves . . ." One can almost imagine many an animal species having such "stories" as this, if they could have stories at all, but not what this girl liked by the age of four. By now her favorite book was a fairy-tale treasury, with stories like "The Emperor's New Clothes":

> The emperor thought he would like to see it while it was still on the loom. So, accompanied by a number of selected court-iers, among whom were the two faithful officials who had already seen the imaginary stuff, he went to visit the crafty impostors, who were working away as hard as ever they could at the empty loom . . .[17]

Two-year-olds tend to see stories as interactive exchanges, in which they can test and confirm their knowledge with an older storyteller or enjoy the dramatization and the prompting and sharing of re-sponse. A two-year-old will often balk at being placed in the role of storyteller, and may not want to go beyond simple naming: "Once upon a time there lived a horse and that's the end."[18] Children's spon-taneous presleep monologues at this age also focus on the famil-iar rather than the novel. Instead of inventing the marvelous, they recount the routines of the day, as if to fix in their minds standard situations or "scripts" they need to master, "the canonical events of [their] world from which other narrative types emerged."[19]

Yet even in early pretense, children test the implications of their imagined contexts without confusing them with the real. They ap-preciate counterfactuality—they do not eat their mud pies—but they can act out situations as if they were actual.[20] A cup that has been pretend-filled by a pretend-pour from an actually empty teapot will spill its pretend contents if knocked over, and children will refill only the "spilled" cup, not the others, even if all are in fact empty.[21] They can run similar inferences with toys that represent animals or people or with purely imaginary characters. Between the ages of three and ten, many children—perhaps 50 percent—develop sus-tained and complex relationships with "imaginary companions." They do not confuse their fantasy and reality, and from the age of three those who do have imaginary companions perform better than others on tests of the ability to differentiate the real and the imagined. They often outperform others "at tasks that require a subtle use of intuitive psychology. They seem to have a firmer grasp of the difference between their own and other people's perspectives on a given situation and are better at construing other people's mental states and emotions." Imaginary companions offer "training for the social mind," as pretend play does in general for many kinds of behavior.[22]

But as children gain confidence in identifying and distinguishing objects and situations, they become interested in testing and overturning expectations, like so many toy blocks to be toppled in order to rouse a reaction. Even at twenty-one months Emily had a different voice for play with her toys and for telling a story. For a story, she raised her vocal pitch and heightened her vocal contours, as if she realized the need to secure attention.[23]

Although young children tend not to invent stories to tell others, as opposed to engaging in pretend play alone or in company, or trading rhymes and riddles, their predisposition to pretend play makes them ready enough to tell stories when prompted.[24] In their third year, before they have much command of story structure, they appear to grasp the crucial role of pattern and of holding the attention of listeners. At this age their stories are almost as much poems as narratives, focusing on striking characters and effects that violate expectations, but in a structure that resembles theme and variation rather than the event continuity that adults expect of stories:

> The monkeys
> They went up sky
> They fall down
> Choo choo train in the sky
> The train fell down in the sky
> I fell down in the sky in the water
> I got on my boat and my legs hurt
> Daddy fall down in the sky.[25]

The two-and-a-half-year-old boy making up this "story" has no idea yet that stories incorporate not just characters, events, and locations but also aims, goals, actions, and outcomes. He cannot *develop* a story but seems to intuit the need to surprise, with his unusual characters in unusual places defying the principles of gravity that he began to understand before he was three months old.[26] Repetition is the simplest form of elaboration, but since pure repetition diminishes interest, repetition of a bold idea with variation offers him the best prospects of holding attention with the imaginative resources he has.

Even infants, like the rest of us, lose interest at mere repetition and suddenly look longer and harder at anything that strikes them as different: "Surprise, pleasant or unpleasant, is more informative than predictability."[27] Neurons in the substantia nigra and the ventral tegmental areas of the brain secrete dopamine in reaction to the surprising but not to the expected.[28] Before they can understand or relate events in anything like an adult way, children seem to understand the value of the rewards of surprise and how to produce them in their first attempts at story—as in the sudden shifts of Jennifer and Muhammed's duck and dinosaur play.

In another game, Sawyer's field notes record Jennifer, now with the dinosaur, playing with two unnamed boys who "have blocks, which are now 'pepper shakers' which shake 'poison pepper.' When Jennifer's dinosaur becomes a little too threatening, they chant 'Pepper, Pepper,' and shake at the dinosaur. Jennifer doesn't want her dinosaur to die; she says 'Pretend he just sneezes from the pepper, he doesn't die.' The boys agree."[29] Against the attention-catchingly extravagant force of the dinosaur they invent the equally extravagant "poison pepper," yet when the sheer power of their weapons threatens to end the game, they easily move out of the pretense, for a moment, to negotiate a sudden new pretense to keep their play going.

By the time children reach three or four, their pretend play and their early stories aim not at realism but at catching the attention through fantastic characters and actions that violate expectations.[30] Since destruction is more arresting than more gradual change, they tend to focus on disaster, often without resolution.[31] The evidence shows that they recall "breach stories," stories that counter expectations, better than those that do not.[32]

As part of his challenge to romanticized notions of play, leading play scholar Brian Sutton-Smith posits that "the delight the young have in destroying things and in general being bound to themes of violence and mayhem is, as [Mihai] Spariosu said, that they have perceived the generally chaotic nature of the universe and wish to participate in it."[33] Surely not. A less far-fetched explanation is that children have discovered they can make more of a difference to their world more quickly by destruction. It can take several minutes of painstaking concentration to build a tall pile of blocks, but less than a second, and one exuberant push, to make the whole thing clatter down. In story as in play, destruction is a way of causing maximum impact for minimum effort. A four-year-old boy made up this story:

> Once there was a dragon who went poo poo on a house
> > and the house broke
> then when the house broke the people died
> and when the people died their bones came out and
> > broke and got together again and turned into a
> > skeleton
> and then the skeletons came along and scared the people
> > out of the town
> and then when all the people got scared out of the town
> > then skeleton babies were born
> and then everyone called it skeleton town
> and when they called it skeleton town the people came
> > back and then they got scared away again
> and then when they all got scared away again the
> > skeletons died
> no one came to the town
> so there was no people ever in that town ever again.[34]

This and other young children's stories are not plotless, but unplanned and episodic, a series of opportunistic riffs, each aimed at catching the attention, from the dragon as a conventional category-breaching monster, to the decorum-breaching "poo poo" on the house, to the house's collapse and killing its inhabitants, to their bones emerging from their bodies, breaking and coming together again, and somehow giving rise to skeleton babies.

As theory of mind continues to develop in middle and late childhood, children learn subtler ways of extending and linking events by incorporating aims and intentions, then feelings, then even beliefs, at first implicitly, then explicitly, into their play, their own stories, and their responses to questions about stories they read. Stories increasingly have a large-scale structure with distinct and well-formed episodes and continuity between the episodes, especially in terms of characters' goals. In pursuit of their goals, characters now resist the calamities that tend to befall them without comeback in younger children's stories, and become more active, resourceful, and resolute.[35]

Pretend play does not arise out of an increasingly sophisticated grasp of story structure; all it takes is the names of a few creatures and a few basic expectations of each of them, and an awareness of actions or of other creatures that would breach expectations. Although pretend play can help children deploy or test what they know of the real, it is more important for them to engage their own and their playmates' attention through impact, surprise, the breach of normal expectations. Children need to be able to grasp the real, to work out what to expect of inanimate and animate, of different kinds of things and animals, of different behaviors and situations; but they quickly become interested not merely in recombining the normal but also in transgressing its boundaries. Holding their own attention or that of their playmates by surprise and intensity matters more than, and precedes, mastering the richness of the real. A choo-choo train falling down in the sky, a dinosaur dying or only sneezing from poison pepper, a dragon poo-pooing on a house until it collapses do little to instruct children about reality, but such stories keep them engaged and help the children explore alone and together the possible around the real.

Although many animal species can understand the actions of other members of their species in subtle ways that *we* have difficulty in deciphering, we understand the behavior of other humans in ways more sophisticated and precise than any other species can understand its own kind. But as young children, before we understand events with any subtlety, before we understand what others can believe about a situation and therefore what they might feel, intend, and do, we are already compulsive creators of pretend events. And these pretend events not only recombine elements of the real but readily step over the actual into the possible and the impossible. Why is so much of early human play *pretend* play? Why is the pretense so often so extravagant? And why, when we master language and theory of mind so that we can use all our special human understanding of social events to tell stories fluently, without needing to act them out, do we still find the "pretend" stories of fiction so engrossing?

13

FICTION AS ADAPTATION

IN THE 1989 TV MOVIE *The Naked Lie* the unpleasant and self-centered Webster shows no sympathy for a prostitute who has been killed. When Victoria asks him, "What if it were your sister?" he sneers: "I don't have a sister, but if I did, she wouldn't be a hooker." Later in the movie Victoria muses to another character: "You know that sister Webster doesn't have? Well, she doesn't know how lucky she is."[1] We easily follow Victoria's initial counterfactual, Webster's counterfactual refutation of her condition, and Victoria's comically contradictory counterfactual consequence, the sister who, because she does not exist, cannot know how lucky she is not to do so if she has to suffer Webster as her brother. Stories help train us to explore possibility as well as actuality, effortlessly and even playfully, and that capacity makes all the difference.

excellent

I forgot
that I
was reading
this book

FICTION: EVIDENCE FOR ADAPTATION

Since, as we have seen, much of our ability and inclination to understand, recall, represent, and invent events precedes language, our storytelling instinct needs a biocultural explanation. To explain how we understand and represent events poses no untoward biological challenge, since the advantages of both are so apparent, and since we can find such clear precursors in other animals. But fiction, *inventing* events, presents a more difficult—but not insoluble—problem.

First, we need to see fiction as an art, and as emerging later than other arts, music, dance, and the first visual arts, like body decoration and the fancy finishing of stone hand-axes. We already had other arts as adaptations—and a strong predisposition to play with pattern—before we began to make up stories.

Second, we need to see the novelty of fiction. We have good reasons to think that it constitutes an adaptation in its own right. Telling fictional stories is a human universal. It develops spontaneously and without training in childhood in the form of pretend play. And even as adults we can and want to tell and listen to fiction.

One sign of a cognitive adaptation is that limited perceptual input yields rich conceptual output: the mind automatically processes information in elaborate ways. In fiction we repeatedly make inferences that far outstrip evidence. Seeing the graffito "Ralph, come back, it was only a rash," we naturally understand it first as one lover's appeal to the other to return, because what had seemed like a sexually transmitted disease was only an innocent rash; then as a joke pretending to be such an appeal. From minuscule data we leap to complex inferences. The difficulty of programming computers to understand natural-language narratives has made plain how richly our minds process the world so that we can understand events and stories from the sparest information.

Another sign of a cognitive adaptation is the inability to suppress a response. If we see a plaster cast of a face lit from above or below, we will think it convex one way, concave another, because our visual system expects the main source of illumination to come from above and not below. We cannot suppress this tendency even when the light comes from below, and we therefore read the highlights and shadows with precisely inverse values, as protruding where the cast recedes, and vice versa. In the same way we are unable *not* to imagine and respond to the characters and events of a well-told story, even if the storyteller invokes fantasy with a time-honored "Once upon a time" or a new twist like Louise Erdrich's "You don't have to believe this, I'm not asking you to."[2] We may know that the story consists of mere words, words with no pretense to report real events, or of contrived images, projected onto a flat screen in pixellated form, of artificially costumed, made-up, and illuminated actors, whose real identities and lives may be quite familiar to us, and whom we may have seen playing very different parts—yet whether on page or screen we cannot stop conjuring up and responding to the story's invented people

[margin handwriting: we respond deeply to stories]

189

and predicaments, and even, if occasion prompts, weeping tears at characters' fates.

Failure in a particular cognitive activity, as a result of genetic defect, disease, or accident, also indicates the likelihood of specialized cognitive subroutines involved in a particular adaptation. Theory of mind is central to the human capacity to understand others as richly as we do and to our understanding of metarepresentation. Autistic individuals have very poorly developed theory of mind, often despite intelligence in other areas, and difficulties in story comprehension, and do not engage spontaneously in pretend play.[3] Williams syndrome, with almost exactly the opposite symptoms—high sociality and playfulness, good theory of mind, linguistic and quasi-narrative fluency, but severely impaired intelligence—confirms that theory of mind is not simply a feature of general-purpose intelligence.

Adaptations show *design* for some *function*. In the preceding three chapters I have explained the design for fiction, and its origins in other cognitive adaptations: evolved systems of event comprehension, our theories of things, kinds, and minds and of event storage, in episodic memory; evolved systems of event representation, whether in action, word, image, or object, and human joint attention both to events and to others representing events; and the reliable emergence of pretend play and story as a way of exercising our capacities for handling social information. A crucial question remains: What biological function(s) has fiction? What difference can it make in terms of human survival and reproduction?

FICTION'S FUNCTIONS

I have suggested that we see art as cognitive play with information-rich pattern. Ordinary play allows animals to extend and refine their competence in standard species behaviors to the point where their skills offer a new freedom that may be crucial in situations like attack, defense, or rearing offspring. The special cognitive play of art, I propose, allows humans to extend and refine key cognitive competences.

Only recently has research begun to confirm the power of play and art to recalibrate minds. Evidence has emerged from ethology and neurology that play of all kinds affects performance, as the intense pleasure it produces motivates repeat practice in secure situations and thereby strengthens synaptic connections. As Marc Bekoff notes, play "provides important nourishment for brain growth; it actually helps to rewire the brain, increasing the connections between neurons in the cerebral cortex. Play also hones cognitive skills."[4] Just as girls enjoy pretend play with dolls, female primates find babies, the younger the better, fascinating; those who have not yet had infants of their own are usually highly motivated to play at mothering the infants of others, while those who have not played the role of mother perform less well when they themselves become mothers.[5] Or, to jump to our own species and the present: computer games can improve the speed and accuracy of visual recognition and manual dexterity. Regular players have improved visual attention and can shift attention more quickly. The best predictor of surgeons' skill at keyhole surgery turns out not to be years of training or number of procedures performed, but time spent at computer games.[6]

Similar benefits apply to art. "Learning requires the assimilation and consolidation of information in neural tissue. The more experiences we have with something, the stronger the memory/learning trace for that experience becomes."[7] Although many people say that their early music lessons didn't "take," cognitive neuroscientists have found otherwise: "Even just a small exposure to music lessons as a child creates neural circuits for music processing that are enhanced and more efficient than for those who lack the training."[8] But because social information matters so much to us, we never need to sign up for story lessons, and we never complain that stories didn't "take": as children and adults we happily succumb to the training in social cognition that pretend play and fiction reliably yield. Children have to hone their capacity to direct and shift attention.[9] The high intensity of pretend play and fiction and their rapid switches of place, time, and perspective must make social cognition, like any other well-learned and much-practiced skill, faster, more efficient, and more accurate, and speed up the capacity to guide and redirect social attention.[10]

Evidence has also begun to arrive that even vicarious and virtual

191

experience activates the mind in ways that partly mimic direct action. Witnessing or even hearing about activities can cause neurons involved in producing the action to fire. Social neuroscience has begun to discover how our minds can be affected by emotional contagion, by responding, even without registering consciously, to cues of specific actions or emotions in others.[11] As we have seen, mirror neurons fire when we perceive someone performing an action, as they would if we were performing it ourselves. Electromyographic studies show that our faces automatically make invisibly small muscle contractions in response to pictures of human facial expressions, even when we are unaware of what we have seen.[12] Recent research also shows that merely encountering an action word like "grasp" can trigger neurons to fire in the area of the motor cortex that would be activated when actually grasping.[13]

Much of this research into neural plasticity, reinforcement, and indirect activation has only recently become possible. It seems likely that evidence will continue to arrive that art in general, through its capacity to compel our attention, improves specific kinds of pattern detection: in the case of fiction, our ability to detect social and agential patterns.

Apart from immediate danger, nothing captures our attention like the actions of others around us, and most of us therefore find fiction even more compelling than music or visual art, and able to hold our interest for hours at a stretch. Fiction allows us to extend and refine our capacity to process social information, especially the key information of character and event—individuals and associates, allies and enemies, goals, obstacles, actions, and outcomes—and to metarepresent, to see social information from the perspective of other individuals or other times, places, or conditions, as Jenny and Muhammed do with dinosaur and duck, or Victoria and Webster do with Webster's nonexistent sister.

Much more than other species, "we refine our resources by incessant rehearsal and tinkering."[14] Fiction's appeal to our appetite for rich patterns of social information engages our attention from infant pretend play to adulthood. Because it entices us again and again to immerse ourselves in story, it helps us over time to rehearse and refine our apprehension of events. Fiction, I propose, does not *estab-*

lish but does *improve* our capacity to interpret events.[15] It preselects information of relevance, prefocuses attention on what is strategically important, and thereby simplifies the cognitive task of comprehension. At the same time it keeps strategic information flowing at a much more rapid pace than normal in real life, and allows a comparatively disengaged attitude to the events unfolding. It trains us to make inferences quickly, to shift mentally to new characters, times, and perspectives. Fiction aids our rapid understanding of real-life social situations, activating and maintaining this capacity at high intensity and low cost.[16]

Fiction also increases the range of our vicarious experience and behavioral options. Like play, it allows us to learn possible opportunities and risks, and the stratagems and emotional resources needed to cope with inevitable setbacks, without subjecting ourselves to actual risk. It does so efficiently because it acts as a superstimulus by focusing on intense experience and concentrated change.[17] These not only hook attention but rouse emotion, which in turn amplifies memory.

By inventing characters and events, by selecting them to shape emotions toward defined conclusions, fiction can stock memory with compact and compelling examples, not usually of scenarios we are likely to relive closely, but of situations with enough emotional and ethical similarities to those we do experience to provide a basis for our thinking. Fiction can *design* events and characters to provoke us to reflect on, say, generosity, or threat, or deception and counter-deception. And it efficiently evokes our intense emotional engagement without requiring our belief.[18]

Many have proposed that a major function of fiction is its ability to provide scenarios or models that we can draw on in planning our own actions and making our own decisions. Others object, arguing that obviously false situations can play no realistic role in planning.[19] The appeal of the surprising in fiction, from pretend play to Shakespeare, does mean that much story, from a dragon poo-pooing to a fairy queen falling in love with a workman transformed into an ass, seems designed to arrest attention rather than to offer scenarios likely to recur in the lives of audiences. If the prime function of fiction were to provide *literal* models of situations we might encounter, then scenarios of an everyday kind, such as we might find in language-learning workbooks, would be the acme of fiction. But fictional situations need be neither just the same as those we encounter in real life, nor realistic, in order to aid our thinking about action.

Thought experiments, Richard Dawkins observes, "are not supposed to be realistic. They are supposed to clarify our thinking about reality."[20] The thought experiments of fiction may opt *for* realism, like Christ's parable of the Good Samaritan, or *against* it, like Aesop's animal fables. We do not need to be a Samaritan or to travel from Jerusalem to Jericho or to encounter a wayfarer robbed by the roadside to learn from the example of the Samaritan's charity. The very brevity of the parable (it takes only a minute to tell, in the version in Luke 10) and yet the clarity of its structure and sense, *because* it is fiction (the threefold repetition, the pointed contrast of uncharitable priest and Levite and charitable Samaritan), are exactly what make this exemplary tale such a powerful basis for reflecting on how we should help others.

This particular parable works though brevity, clarity, patterned structure, and realism: encountering someone in need, even a stranger in need, *is* a situation we are likely to face. Aesop's fables succeed by defying our expectations of the real (animals that talk) and by simplifying character, appealing to our sense of different species' "personalities" (sly foxes, industrious ants). Other stories may entrance us not by concision and schematic concentration but by duration and elaborated verisimilitude. We may never be shipwrecked on a desert island like Robinson Crusoe, but we can learn from the example of his fortitude, resolution, and ingenuity.[21] The very unusualness of the situation catches our attention and animates our imaginations, and our engagement with Crusoe in his predicaments makes the lesson "emotionally saturated."[22] We have the time to see and feel with Crusoe in his plight.[23]

In situations calling for charity, or alertness to flattery, or resourcefulness and resolution, the very conspicuousness of the examples makes it probable that stories like the Good Samaritan, Aesop's fable of the crow and the fox, or Robinson Crusoe might come to mind. Detailed similarity in the scenarios is not required.

Minds generate future: they guide action by trying to predict what will follow. In the high volatility, variability, and flexible responsiveness of human social interaction, any improvement in interpreting situations and testing possible scenarios, actions, or reactions, using not only personal or reported experience but also the thought experiments of pretend play and fiction, offers a telling advantage.

Like other arts, fiction establishes itself as a human compulsion because it improves one key mental mode: in this case, social cognition. But fiction has subsidiary functions arising from that fact.

Art depends on the unique human capacity to share attention. Other species, as they become more social, learn to attend to the attention of others, and individuals can command attention according to their status—which correlates with survival and reproductive success.[24] "While telling a tale a person is, for the time being, dominant over his listeners,"[25] and in the world of specialized storytellers, from a tribal storyteller or Homer to Shakespeare or Tolkien, status proves still more enduring.

Males have more reproductive variance than females: few females do not become mothers, and none can have very large numbers of offspring; unsuccessful males may father no children, and the most successful can father many. Males therefore have on average a stronger drive than females to earn status, since it can gain them better access to females, and hence have more reason to engage in extreme behavior in order to secure status.[26] Males therefore are overrepresented at both extremes, success or genius, and failure, crime, mental illness, or drug dependency. Despite Murasaki, Jane Austen, and J. K. Rowling, males outnumber females as classic and even popular storytellers (including today's leading popular storytellers, filmmakers), while at the other end of the spectrum, males outnumber females by more than four to one in autism, which correlates highly with poor performance in social cognition and pretend play. But females, while on average they do not seek status as urgently as males, also on average invest more in childrearing, and are the principal tellers of fictional stories, of folk tales and nursery rhymes, to their children.

Stories' power to shape reactions can lead to manipulation, and often has.[27] Mythic histories extol rulers from more or less "divine" chieftains or emperors to modern autocrats like Stalin or Kim Jong-il. Folk tales relayed by mothers tell children of the evils of stepmothers.[28] But audiences listen strategically and learn to resist obvious manipulation. Storytellers' appeals to values genuinely shared will meet with much easier acceptance and will help entrench what the community esteems and welcomes: individual values that are also valuable for the group, like courage, resolution, resourcefulness, and cooperative values like sympathy, kindness, generosity, loyalty, and honesty.* Psychology has recently begun to study the emotion of elevation, a feeling that we all know but that science has hitherto spurned: the admiration for moral beauty, and the desire to emulate it, that we experience when we witness a singular act of kindness, gratitude, fortitude, or the like—and that we feel no less intensely when we encounter it in fiction.[29]

Through appeals to the moral and social emotions, fiction can contribute to solving the problems of cooperation that become more acute the larger the population grows. As we have seen, cooperation can become more widespread with effective punishment for noncooperation, and even more robust with the punishment of those who fail to punish noncooperators.[30] With such second-order punishment operating, the costs of maintaining systems of punishment, and therefore habits of cooperation, decrease rapidly. Stories can help motivate cooperation by arousing punitive or retributive emotions like indignation and contempt, in tragedy and satire, or by stressing the need for vigilance against deception and manipulation. Experiments show that "we feel disgust (as evidenced by significant activation of the anterior insula) when faced with the behavior of cheaters, and very real satisfaction (that is, activation of the caudate nucleus) when we punish those cheaters"[31]—feelings activated by fiction as well as by fact. And stories can help solve the problem of common knowledge:[32] traditional stories ensure that *all* know and react to, and know that others know, the core values of the group.

The very nature of fiction makes it likely that storytellers earn least

---

* Occasionally, of course, negative fictive models will also inspire imitators, even though negative behavior in fiction is usually emotionally marked, through the reactions of other characters and the tenor of the story, as behavior to avoid.

audience resistance and most admiration—the highest status—if they tell stories that appeal to values shared by the audience. Dickens' appeals to the generous hearts of his readers, whether overtly in *A Christmas Carol* or more obliquely in *Our Mutual Friend*, did not hurt either his massive sales figures or his stature in his time and since, and may have helped alter the moral climate of English society.

Another feature of fiction—but not of fact—also encourages the development of a moral sense. Story by its nature invites us to shift from our own perspective to that of another, and perhaps then another and another. Stories come most alive when *all* the principal characters have their own vivid life, especially when not only their actions but also their speech and thought are fully realized. Simply in order to animate the story, fiction lets us hear characters' speech and even access their thoughts as we cannot do in life. Merely in following its own bent, fiction cultivates our sympathetic imagination by prompting us to see from the perspective of character after character.

In factual narrative, history tends to be written by the winners, who feed reports of victorious campaigns to their chroniclers. In fiction the story lives the more *everyone* comes to life, the more each character seems to exist in his or her own right. The twelfth-century Persian poet Nizāmī reports his hero's feelings: "While each warrior thought of nothing but to kill the enemy and to defend himself, the poet was sharing the sufferings of both sides"[33]—a remark that could equally characterize Homer in the *Iliad* or Tim O'Brien in his novels of the Vietnam War.

Fiction enormously enhances our creativity. It offers us incentives for and practice in thinking beyond the here and now, so that we can use the whole of possibility space to take new vantage points on actuality and on ways in which it might be transformed.

The ability to imagine the world as other than it is underpins pretend play, and the ability to conceive of alternatives underpins all modeling.[34] Free thought needs alternatives and counterfactuals. A mental architecture that processes only true information remains severely constricted. Most discovery involves supposition.[35] We cannot even think seriously about cause without counterfactuals: if *this* had not happened, would *that* still have occurred?[36]

Nature did not design us to think in the abstract. Minds evolved to respond to their immediate surroundings and could hardly do otherwise. Yet nature *has* prepared minds to attend to, and to act in response to, other agents and actions, and has shaped *human* minds to play, through their long childhood, with models and ideas of agents and actions and reactions. As we grow, we learn to dispense with physical props and to rely more and more readily on the cultural props of our narrative heritage, always on call in memory, in order to think in sustained ways beyond the here and now.

Most of us can easily imagine concrete but complicated scenarios of action and interaction, but have much more difficulty mentally manipulating more than a few highly abstract terms in, say, logic or mathematics. A perfectly rational being, an emotionless Mr. Spock, might see this as a shortcoming—even if, most of the time, we rather like our predisposition to see things in agential terms. That we can think so effortlessly beyond the present in narrative but not, without special training, in abstract realms smacks of excellent but imperfect design for intelligence. Such a telltale combination of the ingenious and the inept itself shows evolutionary adaptation at work, like the brilliant engineering of our stereoscopic and trichromatic vision—but with a blind spot near the middle of each eye—or the advantages of bipedal locomotion and forelimb freedom—but with the curses of bad backs and hard births.[37] Nature cannot design from scratch, but it could let organisms move and act and care about other organisms that also move and act. It was only from there, from beginnings already in place, that evolution could build the higher stages of human creative intelligence.

Because fiction extends our imaginative reach, we are not confined to our here and now or dominated by automatic responses. We can think in terms of hidden causes, of inspiring or admonitory examples from the past, fictional or real, of utopian or dystopian models, of probable scenarios or consequences, or of counterfactuals whose very absurdity clarifies our thought. We know that there are always other spaces of the possible we can explore. And much of the indefinite enormity of possibility space has been made concrete and particular through the examples of story.

This may be the most important function of pure fiction. By appealing to our fascination with agents and actions, fiction trains us to reflect freely beyond the immediate and to revolve things in our minds within a vast and vividly populated world of the possible.

FICTION AND RELIGION

This whole section has a very loose connection *k o[r]*

To explain fully what difference storytelling has made to human survival and reproduction, we need to consider the impact not just of stories recognized as fictions, but also of invented stories that people take as true: of religion.

We can think beyond the actual in narrative because of our sophisticated social cognition and our predisposition to develop it through pretend play and story. This capacity has both an innovative, creative, protoscientific side and a traditionalist and conservative side. Storytelling, like other arts, can open up new possibilities, but it can also be commandeered by religion's power to enhance within-group social cohesion.

First, the novel and exploratory. Explanation moves from the known to the unknown, from the seen to some unseen cause. With other intelligent animals, humans share an understanding of intuitive physics, biology, and psychology—of things, kinds, and minds—that allows us to act flexibly within our world. But for us, uniquely, that seems insufficient.

Because we also develop, as part of our intuitive psychology, an advanced theory of mind, we acquire an intuitive awareness of the possibility and consequences of holding false beliefs: we see how we may misconstrue situations through not knowing what another may know. Other animals, especially omnivorous ones, exhibit curiosity, but unlike them, we also fret about what we may *not* know. Because we know we can know too little, we crave extra information and deeper explanation. But how do we get it? Explanation does not come easily.

Our intuitive understanding of things, kinds, and minds lets us leap to rich conclusions from slight evidence. Our theory of mind, our most powerful intuitive ontology, allows us to infer inner impulses like beliefs, desires, and intentions driving outward actions. Reasoning from ourselves, we see agency as a prototype of causality: unless I intend to strike with this club, it will not swing down upon my enemy. But the powerful explanatory force of agency depends on invisible force: my hands and arms may actually move the club, but something unseen within me determines how my limbs move. And whereas physical causation requires contact, psychological cause can work at a distance: I can make you move without touching you, merely by saying the right words, pointing to the right thing, or looking at you or moving toward you in a certain way.

Not only do we see agency as a prototype of cause; we also overattribute it, since erring in that direction is almost always safer than erring in the other. If we discount a rustle as no more than wind, we could find ourselves surprised by a snake or a spy. We know the potentially dire consequences of *not* detecting unseen agents, and we assume that if we *did* know of agents affecting our lives in ways we cannot see, we could take answering action.

We also know other creatures with powers different from ours: animals that can see in the dark better than we can, or hear or smell what we cannot, or run faster, or exert more force with jaws or claws. Even in modern cities children remain fascinated by animals, and in their play learn to reason about different characters, perspectives, intentions, and powers through the sharp distinctions that animals provide. The preoccupation with animal versus human powers has long been with us. Paintings of aurochs and mammoths adorn Paleolithic cave walls. Totemism pervades preliterate cultures. And animals whose powers challenge or question those of humans feature in stories for children or in classics from *The Epic of Gilgamesh* and the *Mahabharata* to Ionesco, Cortázar, and Auster.

We desire deeper explanations. We see cause in terms of agency, and recognize the special characteristics of psychological or "spiritual" rather than physical causation. We recognize other creatures' different powers. We readily invent, recall, and retell stories involving agents that violate expectations. Across humankind we have therefore repeatedly offered (1) deep causal explanations in terms of (2) beings with powers different from ours, (3) understood in terms of mind or spirit, moved like us by beliefs, desires, and intentions but (4) somehow violating our expectations of things or kinds, especially by transgressing normal physical limits—perhaps by being invisible, or existing in more than one place at a time, or being able to change shape or pass through solid obstacles or live forever.

Explanations in terms of spiritual cause do not dispense with explanation by physical cause. Omitting *that* would be dangerous in the extreme. Instead, supernatural stories attempt to offer an additional level of explanation. A hut may collapse because termites have eaten it, but why *this* hut, at *this* time, with *this* woman inside? What spirit or witch has she offended?[38] Such thinking always searches for a concealed agential cause to supplement the evident physical cause.[39]

As we should expect, young children naturally first understand the world in purely natural ways that reflect human intuitive ontologies, but then begin to learn and subscribe to the "deeper" explanations of local culture.[40] In Manu society in Papua New Guinea, children would first interpret events, such as a canoe's going adrift, in terms of pure physical causality—the loosening of a rope, water currents—whereas their elders would add an additional layer of explanation in terms of ghosts and evil intents. Explanations for dreams are similar in young American and Atayal (indigenous Taiwanese) children, but by the age of six or seven Atayal children, like their elders, come to view dreams not as inner representations of the sleeping mind but as real, visible external manifestations.[41]

Supernatural stories need not seek to explain the world in general[42]—although some religions and philosophies eventually attempt such ambitious systematization, once intellectual specialization and speculation take firm root—but always aim to identify unseen forces that supposedly impact our lives. We can envisage possible futures and their dangers. We know we can often be at risk from *not* knowing the full truth of the present, let alone the future. We naturally welcome what seems a deeper explanation, an unseen cause in invisible agents whom we can appease or supplicate so as to gain some measure of control over the unknown.[43]

All storytellers can earn status. But those whose stories promised an apparently deeper explanation for what happened, and therefore a better chance of reducing future uncertainties, could earn a still higher status—as those who elaborated mythologies once did, as scientists do today. To modern science, explanations in terms of unseen spiritual forces are mere fictions, invented stories without real referents. Yet despite being untrue, religious myths have provided the standard form of apparently deeper explanation for most humans since we emerged fully into culture. Such stories, arising readily from our quest for extra explanation, our overreading of agency, and our memory for the exceptional,[44] prove hard to dislodge despite their detachment from fact.

Our intuitive psychology efficiently explains a great deal in the human affairs that matter so much to us. Agential explanation has therefore seemed to us natural for so long that the idea of explaining things *without* agents can still seem hard to accept, even after science's successes. Newton's account of planetary motion in terms of gravity delighted many but also repelled others who felt that it allowed room for a godless universe or a remote and impersonal God, a cosmic clockmaker no longer actively intervening in our world. The Darwinian notion of design by natural selection—design operating retrospectively and mindlessly, merely by culling the weakest, cycle after cycle, rather than consciously and prospectively aiming for the strongest—remains still too contrary to ingrained human predispositions for many to comprehend.

Our capacity to think beyond the here and now has made our species distinctively different. It has led to attempted explanations at a level deeper than the visible. Until the emergence of science, such superhuman explanations have been all too human and flawed. Even scientific hypotheses are fictions of a kind, but *science's* invented stories offer explanation without what Daniel Dennett calls "skyhooks"—without top-down forces, without mental or spiritual agents—and restrict themselves to what he calls "cranes," explanations building slowly from the ground up, from the simple to the complex, rather than hanging answers from the heavens.[45]

Science has improved immensely on the fictive agential explanations of the past—although even scientists find they cannot help anthropomorphizing causal factors; but science could not have begun without our persistent inclination and ability to think beyond the here and now, to invent agents and scenarios not limited to the actual or the probable but exploring also the merely possible or the eerily improbable.

We forget that a fiction was invented once we start to rely on its apparent explanatory power—as we do with scientific "fictions" once we think *they* work.

Science tests its stories hard, and some stand the test. But how can belief in stories that lack any foundation in fact ever have benefited humans? Why has fiction so often been assimilated to belief? If our senses fed us untrue data, if our minds continually made incorrect inferences, we would not last long. We need true information to survive. Why, then, did we not evolve a reluctance to believe the fictions of religion?

Operating in real time with limited resources, minds have to have stopping rules. We look for the first explanation that surfaces to consciousness that seems to fit. The uncertainty that theory of mind arouses, that we do not know the complete situation, will often be appeased by the first ready-made agential explanation that comes to mind.[46]

Since it pays to detect patterns and therefore to recognize recurrence, we also have a confirmatory bias in our thinking: we welcome apparent confirming instances and often shun falsifying evidence.[47] We have a conformist bias, which inclines us not to want to think in ways too unlike others.[48] We do not readily seek disproof, and would not easily know where to start looking for it. It took millennia of civilization before we recognized the value of systematically and rigorously testing explanatory ideas, and even after four centuries of scientific method, the notion of searching for evidence that could count *against* our best theories still seems counterintuitive and disconcerting. Our forebears had neither the inclination to suspect nor the methods to test the existence of invisible beings who could perhaps cause them harm. Even today the cost of testing a belief in unseen forces is extremely high—as we know from the modern quest for neutrinos or the Higgs boson.

Our supernatural ideas seem natural to us. We fixate on spirit beings because of their difference from ordinary agents, through their invisibility, immateriality, immortality, and the like. But evolutionary anthropologists of religion show that we actually think about supernatural agents in ways that depend less on these attention-catching properties than on our intuitive ontologies, our folk beliefs about how ordinary minds work. Experiments on adults in the United States and India show what this means:

> When asked to describe their deities, subjects in both cultures produced abstract and consensual theological descriptions of gods as being able to do anything, anticipate and react to everything at once, always know the right thing to do, and be able to dispense entirely with perceptual information and calculation. When asked to respond to narratives about these same gods, the same subjects described the deities as being in only one place at a time, puzzling over alternative courses of action, and looking for evidence to decide what to do.[49]

Religious convictions derive less from doctrine than from story: "Public expressions of supernatural beliefs rarely, if ever, take the form of generalized or universally quantified statements, such as 'God reasons,' 'All spirits fly,' and 'Every frog turns into a prince.' Instead, supernatural phenomena are usually embedded in explicitly contextualized episodes; for example, 'The *arux* [an Itza' Maya forest spirit] came as a child to steal my moccasins but disappeared as a gust of wind before I could catch it.'"[50] Such episodes usually focus on supernatural spirits' violations of the ordinary, like their being born of a virgin or walking on water or rising from the dead. Religion consists less of a coherent body of dogmas or explanatory systems than of these memorably surprising stories that we then flesh out through our common intuitive ontologies.

The spiritual beings that actively matter for believers all have unrestricted access to strategic social information.[51] We ourselves attend closely to such information, and we know that because it continues to accrue wherever people act and interact, no mortal eye can see it all. We watch others assiduously when we can, as we know they watch us. Sometimes we may wish to act unobserved, but our fear that others might notice us undetected may make us wary of committing even apparently unseen actions that could earn others' disapproval and reprimand. But belief in invisible spirits who in *any* situation can observe us undetected will incline *all* of us to behave more often in the ways generally approved of by all—in ways that aid cooperation within the group rather than short-term individual advantage. Let me quote again Malidoma Patrice Somé's echo of Dagara elders: the "real police in the village is Spirit that sees everybody. To do wrong is to insult the spirit realm. Whoever does this is immediately punished by Spirit."[52]

Evolution, biologist D. S. Wilson observes, places no premium on truth. I may need to see where a rabbit is to hit it with a stick, but "there are many, many other situations in which it can be adaptive to distort reality . . . Even massively fictitious beliefs can be adaptive, as long as they motivate behaviors that are adaptive in the real world."[53]

Fictions of course can motivate without belief: we can heed fables without thinking them literally true. But stories that start as invention but persist as explanation, as meriting belief, are considerably more powerful than mere illustrative examples. All the more so if they are designed to encourage cooperative behavior. Wilson writes:

> Emotions are evolved mechanisms for motivating adaptive behavior that are far more ancient than the cognitive processes typically associated with scientific thought. We might therefore . . . expect stories, music, and rituals to be at least as important as logical arguments in orchestrating the behavior of groups. Supernatural agents and events that never happened can provide blueprints for action that far surpass factual accounts of the natural world in clarity and motivating power.[54]

Evolution will favor belief in a falsehood if it motivates adaptive behavior better than belief in a truth. (In any case, truths other than those supplied by the senses and intuitive inferences can be costly to discover.) It will favor beliefs in spirits that monitor human actions and that need to be appeased, if such beliefs reinforce group cohesion. Of course they may do so at the cost of exacerbating external strife. But if our group gains in cohesion through a sense of divine support, it is likely, other things being equal, to be able to overcome enemies who lack such a compelling belief.

We have a desire for deeper knowledge and greater control; a strong inclination to read and overread the world in agential terms; a lively interest in, and a keen memory for, what violates expectations; a predisposition to learn from others; an instinct for conformity and fear of exclusion; and an inclination to seek confirmation of what we think we know. It is hardly surprising, therefore, that we progress from explanation in terms of intuitive ontologies to the apparently more profound explanations of the supernatural stories that our cultures devise. And if we see belief as a mark of willingness to abide by the collective practices of the group, and a powerful prop for cooperation, we may even see reluctance to believe as a challenge to group unity and as tantamount to treason.

VARIETIES OF FICTIONAL EXPERIENCE

To consider the functions of fiction, we have had to consider not only "pure" fiction, but all storytelling in any medium with invented elements, including stories that begin at least in part in imagination but are relayed at some stage as true.

In the evolution of biological adaptations, different functions may dominate at different times. Bird wings apparently evolved first as thermoregulatory flaps, and can still be used that way, but they much more centrally serve what is now their main function and what has shaped their recent evolution: flight. In the same way, our predisposition to fiction has served different functions at different stages of cultural evolution. From the start, it has sharpened social cognition and extended our capacity to think beyond the here and now, to entertain other possibilities and not simply accept the given. In most societies and for much of human history, however, that storytelling power has often been commandeered by the apparent promise of explanation, cohesion, conformity, and control offered by fiction that has hardened into mythological and religious belief.[55]

Yet even in societies saturated in religion or shored up by ideological fictions, other stories persist alongside official myths. *The Arabian Nights* reached their fullest form in Islam's most expansive years. Shakespeare conjured up the fairies of *A Midsummer Night's Dream* and the enchanted isle of Prospero despite an official Anglicanism that often sent religious dissenters to their deaths. Anti-Stalin jokes circulated in whispers in the stalls of Red Star cinemas screening party propaganda films. Or to take a more traditional example: among the Kalabari in Africa,

> serious myths intended to throw light on the part played by the gods in founding social institutions shade into tales which, although their characters are also gods, are told for sheer entertainment. And although the Kalabari do make a distinction between serious myth and light tale, there are many pieces which they themselves hesitate to place on one side or other. Belief shades through half-belief into suspended belief. In ritual, dramatic representations of gods carried out in order to dispose them favourably and secure the benefits which, as cosmic forces, they control, are usually found highly enjoyable in themselves. And they shade off into representations carried out almost solely for their aesthetic appeal.[56]

Story emerges out of our focus on one another and other animal agents, and out of the play that helps us learn to imagine by way of actions and agents. Although story can be requisitioned to the serious and even solemn purposes of religion, play remains, on the fringes or right back in the center.

We have, and have long had, stories of low cost but high long-term benefit, like proverbs, parables, or fables that still provide moral rules of thumb. We have stories of low cost and high immediate benefit for teller and audience, like jokes, which—being brief, portable, and steeped in the pleasures of surprise—act as an efficient social lubricant, like all social play.[57] We have stories of high cost and high immediate benefit, as in the screen and print fiction produced especially to earn short-lived attention. And we have stories of high cost and high long-term benefit, like the serious stories that provoke us to reconsider what it is to be human.

In modern secular societies, with an unprecedented degree of specialization of labor and mechanization of print and other media, fictional storytelling again serves the purpose it has usually served, whenever not appropriated by faith or power. It helps us to understand ourselves, to think—emotionally, imaginatively, reflectively—about human behavior, and to step outside the immediate pressures and the automatic reactions of the moment. From pretend play and jokes to Homer, Murasaki, or James Joyce, fiction taps into the swift efficiency of our understanding of agents and actions. Old and new stories and characters open up and populate possibility space. All these fictions make us the one species not restricted to the here and now, even if that must be where we act and feel—and imagine.

BOOK II

FROM ZEUS TO SEUSS: ORIGINS OF STORIES

> The Boston flight had only just landed and I guessed I had a half-hour wait. If one ever wanted proof of Darwin's contention that the many expressions of emotion in humans are universal, genetically inscribed, then a few minutes by the arrivals gate in Heathrow's Terminal Four should suffice. I saw the same joy, the same uncontrollable smile, in the faces of a Nigerian earth mama, a thin-lipped Scottish granny and a pale, correct Japanese businessman as they wheeled their trolleys in and recognised a figure in the expectant crowd. Observing human variety can give pleasure, but so too can human sameness.
>
> Ian McEwan, *Enduring Love* (1997)

IN BOOK II OUTLINED a naturalistic account of art and fiction and argued that both are adaptations: we have evolved to engage in art and in storytelling because of the survival advantages they offer our species. Art prepares minds for open-ended learning and creativity; fiction specifically improves our social cognition and our thinking beyond the here and now. Both invite and hold our attention strongly enough to engage and reengage our minds, altering synaptic strengths a little at a time, over many encounters, by exposing us to the supernormally intense patterns of art.

These hypotheses need to be tested empirically against alternative explanations. But an evolutionary approach to literature does not depend on them or on any other hypotheses that claim art, literature, or fiction as adaptive. A biocultural approach to literature simply requires that we take seriously that evolution has powerfully shaped not just our bodies but also our minds and behavior. We can do that whether literature or fiction is an adaptation, byproduct, or some combination of the two.

Evolutionary literary criticism will be worth the detour into biology and psychology only if it deepens our understanding and appreciation of literature. Book II offers two contrasting case studies—Homer's *Odyssey* and Dr. Seuss's *Horton Hears a Who!,* ancient and modern, adult and children's, serious and comic, massive and minuscule—to show that it can.

Part 4 explores in depth the most vital classic of antiquity, the *Odyssey.* It tries to suggest what an evolutionary perspective allows us to see afresh even in a work astutely analyzed for two and a half millennia. Of course readers have already seen much in the *Odyssey* or it would not have had such an audience for a hundred generations. An evolutionary approach will not reject all that others have seen but will also suggest new directions in which to look.

The six chapters fall into three pairs, focused in turn on attention, intelligence, and cooperation. The differences between each pair, and each of the six chapters, should suggest the variety of possibilities that evolutionary approaches can open. They certainly do not limit the evolutionary approaches to a work as rich as the *Odyssey*.[1]

Part 4 has a second principal purpose. It introduces core elements of fiction: character, plot, structure, dramatic irony, and theme. It also attempts to explain these familiar narrative features from first principles by treating fiction as a human activity arising naturally out of other human and animal behavior, and to consider them in terms of a particular work's unique problems and solutions.

Part 4 also has other aims: (1) to show the sophistication of human thought and art close to the origin of recorded stories, almost 3,000 years ago, before many of the storytelling devices we take for granted had been discovered; (2) to offer an example of the way in which creativity in art precedes and prepares for creativity in the more resistant nonhuman world, in this case the achievements of Greek philosophy and science; (3) to show the necessary link between problems of historical knowledge and universal aspects of human nature. An evolutionary approach to literature needs to be biocultural: it needs to take into account historical, textual, and cultural scholarship; (4) to demonstrate the power of a problem-solution model, emerging with the emergence of life and fruitfully applied to art by art historian Ernst Gombrich and film scholar David Bordwell;[2] (5) to show that although art as a behavior has been designed in part by evolution, its power also depends very much on traditions and individuals; (6) to demonstrate that while all fictions offer intense doses of pattern, classics offer concentrations of pattern high enough to reward continually renewed attention.

Part 5 focuses on Dr. Seuss's 1954 children's picture story *Horton Hears a Who!* Whereas Part 4 explores near the *historical* origin of story, where story remains closely linked to tradition and religion, Part 5 selects a story near the *individual* origin of story, once designed for young children yet also in its different way aiming, as Homer does, at classic status, but composed in a world in which tradition and religion have ceded ground to innovation and creativity.

Just as Part 4 introduces the elements of fiction and the particular case of the *Odyssey,* so Part 5 introduces different levels of reading or literary explanation—universal, local, individual, and particular levels, and the relationships between story and meanings—as it examines the special circumstances of a single story.

Standing near the origins of recorded story, the *Odyssey* offers us both a complete and highly reliable text and a case study in the problems caused by the lacunae of early historical knowledge. *Horton Hears a Who!* by contrast shows the opportunities provided by historical knowledge—Dr. Seuss's prolific work in many media, his working processes, his sense of his own aims, and his immediate contexts—and the fertile interrelationship of the immediate and the wider evolutionary contexts. Mystery shrouds Homer's mode of composition and his aims. Dr. Seuss's methods and aims can be readily retrieved, and show how creativity operates as a Darwin machine, through repeated cycles of variation and selection.

PART 4

PHYLOGENY: THE *ODYSSEY*

poetry . . . in the noblest nations and languages that are known, hath been the first light-giver to ignorance, and first nurse, whose milk by little and little enabled them to feed afterwards of tougher knowledge . . . the poet . . . cometh unto you, with a tale which holdeth children from play, and old men from the chimney corner.

Sir Philip Sidney, *A Defence of Poesy* (1595)

14

EARNING ATTENTION (1):

NATURAL PATTERNS: CHARACTER AND PLOT

FOR NEARLY 3,000 YEARS, audiences and readers have enjoyed the *Odyssey* as "one of the best tales ever told."[1] It stands with its predecessor and companion, the *Iliad*, as the oldest masterpiece of ancient story to survive, more accomplished than the earlier and incomplete *Epic of Gilgamesh*, more concentrated than the later and over-replete *Mahabharata*. Though not as mighty as the *Iliad*, the *Odyssey* has earned a larger audience through its hero, its story, and its narrative art.*

What can an evolutionary perspective offer on a story subjected to so much attention from audiences and critics for over two and a half millennia? For one thing, it can stress the importance of attention itself, so often taken for granted in literary criticism, or dismissed as the mere courting of popularity, but in fact a sine qua non of all art. Art can affect minds over time because it so compulsively engages attention.

An evolutionary account of the *Odyssey* will not ignore either the criticism or the critical skills that have developed over the centuries. After all, these skills start with our evolved capacities to understand events and minds and develop via the social learning that our species has also evolved to rely on. Indeed, I will take the opportunity provided by Homer's mastery of character and plot to explain these core

---

*Unless otherwise noted, I quote from Richmond Lattimore's translations of *The Odyssey* and *The Iliad*, as offering the best combination of accuracy and poetry, and follow his spellings of proper names.

elements of story in terms of an author's shaping audiences' attention by appealing to their evolved cognitive predisposition to foreground and respond to—automatically, amid all the possible patterns in a story—the patterns of character and plot. A biocultural analysis does not ignore the common ground, but as we will see increasingly over the next six chapters, it can gradually deepen the perspective—*and* our sense of a storyteller's art.

Homeric Problems

Despite millennia of criticism, much about the *Odyssey* remains unsettled. Who composed the poem, and where, and when? Was it the work of a single author? If so, was it the "Homer" who composed the *Iliad*? Or did both epics emerge from a long tradition of oral storytelling that slowly crystallized into the texts we now have? Was the *Odyssey* composed orally and written down or dictated by the poet, or handed down orally in roughly the form we now have, or continuously modified before finding written form centuries later? Does it reflect the society of its time, whatever that time was, or of the Mycenaean age, in which something like the Trojan War may have occurred or been imagined, or of the so-called Dark Ages of Greece, between the collapse of Mycenaean culture and the time of composition, or some temporal blend? Did its first audiences regard it as history or as fiction?

Debate rages over all these issues and more, and just enough information survives to ensure that any answers remain problematic. Nevertheless, since the work of Milman Parry early last century, scholars accept that both the *Iliad* and the *Odyssey* derive from a long tradition of oral epic storytelling and that their outlines and many details can be explained by this tradition.

Although contention persists, consensus also exists: "it is now universally agreed that our epics present narratives of such coherence and sophistication that they cannot represent a mere accumulation of material over several centuries: a major creative effort must have gone into their final composition."[2] Many agree that each epic is so internally intricate as to suggest strongly the work of a single author, though not necessarily the *same* author, who is "not subordinated to his tradition, but master of it,"[3] able to use, ignore, violate, or transcend "his inherited techniques ... as he wishes."[4] Indeed many think that the composition and preservation of epics of such unprecedented scope and scale (15,000 lines in the *Iliad*, 12,000 in the *Odyssey*) required "the special ability, aims, imagination and reputation of a particular singer," whose prestige enabled him "to outstrip normal performances, normal audiences and normal occasions ... [and] *impose* his own will and his own vast conception"[5] on both his own time and posterity. Many also think that although Homer grew up in the oral tradition, writing (using the newly invented alphabet) may have played a part in his composing the poems, probably in the late eighth century B.C.E. for the *Iliad*, and perhaps thirty years later for the *Odyssey*.

While all agree that the *Iliad* precedes the *Odyssey*, not just in story line but in composition, only some of those who accept solo authorship for the *Iliad* believe that the same author composed the *Odyssey*. The two epics differ not only in setting—war versus peace—but also in tone, trajectory, and the role of the gods. In the *Iliad*, the gods bicker; mortals on both sides gain and lose from divine interventions; no side consistently has a moral advantage. In the *Odyssey*, all the gods except Poseidon back Odysseus; none backs Penelope's suitors; Odysseus, though not flawless or idealized, easily outstrips his opponents physically, intellectually, and morally. For that reason, and because Zeus and Athene support him, we anticipate and eventually witness victory for Odysseus. The *Iliad*, by contrast, has no clear-cut moral lines, no winners, and many losers.

Yet if nothing of Shakespeare survived but anonymous texts of, say, *A Midsummer Night's Dream* and *King Lear*, could we imagine these two plays as products of the same mind?[6] For all their differences in tone and tenor, much links the *Iliad* to the *Odyssey*, and much distinguishes them from the surviving glimpses of the Epic Cycle tales composed over the next two centuries. There are no significant differences in diction between the two Homeric epics,[7] and although in *modern* oral tales each singer has his own characteristic standard forms for each type-scene, the type-scene templates in the *Iliad* and the *Odyssey* are indistinguishable.[8] Both poems have a high proportion—almost two-thirds—of dialogue, and "the little we know of other ancient Greek epics supports Aristotle's statement that this ratio is unique."[9] Both Homeric epics focus intensely on a small part of their overall stories, unlike the Epic Cycle stories, and both downplay the fantastic and avoid the proliferation of intrigues characteristic of the Cycle.[10]

Except for guesses based on Homer's own, perhaps very stylized, depiction of the bards in the *Odyssey*, we know nothing of the circumstances of the poem's composition or performance. But across cultures patrons often play a role in producing oral epic.[11] It would be surprising if a work as extraordinary as the *Iliad* had not prompted patrons to request another, similar story. Such differences as exist between the two poems could easily be explained by the catalyst of a patron seeking a story with Odysseus rather than Achilleus as hero, a positive rather than a tragic end, and a more positive view of the gods. Of course such shifts could also easily have arisen from the poet's own reflections rather than a patron's promptings, but in either case the differences between the *Iliad* and the *Odyssey* occur at this level, easily subject to conscious redirection by an author serving a slightly altered artistic purpose, and not at the level of unconscious lexical or scenic habits that authors cannot readily alter by choice and that usually differ markedly from author to author—but not from the *Iliad* to the *Odyssey*.

Whether Homer is one, two, or many, I will refer to "Homer" and assume a consistency of artistic purpose. This allows us to discern qualities in the *Odyssey* obscured for those who see the Homeric epics as poorly cobbled together, "a miserable piece of patchwork" (Wilamowitz on the *Iliad*), "a crime against human intelligence" (Fick on the *Odyssey*).[12]

Strategies for Attention

An evolutionary approach to literature, as to life, sees organisms as seeking strategically to solve their particular problems, immediate and longer-term. Both authors and audience members will be strategists. Authors pay steeply in composition costs in return for engaging the attention of audiences and directing their response as much as they can without compromising their engagement. Audiences pay costs in time and effort and perhaps more (admission or book fees, for instance) and will seek to recoup as much benefit—as much cognitive and emotional stimulus—as possible. They will be particularly wary of attempts to persuade them to attitudes contrary to their sense of their own interests. By stressing explicitly the costs and benefits of earning and paying attention, an evolutionary approach to literature can focus rather than blur our appreciation of artistry.

As a singer of tales, Homer needs first to catch, hold, and move his listeners: if he bores them, he will not be invited to dinner again.[13] But how could he shape his story to appeal not only to his original audience, whoever they were, but also to an audience who cannot believe in Greek gods, the right of powerful males to whatever females they can obtain by force, or the justice of massive revenge? How can a classic remain alive over millennia of change?

One part of the evolutionary answer is that we all share a great deal as humans, however great our cultural differences. Another is that classics make multiple appeals to our attention, and that while some of these appeals fade over time, other attractions become more vivid as we uncover the insights and implications of genius. I start therefore with what almost all audiences agree on: the first reasons for enjoying the *Odyssey,* for thousands of years, some depending on the special place that Odysseus and the Trojan War have in Greek legend, but most involving features of character and plot at the heart of all story.

As we have seen in Part 3, storytelling can command the attention of others by delivering high-intensity social information. Even expert storytellers have to compete against other demands on potential audiences, including their attention to the immediate environment, to everyday needs, and to other artistic appeals for attention. Strategizing bards should aim to reduce composition costs while earning the maximum rewards of attention.

Outstanding storytellers tend to choose the forms most valued in their culture. The Greeks canonized epic form and epic content, their tales of gods, heroes, and mortals, but only Homer survives of the storytellers of his time—and, Aristotle suggests, only his epics *de-*

served to survive. How did Homer appeal not just to his own time but beyond?

Studying storytelling in black inner-city American neighborhoods, sociolinguist William Labov identified the importance of "evaluation," of making clear to a possibly restive audience the *point* of the particular story you wish to tell.[14] In a tradition such as Greek epic, the most prestigious narrative form of Homer's time, the epic matter and manner themselves provide a guarantee: if you are Greek,* this concerns you, this is your story. But Homer took it further. He begins his two epics with proems, poetic promos encapsulating the subject of the story to follow (the wrath of Achilleus, the return of Odysseus) and promising a concentration that to judge by Aristotle was far from the norm for epic. Homer promises mighty material, and minimal distraction—promises that in a world of competing demands appealed then and still appeal now.

TRADITION AND INNOVATION

Competition for attention has always been a feature of conscious life. The epic mode may have guaranteed at least initial attention from a Greek audience, but even the best-disposed audience can drift. How does Homer hold his audience while telling a story already familiar?

Tradition gives his subject the stamp of authority, but how much did he depend on the prestige of the old stories and styles? How much did he fill in the existing outline of legend? No documents prior to the *Iliad* or the *Odyssey* survive to suggest how much he drew from tradition. But Aristotle, centuries later, expected no rigid adherence to the past.[15] The Athenian tragedians allowed themselves considerable narrative freedom, although by then, three centuries after Homer, tradition was much more open to challenge than in Homer's own time. Nevertheless in the *Odyssey* itself Telemachos

---

* An anachronism here. Homer thought in terms of local populations—Ionian, Achaian, Spartan, and so on. But he used the terms "Achaians," "Argives," and "Danaans" almost interchangeably to refer to the whole "Greek" force massed against Troy, and through both this sense of a common purpose and the prestige of his achievement his work played an important part in the creation of a sense of a panhellenic culture.

observes, when the bard Phemius sings his tale: "People . . . always give more applause to that song which is the latest to circulate among the listeners" (1.351–352). Homer's characters report the past freely, compressing, expanding, altering emphases, and we can infer that Homer likewise "felt free to expand, adapt, or complement his story (provided he did not change its core, which was quite small, presumably," suggests Irene de Jong: "in the case of Odysseus, it may have comprised no more than the hero's successful return home after twenty years)."[16]

This may overstate the case—slightly. But other than these elements—Odysseus as the last to return from Troy, reaching Ithaka after many delays and adventures, including the challenge posed by Polyphemos, and finding a wife still faithful despite the importunate attentions of suitors—the story of Odysseus' return may not have been much more elaborated within the larger story of the Trojan War.

CHARACTER

How does Homer engross his audience in Odysseus' return? One sure route to attention is to focus on highly memorable characters. We heed others, we distinguish one from another, and we take particular note of those of high status or exceptional powers or both. Epics from *Gilgamesh* to *Beowulf* focus on redoubtable heroes, especially those like Odysseus who have already acquired local cultural capital. Odysseus finds that even among the remote Phaiakians his exploits at Troy are already a standard theme of their song.

Odysseus is the king of Ithaka; a warrior hero, second only to Achilleus among the Greek fighters at Troy; the man whose ruse of the Wooden Horse led to the sack of Troy and the recapture of Helen after the impasse of a ten-year siege; the most intelligent of mortals, a favorite of Zeus, ruler of the gods, and of Athene, warrior goddess of wisdom.

Tradition already defines him: in the *Iliad* he is "like Zeus in counsel" (2.636) and "his mind is best at devices" (10.241), although the most famous of his devices, the Wooden Horse, plays its part after the events recounted in the *Iliad*. But Homer makes him even more remarkable than legend recorded. Three times in the *Odyssey* he tells the story of the Wooden Horse, in circumstances more dramatic each time. Odysseus' reputation at Troy has already spread across the Mediterranean, and he is on the minds of all, from Olympos to Ithaka. Homer makes the most of Athene's unique relationship to Odysseus, not as a mortal lover, as other goddesses have, but uniquely as a kindred spirit, an intellectual comrade, a representative among mortals of the things *she* stands for among the gods.

Even with Athene's support, Odysseus also arouses our interest and sympathy because he has suffered more than anyone else returning from Troy, yet still endures. Although the victory at Troy owed more to him than to anyone else, he has had to face years of grief as he pines for home. We first see "great-hearted Odysseus," "wretched above all the other men," weeping for a way home, with "the sweet lifetime . . . draining out of him" (5.150, 105, 152). His absence casts a pall over all who fondly recall him.

In each epic Homer focuses intently on a single hero, although Achilleus withdraws from the action for much of the *Iliad* and Odysseus remains absent or unrecognized through most of the *Odyssey*. In each, he builds up the hero as a nonpareil. The structure of each poem magnifies the hero, in a different way each time, yet shows him to be not only the greatest warrior but something else: the most compassionate, in Achilleus' case, for all his fury on the battlefield; in Odysseus' case, the hero who can achieve his most glorious victory, against more than a hundred suitors, by playing the part of a lowly beggar abused by all.

We keep close track of the personality of others, because doing so helps us to predict their behavior and to plan our own. Since personality remains relatively stable, it pays us in life to monitor it, especially if a person is complex enough not to be entirely predictable. Story offers us intense concentrations of character information, and never more than in the case of Odysseus. He is not only the shrewdest strategist and the most dauntless warrior in the *Odyssey*, but the most multifarious character in ancient literature, and perhaps even since. James Joyce asked a friend whom he considered the most many-sided of all literary characters. He knew his own answer:

Odysseus—or Ulysses, in the Latin version of his name.[17] His admiration spurred him to the challenge of composing his own *Ulysses* with its own all-encompassing hero.

Odysseus is "king and rogue, hero and trickster, gentleman and knave, family man and wanderer, faithful husband and roving lover; egotistic and considerate, ruthless and tender, prudent and foolhardy."[18] He is inexhaustible, and an unexampled hero, yet never merely idealized. The greatest triumph on his return voyage, as we will see, is also his greatest disaster, entirely preventable, entirely his own fault, earning him the wrath of the god Poseidon for his own folly (*and* his resourcefulness), and ultimately costing him the lives of his companions. He stands out from all men, and yet he also seems to undergo the whole range of human experience.

PLOT

Episode

Character provides one of the surest means for a story to earn attention, but in his *Poetics* Aristotle places plot first. Surprising incident, especially if stakes are high, offers the simplest motive for story. The most famous parts of the *Odyssey*, even for those who have never read it, are Odysseus' fabulous adventures beyond the fringes of the familiar, especially his ordeal in the cave of the one-eyed Cyclops, his encounter with the sorceress Circe, his sailing between the cliff-dwelling monster Skylla and the deadly whirlpool Charybdis. The Cyclops and Circe stories, "more folk tales than epic episodes,"[19] seem to have circulated widely before Homer composed his story.

But Homer allows only one-sixth of the poem to the wanderings, and even then he has Odysseus himself tell the story to his Phaiakian hosts, as if both to provide eyewitness authentication and to minimize his own narrative responsibility for these wild stories. Mere incident can become repetitive, predictable, disjointed: one damned thing after another, with each adventure having to top the last. As a master storyteller, Homer recognizes this problem and solves it in multiple ways. Surprising fates can engage us even more than surprising incidents. Homer makes Odysseus' extraordinary destiny—

his regaining his wife and home after twenty years—the key to every door of his many-chambered story.

Unity and Diversity: Goals

Aristotle critiqued the episodic as the worst kind of plot. The summaries of other Epic Cycle stories show that although they concentrated on characters like Herakles or Theseus, still weightier in Greek mythology than Odysseus, they had no other focus, merely running their heroes through obstacle courses of traditional adventures and cheaply contrived new surprises. In contrast to these stories Aristotle singled out Homer's epics and their unity of action. Homer tells the story not of Achilleus' entire life, but only of his wrath and its consequences in a forty-day stretch in the Trojan War; he does not tell Odysseus' whole story, but only his return home.

Aristotle did not explain why or how "unity of action" matters in story. It matters because it focuses our attention and promises an overall pattern into which every event fits like a voussoir into an arch. In any situation, we look first for information relevant to our current concerns and resist the irrelevant. In story, the protagonists' goals matter and *become* our criterion for relevance. From infancy we can identify others' goals, and we have a default sympathy with their pursuing them—unless they are at odds with ours. In the case of Odysseus, we, the gods, Odysseus, and his family and friends all desire his return, so that anything relevant to that overarching goal instantly counts.

Intelligence evolved out of movement, to guide organisms away from threats and toward opportunities. All languages equate movement toward with approaching or attaining a purpose.[20] Goals have shaped stories since before the *Epic of Gilgamesh,* and quest narratives like *The Lord of the Rings* still flourish. Few of these stories make the terms of the quest sharper than the *Odyssey,* where Homer establishes from the first Odysseus' frustration with his confinement to Kalypso's island and his desire to return home. The Olympian gods support his goal; so do all those we admire in Ithaka and on the Greek mainland; and so do we.

Odysseus' motivation has not just simplicity and clarity but emotional amplitude. We notice an exceptional hero, but our sympathy can be engaged when *anyone* pursues a goal central to life. Odysseus' goals could not be more fundamental: survival and security, his own and his family's; and reproduction, including mate selection and retention, and the defense of offspring and the resources to maintain them.

Odysseus enters the poem in what might seem a male fantasy come true,[21] sharing a bed with a goddess who can surround him with love and luxury and even the promise of immortality; yet we find him "unwilling beside the willing nymph" (5.153),[22] weeping to return to home and to family. In Greek culture, *philia* sums up "that attachment to one's normal and natural social environment which underlies so much of Greek happiness,"[23] but under other terms *philia* is a human universal. As research inspired by the attachment theory of John Bowlby has shown, human and other infants need a secure base, especially a mother or other primary caregiver to return to, before they have the confidence to venture out into the world.[24] Odysseus is both the ultimate explorer, who in that sense represents all human curiosity—and we are the most curious species on Earth —and the personification of the desire to return home, which drives the whole narrative of the *Odyssey.*

The attachment of parents and children preceded in evolutionary terms the more or less monogamous attachment of male and female partners. The link between the two is revealed by the fact that in mammals attachment both between parents and children and between partners is expressed neurochemically through oxytocin. A central focus of story across the world is falling in love,[25] or, in unromantic biological terms, "mate selection." The *Odyssey* tells us nothing about how Odysseus and Penelope fell in love, although an *Odyssey* as linear and loose as the *Herakliad* or *Theseid* that Aristotle mocks could have told such a story. But although falling in love provides the climax of many a tale, it could never have been the climax of the *Odyssey,* whose story is the hero's return after twenty years to a faithful Penelope. Homer holds our attention by eschewing the endorphin high of Odysseus and Penelope falling in love—which would have made his story sag in its long middle—and moving straight to the point where he can build steadily toward the climax of their reunion.

Nevertheless he evokes aspects of mate selection. He offers us an ideal opportunity for Odysseus: the virginal young princess, Nausikaa, who realizes that she is ripe for marriage and demurely but unmistakably signals that she would be happy to be wooed by Odysseus. Though charmed, Odysseus wants nothing to deflect him from his return home. But Penelope at home shows another side of the process of mate selection. More than a hundred males compete against one another for her, Ithaka's prize female, but she resists them all because she has already selected her prize male.

In monogamous relationships, in which both partners help to raise children, mate retention is a problem. Homer intensifies this problem to the extreme, not only making Odysseus the last to return from the Trojan War, but having him return after twenty years, double the length of the war itself, and making him face a wife besieged by a swarm of suitors. While the need to defeat the reckless and murderous suitors determines Odysseus' strategy over the last half of his tale, its climax comes not with the bloody defeat of the suitors but with Penelope's at last recognizing Odysseus in a way that shows their unique compatibility. Odysseus enters the *Odyssey* in a situation that seems a male fantasy, as the paramour of a goddess; he ends the poem realizing a still greater fantasy, returning home after twenty years away to find Penelope still weeping for him every day, still enchanting enough to have 108 suitors in hot pursuit, only for him to be able to establish his absolute mental and physical superiority to all his rivals and his unmatched, unbroken bond with his wife.

Unity and Diversity: Obstacles

Odysseus' goals are decisively clear, central to life, emphatically intense, and emotionally rousing for the gods, the mortal characters, and us. But goals by themselves rarely make a story. "Odysseus wants to return home, so he does" does not promise much of an epic. Story is not only the pursuit of goals, but the interaction of goals and obstacles—often, obstacles arising from the competing goals of others. We know that the world does not depend on our desires and often frustrates them, so that to achieve them we need resolution and inventiveness. Odysseus has both, and to an unprecedented degree.

He faces dire obstacles: raging elements, daunting distances, angry gods, monsters and mortals galore, even amorous women, both human and divine. The giants and goddesses on his travels provide the spectacle in his story, but from the first Homer establishes the suitors as his ultimate obstacle, not monsters but mortals, an army of reckless insolence installed for years in his own home. None of Odysseus' adventures in his wanderings take as much as a single book, but the challenge that faces him on his return dominates the first four and the last twelve of the poem's twenty-four books.

Revenge elicits intense emotions, from gnawing curiosity to moralistic anger to acute apprehension as the rewards and risks of retribution loom. Until it is satisfied, moral anger can preoccupy us like little else. Since revenge can whet an appetite at the start of a story and be amply satisfied at the end, it is no surprise that it has long been and still remains a staple of narrative, from the *Oresteia* to *Hamlet* and *The Revenge of the Jedi*.

Since we became human long before we invented legal institutions to break cycles of offense and retaliation, the urge to avenge affronts and injuries remains deep in our psyche. We may no longer share Odysseus' or Athene's sense that the suitors and the serving-women who consorted with them all deserve death, but Homer summons all his craft to ensure that his audience—including those living under conditions he could not readily imagine—feels a moral outrage against the suitors for their indifference to all responsibility, whether to the gods or their own social superiors or inferiors.

He establishes two interlinked problems from the start. The first—how will Odysseus return to Ithaka?—we, the gods, and Odysseus share from the outset. The second—how will the suitors be dealt with?—Odysseus will be the last to ponder. Homer establishes the problem of the suitors' behavior not as his hero's narrowly personal motive but as something that outrages us, and the gods, and Odysseus' family, his peers, and his trusty servants. Long before he returns to find the suitors wooing his wife, we all recognize the moral problem they pose and Odysseus' return and revenge as the only solution.

In the *Odyssey* the unity of action that Aristotle extols in Homer is therefore a double unity, twining through the poem in two sturdy, mutually reinforcing strands. First, the return, a problem and a goal we share with Odysseus from the first. Then, the revenge against the suitors, a problem that we know lies ahead before Odysseus does, although we look forward eagerly to him as the solution to the problem. Homer solves his own problem, as so often, in several ways. He finds a goal that can unify the poem's action and encompass the traditional adventures of Odysseus, and more; he then invents a subordinate goal that he establishes for us but does not disclose to Odysseus until the final phase of his story, and that he treats very differently, in a much more detailed and psychologically strategic way. He creates both a complex unity of action through his hero's nested hierarchy of goals, and a complex diversity of action. He offers not just a higgledy-piggledy succession of adventures but a series of changes in perspective, pace, and tactics. He keeps the principal obstacle before our eyes, but hidden from his hero's, through the interim goals of the first half of the story, then transforms Odysseus from someone buffeted by fate, and apparently without support, to someone resolutely in command as he plans his revenge, confident of Athene's eager support.[26] And he makes his hero, enduring, resourceful Odysseus, someone uniquely endowed to withstand and overcome all the obstacles he must face.

Outcome

Vividness of incident captures immediate attention but runs the risk of lapsing into a trite succession of spectacular moments. Unity of action, by integrating goals and obstacles, can hold an audience for far longer, by making every event relevant to an end we either dread or desire, part of a pattern that connects and sustains disparate details. And this leads to a third major factor in plot. The likelihood of change in the fortunes of characters we care for raises our expectations and rouses our emotions.

Feelings evolved as a guide to behavior: an appraisal of the positive or negative value of current conditions and of actions to be taken in response to them. In real life we naturally prefer a positive emotional outcome, indicating that a course of action has been of benefit. But the negative emotions, like fear and anger, are more powerful, because more critical: we need to avoid being eaten much more urgently than we need to eat. Hence tragedy and thrillers can engage emotions at their most intense. Yet most of us naturally prefer happy endings.[27]

Homer simplifies his story through the clarity of Odysseus' goals, and our sympathy with them, while he amplifies it through the range of experience these goals involve, without the story's ever losing its strong sense of direction. He makes the most of the intensity of negative emotions through the mortal dangers Odysseus confronts and the outrage the suitors' behavior provokes. But he also holds out the promise of ultimate pleasure. He maximizes the distance between Odysseus' initial trials and his ultimate triumphs, and allows us to anticipate those triumphs even when no other mortal characters can.

More than anyone else, Odysseus made possible the Greek victory at Troy. Of all the returning warriors, he deserves most, yet his sufferings in the wake of the war continue longest of all. He loses his companions, lives for seven years without mortal company, and we first see him weeping in despair of returning to home and humankind. But because we can see he has the support of all the gods but one, we can also foresee that his fortunes will change radically, transforming him from the most unfortunate to the most fortunate of homecomers. Physically he has been the most direly threatened of all the heroes at Troy. Socially his trials are no less extreme: back in Ithaka at last, he will be reduced to a level more abject than any other Greek hero could endure, a beggar abused by another beggar, by servants, by the suitors despoiling his home and wooing his wife. But he reestablishes himself as the father of staunch Telemachos, the victor over the suitors, the husband of faithful Penelope, the lord of a newly pacified Ithaka, the upholder of the line of Laertes. Priam, king of Troy, had had fifty sons, but Zeus decrees he shall lose them all, and his kingdom, along with his life. Odysseus by contrast is his father's only son, and Telemachos *his* only son, but together he and his son defeat the army of the suitors, and together with both his father and his son he stands resolute at the end of the poem against those who seek to avenge the suitors' death.

As the contrast with Priam indicates, the scale of Odysseus' triumph is all the greater through the contrasts Homer repeatedly establishes between Odysseus and other heroes of the Trojan saga: Menelaos, reunited with his fickle Helen, the *cause* of the whole war; Agamemnon, who returned home only to be killed by his faithless wife and her lover, the grimmest *consequence* of the war's hollow victory;[28] and Achilleus, dead and dejected in Hades, unconsoled by the glory he obtained at Troy, his despair relieved only by the news of his son's bravery in battle. Odysseus has already earned glory at Troy as the architect of the city's defeat, and now adds to it as the most heroic of the returnees, as the victor over the suitors, as the husband of the most faithful of wives. And he still lives to enjoy both his renown—*and* the renown of his son—and life itself.

Storytellers need to balance audience benefits against audience costs in time and comprehension effort. Homer ensures the popular appeal of the *Odyssey* not only through the features of character and plot already considered, but also by merely simplifying his plot, dividing the characters the easiest way of all, into "goodies" and "baddies."[29] He focuses overwhelmingly on one hero, the foremost of the goodies, and contrasts the singularity of this hero with the often anonymous rabble of the suitors. In these respects, the strategies of the *Odyssey* stand in stark opposition to those of the *Iliad*. There, there is no division into goodies and baddies—indeed, Homer appears to have gone out of his way to allow much more of the Trojan point of view than the traditional story granted[30]—and a score of highly individualized heroes take turns on center stage. These are major reasons for many critics' preferring the *Iliad*, and for thinking that the author of the *Iliad* did not compose the *Odyssey*. They are also major reasons for the greater popular appeal of the *Odyssey*.

But as we shall see, Homer simplifies the story of the *Odyssey* into goodies and baddies—perhaps at a patron's request, perhaps for his own reasons—without sacrificing moral complexity. Although the suitors are almost all irredeemably offensive, and Odysseus is unrivaled in strength, shrewdness, and steadfastness of purpose, he also has his flaws, even in the midst of his successes. And although the story focuses resolutely on Odysseus, even in his absence, it makes him the most multifaceted of characters, in his roles, his personality, his experiences, his states of mind; it focuses on him in surprising ways, through sophisticated angles and ironies, as we shall see in the next chapter; and it does not ignore the point of view of others, even characters of humble station, like the nursemaid Eurykleia and the swineherd Eumaios.

Homer devises scores of ways to secure his audience's attention as he draws on the strength of a local tradition, the Trojan saga, and the universal impact of character and plot as elements of story. Character and plot are from one angle problems that all storytellers must solve to hold an audience, but Homer easily outdoes the usual solutions. He generates multiple simultaneous solutions, even in these most ordinary and obvious aspects of story. The inventiveness and originality of his quest for his audience's attention and response will take us still further into the art of story.

15

EARNING ATTENTION (2): OPEN-ENDED PATTERNS:
IRONIES OF STRUCTURE

NATURE HAS SHAPED OUR ultrasocial selves to attend to character and event in life and story. Homer not only manages both superbly, offering us an extraordinary hero and shaping events toward
a supremely challenging goal, but he also makes them multifaceted
enough to encompass wide swathes of life without losing focus or
force. But all this is still not enough to explain why the *Odyssey* remains such a classic after 3,000 years.

Authors can appeal to their immediate audience, or to future audiences they can only vaguely imagine, to first-time audiences, or to
repeat audiences. Homer appeals to all four. He sought and found
ways to hold the attention of his time, of posterity, and of newcomers and oldtimers.

Academic literary criticism tends to focus on meaning, on the
themes of traditional critics or the *ideologies* of more recent ones.
But works of art need to attract and arouse audiences before they
"mean."[1] Every detail of a work will affect the moment-by-moment
attention it receives, but not necessarily a meaning abstracted from
the story. Our minds can focus on only a few things at once. To hold
an audience, in a world of competing demands on attention, an author needs to be an inventive intuitive psychologist. Yet criticism has
tended to underplay the "mere" ability to arouse and hold attention.

A biocultural approach to storytelling, by focusing on authors'
appeal to audiences, can draw closer to the design and details of the
text, to authors' strategic decisions, large and small, and to the global

and local responses of readers. It will not ignore meaning—indeed, in the following four chapters I will suggest that it can add to the breadth and depth of literary meanings—but it also will not overlook what first makes us interested enough and moved enough to linger, sometimes, over meanings.

Our minds automatically send information worth further processing up to our highest evolved mental systems, the social cognitive systems that track patterns of agents and actions. Fiction engages these systems to *play* hard so that outside fiction they can *work* harder. But as we have seen in earlier chapters, we also have an open-ended appetite for patterns of new kinds that may yield new implications we may have to work out on a case-by-case basis. Classic literary works amply engage our swift evolved systems of social cognition, but they also abound in other, open-ended patterns, internal structures that lack the popout effect of character and event but may nevertheless alter our expectations even without our conscious recognition. We catch and lose sight of such patterns, in a kind of hide-and-seek surrounding the more immediate and insistent patterns of character and event. The structures the greatest storytellers set up around the more evolutionarily salient effects of character and event can shimmer with implications that invite us back again and again to their stories. Their shifting vistas help keep our attention alive.

I make no claim to have discovered all the patterns and ironies I marshal here, nor do I claim that they can be discovered only by adopting an evolutionary viewpoint. Neither the evolution of life nor the development of art requires organisms to be *aware* of evolution to do what evolution shapes them to do. But an evolutionary appreciation of social behavior nevertheless helps us to highlight authors and audiences as strategists: authors trying to earn the attention of audiences, audiences always ready to redirect their minds to their own concerns should an author's grip on their attention weaken. Authors strategically assess the traditions they operate within or against, and their own relative skills and standing, and the expectations audiences have both of traditions and of themselves as authors. An evolutionary perspective helps us see the *Odyssey*'s subtler ironic patterns as a concerted suite of solutions to problems of maximizing audience attention in the particular situation Homer faced, at his place in oral tradition, and with the *Iliad* already acclaimed, and given his own talents, inclinations, and ambitions.

Tradition and Attention

Greek bards reciting or singing tales about their gods and heroes already belonged to a system of competition for attention. But since traditional stories—a safe bet in a traditional world—were already familiar, how could a gifted storyteller stand out? How does a Homer appeal both to his immediate audience, who know the traditions, and to future audiences, who may or may not know them? How does he excite audiences by *his* version of the tale, the first time they encounter it, yet also invite repeat invitations to perform his particular version?[2] Will his immediate audience not already have heard successively more elaborate versions from childhood, so that the usual narrative elements of suspense, mystery, and surprise are absent?[3]

Ruth Scodel examines bardic traditions cross-culturally, and notes that although audiences may know aspects of a traditional story, they are likely to do so not uniformly and comprehensively, but to various degrees, in various versions, and almost always much less well than the bard. The bard cannot assume even his immediate audience's prior knowledge, except of the story's broadest outlines and most salient moments; but his awareness that their knowledge may be patchy allows him to omit, include, or elaborate old details or invent new ones to create his own effects and emphases.

Scodel cogently explains why Homer's epics are "notoriously odd as possible oral compositions," why they seem designed for a special kind of attention:

> They are very long, too long for most performance situations. But despite their length, the epics seem clearly, in their present form, designed for continuous reception or at least for reception within the context of a whole they themselves define. Both epics are mere episodes in a much longer story, the Troy tale, which is itself incorporated into the overarching legendary history of the Greeks. The epics explicitly locate themselves within this larger frame. They belong to a

multi-story tradition in which cycles of stories are well established. However, they are relatively self-contained. Most of their episodes do not work well if their implicit frame is the Trojan cycle as a whole, instead of the epics themselves. [Although one or two episodes are self-contained, many] completely lose their point in a different context. The Telemachy [the story of Telemachos' voyage in search of his father in the first four books of the *Odyssey*] works only in a richly expanded version of Odysseus' return . . . The *Iliad* obviously wants to be *the* Trojan epic, even though it tells a section of the Troy story that could easily disappear completely without serious consequences for the tale of Troy as a whole; the *Odyssey* seeks to be *the* return story.[4]

Homer makes the most of the prestige of the tradition, yet he also appeals not only to his first audience, but to any later ones, by inserting his own self-contained "tradition" within the tradition. He focuses on Troy and its aftermath, but chooses to omit the outlines of the story already known to all in his own world, like Aphrodite's rewarding the Trojan prince Paris, for his judgment that she is the fairest of goddesses, with the love of Helen, the beautiful wife of the Greek Menelaos, and the lasting enmity against Troy that Paris' judgment arouses in Athene and Hera, the two goddesses he deems less fair.

THE *ILIAD* AND THE TRADITION

To appreciate Homer's originality and ambition in the *Odyssey,* we need first to see how he took tradition by surprise in the *Iliad.* As Aristotle noted, Homer recounts neither the beginning nor the end of the Trojan War,[5] yet the *Iliad*'s scale and scope make it seem *the* epic of the war. Homer limits himself to a single episode, the wrath Achilleus feels for Agamemnon when the Greek leader takes Briseis from him, an episode that in all likelihood Homer invented himself. Yet although the *Iliad* recounts only this dispensable episode and its aftermath, it never feels remotely episodic or minor. The poem has a sustained sublimity of tone, a controlled but capacious architecture, and a high grandeur of subject: the opposition of the leader of the Greeks, Agamemnon, and the Greeks' greatest fighter, Achilleus, and its consequences, especially Achilleus' defeat of Hektor, the greatest fighter and indeed the mainstay of Troy. And although the *Iliad* focuses only on a brief and not particularly decisive phase of the ninth year of a ten-year war, Homer works hard to create a sense that it encompasses the whole campaign.[6]

Homer chooses the story with the most cultural cachet, *the* major war of Greek legend, but to pique his audience's interest recounts a phase of the war that is wholly new. Yet in a society that depends on the authority of tradition, his stance is not to challenge the tradition but to suggest that his story embodies and enshrines it in its most magisterial form. His massively circumstantial account of a previously unknown episode, he implies, testifies to his command of the whole Trojan saga.[7] He offers his first audience both the authority of tradition and the surprise of the new.

By restricting the *Iliad* to a short phase of the war, he can achieve what Aristotle, Horace, and innumerable others have praised him for, the tight unity of action that keeps attention focused. But he also invents this episode to reflect in complex and fascinating ways on the already familiar elements of the Troy saga, so as to invite us back again and again to reconsider the *Iliad*'s own shape and sense and its relation to the larger story.

Deep ironies saturate the *Iliad* and become both more troubling and more satisfying the more we return. Although the Trojan War is for the ancient Greeks *the* war story, Homer begins *his* story with a surprise and an enduring irony, setting not Greek against Trojan but Greek against Greek. He makes this ultimate war story, about the ultimate warrior, a fight about not fighting, a fight between the leader of the Greek fighters and the leading Greek fighter.

The reason for Achilleus' refusal to fight contains another pointed ironic surprise and reversal. Agamemnon has had no choice but to return Chriseis, his latest battle prize, to her father, a Trojan priest of Apollo, in order to end the plague that irate Apollo has brought down on the Greeks. To compensate himself, Agamemnon commandeers Briseis, whom Achilleus has earlier won in battle and now loves as dearly as she loves him. As critics comment, "Not Helen but another woman is the contended prize between two men as the *Iliad* opens."[8] Agamemnon has led a war to recover Helen, the wife Paris took from his brother, Menelaos, but he has no compunction about himself taking the woman Achilleus loves (Achilleus himself notes the irony, at 9.336–345). The *Iliad* begins with the Greek leader utterly undercutting both the Greek *grounds* for the war—their claim to be redressing the injustice of a woman's abduction—and their *power* to fight it, when Achilleus refuses to take to the field.

Although Homer surprises by making *his* war poem a fight about not fighting, that twist does not mean it is *not* a war story. Achilleus' refusal to fight emboldens the Trojans, who come perilously close to defeating the Greeks and sacking *their* encampment. Suddenly the familiar outlines of the Troy story seem much less reassuring, as Homer makes us feel the anxiety and desperation as well as the gruesome cost of war.

He ends the epic with another surprise, another complex and poignant irony. Almost throughout, Achilleus has been at odds with his own leader, but the *Iliad* closes, movingly, with his reconciliation not with Agamemnon but with the *Trojan* leader, Priam, as he offers him a guarantee of peace until Hektor has been laid to rest. The greatest Greek hero proves his greatest heroism not by fighting—although he has easily slain Troy's greatest fighter—but by agreeing not to fight, by offering Priam a truce in recognition of their shared grief: his own for his best friend, Patroklos, Priam's for his favorite son, Hektor, Patroklos' killer. This is the sad but high end of the poem. It began as a fight about not fighting, and now ends with two foes with bitterly personal reasons for enmity agreeing not to fight. But the somber stillness of reconciliation at the end of the poem incorporates a further tragic irony, for, as we know, this is not the end of Troy's sad story. When the truce is over, battle will resume, Achilleus will be killed, and Troy, along with Priam and all his remaining sons, will be destroyed.

All of which makes astonishing the judgment of the eminent critic Georg Lukács: "The way Homer's epics begin in the middle and do not finish at the end is a reflexion of the truly epic mentality's total indifference to any form of architectural construction."[9]

Homer depends on tradition, but transcends it. He takes over the epic form, and the Trojan saga, and the predilection for powerfully ironic parallels in Greek tradition—most famously, Agamemnon's successful campaign against Troy to retrieve his brother's straying wife, only to return to be killed by the lover his own wife has found in his absence. From such elements he invents his own telling irony— Agamemnon's taking Achilleus' beloved Briseis, and Achilleus' implacable wrath in response—to surprise and hold his first audience, but also to complicate and amplify the story's internal tensions so as to appeal to repeat audiences or to later audiences drawn not so much by the Trojan tradition as by the *Iliad*'s self-sufficient energies and ironies.

THE *ODYSSEY* AND THE TRADITION

Although the *Iliad* makes multiple claims to attention, the *Odyssey* outdoes it. Again Homer makes the most of the cachet of tradition, a tradition that by now also includes the *Iliad*. As before, he makes his new epic self-contained, not a mere segment of a longer saga, but also continues and complements his earlier poem by making his new epic *the* story of the return to set alongside *the* story of the war.* Indeed he seems to have sought to create two self-sufficient but mutually reinforcing works of art, *the* exemplary stories of their subject, while keeping attention fresh by avoiding almost all familiar ground. In the *Odyssey* that strategy involves resisting the expectations created both by the old story of Odysseus and by his own recent *Iliad*.

In neither poem does Homer dwell long on the expected core of the traditional story. Instead he chooses unexpected episodes, apparently of his own invention—the wrath of Achilleus and all that follows, Telemachos' voyage in search of his father, and Odysseus' alliance with Telemachos and the swineherd Eumaios—yet in each case offers a comprehensive and even definitive treatment of the tale of Troy. That he does not merely expand episodes already embedded within the traditional stories seems evident from the fact that these

---

* Given that Homer repeats many lines and even blocks of text within the *Odyssey*, as within the *Iliad*, the avoidance in the *Odyssey* of any part of the Troy story covered in the *Iliad*—despite the fact that Nestor, Menelaos, Helen, Demodokos, Odysseus, and Achilleus all report on the events of the war itself—must have been a highly conscious decision.

incidents would have no place except in the poems as he devised them.[10]

As in the *Iliad,* Homer again in the *Odyssey* prefers the intensity of compression to the slackness of mere sequence. Rather than re-counting Odysseus' whole life, or even the ten years of his return, from the sack of Troy to his final restoration to Penelope and to Ithaka, he trims Odysseus' wanderings to fit his Trojan diptych. Con-centrating the action into the four weeks or so between Athene's ap-peal to Zeus to release Odysseus from Kalypso and Odysseus' return and defeat of the suitors,[11] he devotes almost five-sixths of the poem to the last ten days.

The traditional tale appears to have focused on Odysseus' fabled ordeals on the fringes of the Mediterranean. Homer relegates this material to one-sixth of the poem, recounted by his hero during a single night, the last before his return to Ithaka. Already the tradi-tion seems to have contrasted Helen and Penelope, faithless wife and faithful, but Homer places this contrast at the center of his story when he shifts the emphasis from Odysseus' wanderings to the vir-tual siege of Penelope in Ithaka by an army of suitors. Against ex-pectation, Odysseus finds Ithaka not a home and a refuge but the greatest danger of all, almost a war zone. It appears to have been a traditional irony of the stories of the return that many suffered worse losses *after* leaving Troy. Homer tightens the irony: once again Odys-seus must do battle, this time almost single handed, over a woman, this time his own wife.[12] His reception in Ithaka is at once a kind of miniature second Troy, and an escalation and refutation of the hostile welcome awaiting Agamemnon on *his* infamous return from Troy.

For all the shifting fortunes of war, the *Iliad* records a stalemate. Homer builds the *Odyssey* by pointing it always toward a single deci-sive battle. The narrative moves swiftly from its own initial stale-mate—Odysseus far from home, languishing for seven years on Ka-lypso's island—to an emphatic final reversal of his fortunes. In the *Iliad* the gods' sympathies, like our own, are evenly balanced; no wonder no decisive outcome can be reached. In the *Odyssey,* the gods' sympathies, like ours, are all but unanimously for "long-suffering Odysseus." Even before *he* knows it, we know he has the backing of the gods, and we expect his ultimate victory from the first. That expectation makes all the difference to the mood of the *Odyssey* and to its popularity.

Homer also contrasts his two heroes to the hilt. The entire story of the *Iliad* magnifies Achilleus, even in his absence: the Trojans surge forward only when they know that he has withdrawn from the field. In the *Odyssey*, too, the whole story magnifies Odysseus, even when *he* is absent: the suitors run riot because he is presumed dead. But Homer's Achilleus is utterly single-minded, focused intently first on his anger at Agamemnon for taking Briseis, the woman who means most to him, then on his rage against Hektor for killing Patroklos, the man who means most to him. In the later epic, by contrast, Homer makes Odysseus always and insistently multifaceted, circumspect, myriad-minded.

And throughout the story Homer sustains audience attention by choosing unexpected angles and rhythms and ironies for each new phase.

THE TELEMACHY

The *Odyssey* opens with a proem, a pre-poem (1.1–9):

> Tell me, Muse, of the man of many ways, who was driven
>   far journeys, after he had sacked Troy's sacred citadel.
> Many were they whose cities he saw, whose minds he
>   learned of,
> many the pains he suffered in his spirit on the wide sea,
> struggling for his own life and the homecoming of his
>   companions.
> Even so he could not save his companions, hard though
> he strove to; they were destroyed by their own wild
>   recklessness,
> fools, who devoured the oxen of Helios, the Sun God,
> and he took away the day of their homecoming. From
>   some point
> here, goddess, daughter of Zeus, speak, and begin
>   our story.

From the start Odysseus is singled out as major—by tradition, by Homer's appeal to the muse to tell the story of someone whose travels have made him too famous to need naming, and then by the first scene of the poem, where Athene makes the question of Odysseus' return the only item on the agenda in the council of the gods. Zeus's reply to Athene refers to Odysseus' fateful clash with Polyphemos and raises our expectation that we will hear more of his wanderings and his most famous adventure.

But instead Athene takes us with her, not to Odysseus, but to Telemachos in Ithaka. We see nothing of Odysseus for four books, one-sixth of the story, and this comes, "or should come, as a considerable surprise to anyone encountering the poem for the first time."[13]

Homer would seem to risk disappointing us here, by *not* offering us the promised wanderings. But in fact these first four books constantly whet our appetite for the story's conclusion, by means of a series of images—prophecies, wishes, supposedly vain hopes, forlorn fantasies—of Odysseus returning to Ithaka and setting to rights the chaos caused by the suitors. In the first book alone, Odysseus is referred to on twenty-nine separate occasions, or more than once every fifteen lines, and often for several lines at a stretch. He is absent, and *felt* to be absent, both by the characters, who are not expecting to see him, and by us, who *were* expecting him, and he is nevertheless very much present, on everybody's minds, and of paramount importance to all. Homer creates such a strong sense of anticipation throughout these four books that they become the oral equivalent of a page-turner, not through the quiet onstage action but by rousing our keen confidence in Odysseus' return to confront the suitors. And as many have noted, from Goethe onward, although we have a strong inkling of the outcome, we do not know how it will be achieved.[14]

Our pleasure in anticipating Odysseus' homecoming is magnified by our consciousness that the mortal characters in the story often view his return as no more than a hollow hope, forever excluded by the presumed "fact" of his death. We anticipate his return, we enjoy the privilege of knowing that the gods have already initiated the process, and we look forward to the surprised pleasure of those who currently conjure up his return as no more than an impossible dream. At the same time, since we do not know in detail what will happen, we remain curious (how can he possibly deal with all these suitors?), with a tinge of misgiving spicing our confidence. What if things do not turn out quite as Athene seems to expect? What if Poseidon finds out that Odysseus is on his way back to Ithaka, and vents his wrath on him while he is still at sea?

Homer's strategy here also appeals upon our return to the story, knowing how it ends. We can relive the pleasure of our first encounter, recalling that we already knew more than any of the mortal characters, since we were already privy to Athene's and Zeus's acting in support of Odysseus. Yet now we can also enjoy the pleasure of enjoying more than our already privileged earlier selves, since we now know exactly how much each of the long series of predictions or dismissed dreams of Odysseus' return is true or false.

Why else does Homer focus the first four books on Telemachos, rather than on his father? Among other reasons, Telemachos' quest allows Homer to recount the major returns from Troy: Nestor, happily and rapidly reinstalled in his loving family; Agamemnon, murdered by his wife and her lover, and avenged by his son; and Menelaos, back with Helen, the cause of the war, in a relationship that remains brittle, despite all the glitter, because of the long history of her infidelity. Homer can tell these stories not as mere authorial summaries but in dramatic form, saturated with feeling for the tellers as they recount their own returns, and saturated with particular feeling for the young listener straining to learn more about his father and his comrades in war for so many years. Homer sets up each of these returns as a pointed contrast, both for Telemachos and for us, to Odysseus and Penelope: Nestor, quickly recovering his old warm domestic happiness; Agamemnon, grimly betrayed by his own wife after his long attempt to recover his brother's faithless wife; Menelaos, reunited with his peerless Helen, but unable to ignore the old painful differences beneath the dazzling show.

Homer could have told the story without Telemachos' quest for his father. But by starting this way he does more than establish Athene's hands-on help, Telemachos' own stature, and the ironic light cast upon the returns of the other Greek heroes and Odysseus' eventual homecoming. Perhaps most important of all, he also introduces the ultimate goal of Odysseus' own greater quest, Ithaka, yet not the old Ithaka he pines for, but the new one wasted by the "insolent," "arrogant," "reckless" suitors. The Telemachy ends with the suitors at anchor, lying in wait for Odysseus' son, "in their hearts devising sudden death for Telemachos" (4.843). By showing us Telemachos in Ithaka, unable to persuade the suitors to disperse, unable to curb their outrages, Homer discloses not only the goal of Odysseus's quest, but the painful cost of his absence, the scale of the threat awaiting him once he does return, and the son and heir who will help him regain their heritage. More than he could in any other way, Homer also makes us *feel* Odysseus' absence and its cost, makes us long for his return and look forward with edgy excitement to the surprise it will cause for his family, friends, and foes.

Part and Whole

While the *Odyssey* as we have it is too long for a single performance, its parts are not self-sufficient wholes. The Telemachy makes splendid if unexpected sense as the introduction to the story of Odysseus' return, but it cannot stand on its own—and not only because of its cliffhanger ending.

As we now have it, the *Odyssey* divides into twenty-four books. These divisions may well be post-Homeric, yet many agree that the poem naturally falls into four-book units.[15] Here we can see another aspect of Homer's care to secure his audience's attention. Each four-book block would constitute something like a two- to three-hour performance, a dedicated afternoon's or evening's entertainment. Each has its own focus: Telemachos and his quest for his father (1–4); Odysseus' escape from Kalypso's island to the Phaiakians (5–8); his account to the Phaiakians of his great wanderings (9–12); his return to the island of Ithaka, to the hut of his loyal swineherd Eumaios, and his reunion there with Telemachos (13–16); his return to his palace in disguise as a beggar, subject to the abuse of the suitors (17–20); his disclosure of himself as Odysseus to the suitors, to Penelope, to his father (21–24). Each block also marks a clear advance toward the return and restoration of Odysseus and thereby offers a reason to turn up for the next session.

Yet each block also marks a shift in scene or strategy or both. In the first, the Telemachy, we enjoy the irony that Odysseus is alive and about to return, so that we expect a pleasant surprise for those wishing for his return but sure that he never will. In the second block, Odysseus' escape to Phaiakia, the ironies prove quite different, as we shall see shortly. In the third block, the ironies disappear altogether as Odysseus recounts his story directly to the Phaiakians. In the fourth block, irony returns more emphatically and immediately than ever as Odysseus stands before two who love him and do not recognize him, his loyal swineherd and his son. In the fifth block, the ironies multiply again, as the suitors heap abuse on a beggar they do not know as Odysseus and unwittingly stoke his revenge with every insult, while Odysseus also meets Penelope and his old nurse, Eurykleia, in scenes where the insecurity of his disguise creates a new kind of suspense. In the final block, Odysseus springs his deadly, long-prepared-for surprise on the suitors, only for Penelope to spring on us and on Odysseus a surprise that takes us all unawares: her reluctance to recognize him as Odysseus.

Homer builds a strong narrative drive throughout the story, but breaks it into sessions that respect the limits of human attention. Each of his partially self-contained sessions advances the story but also offers often a new scene, a new cast, and a new set of strategies. He devised this complex but lucid scheme for his original audience, but it has the same kind of appeal for modern readers as for old auditors, for newcomers and for rereaders. Although on a repeat encounter we find ourselves back in the familiar world of Odyssean irony, we also find ourselves never settling into the same set of ironies for long.

This scheme has little to do with the meanings of the *Odyssey,* but everything to do with our desire to attend to its story. Millennia later, Tolstoy introduced stream of consciousness into literature late in *Anna Karenina,* but he also divided the novel into eight books, and within each book into the interweaving strands of the story, and within each strand into short chapters easily assimilated in a reading situation subject to interruptions. Like Tolstoy, Homer thinks about human psychology not only in terms of his characters but also in terms of his audience. One requires at least as much ingenuity and insight as the other.

From Kalypso to the Phaiakians

Each four-book block of the *Odyssey* has its own taut relation to others. After the Telemachy shows us the suitors who will not leave the home of the woman they would like to wed, the next block begins with Odysseus unable to leave the home of the goddess who would like to wed him. Like many of the ironies and symmetries of the *Odyssey,* this shapes our response immediately, although we may not enjoy it consciously until a repeat encounter. There is too much else happening.

In the proem to the *Odyssey,* Homer introduces Odysseus as the man who tried to save his companions, although, as Odysseus himself knows, his actions have often cost as well as sometimes saved the lives of his companions. From the first, and especially just before he at last brings Odysseus onstage, Homer stokes our admiration of his hero to ensure our sympathy and our eagerness for his ultimate triumph.

In the first council of the gods, Zeus tells Athene that he has already been planning Odysseus' homecoming. Athene suggests sending Hermes, the messenger of the gods, to Kalypso to announce to her "our absolute purpose, the homecoming of enduring Odysseus, that he shall come back. But I shall make my way to Ithaka, so that I may stir up his son a little . . ." (1.86–89). Immediately she dons her sandals and streaks toward Ithaka and Telemachos. Directly after the Telemachy, book 5 opens with a second council of the gods and Athene again raising the case of Odysseus. Again Zeus says that Odysseus' release is already his plan, *their* plan. He orders Athene to bring Telemachos safely home past the suitors waiting in ambush, then turns to Hermes, commanding him to announce to Kalypso "our absolute purpose, the homecoming of enduring Odysseus, that he shall come back" (5.30–31). Verbal repetitions are common in the Homeric epics, a sign of their oral heritage, yet critics have been puzzled by the overlap of these two councils of the gods. Why are they so similar? Why does Athene suggest dispatching Hermes immediately in the council in book 1, and Zeus order Hermes' dispatch in the same words again in the council in book 5?

Throughout the Telemachy Homer highlights Athene's role in restoring the fortunes of Odysseus and his family, to boost *our* anticipation that he will be restored. She brings Odysseus' case before Zeus, who announces that he already has Odysseus' interests at heart, then promptly takes her own initiative in grooming Telemachos for heroism. Throughout the Telemachy, she appears before Telemachos in this or that mortal form. Not only does she supply specific advice, but he usually recognizes that he has been instructed by a god, and his confidence grows accordingly.

Homer wishes to establish both Athene's support and the long-term prospects for Ithaka's ruling family, yet at the same time he wants us to experience the sufferings of "much-enduring" Odysseus, at his lowest ebb, facing severe trials without any reassurance that he has the aid of the gods. He wants us to look forward keenly to Odysseus' restoration, but also to sympathize with him now, amid all his ordeals.

Athene is a hands-on goddess, and a passionate partisan of Odysseus. Were *she* to travel all the way to Ogygia to announce the gods' decree to Kalypso, she would also visit her favorite mortal. Instead, Homer and Zeus assign her other pressing tasks, in the first council, from which she rushes after Telemachos of her own accord, and in the second council, where Zeus orders her to guide Telemachos safely back to Ithaca. Hermes heads for Ogygia in her place, and talks only to Kalypso. Goddess though she is, Kalypso has no choice but to comply with the Olympian decree, but still hopes to deter Odysseus from leaving. She tells him nothing of the gods' command, as if *she* should have the credit for his release,[16] and seeks to deter him by reminding him of the possible malice of the gods.

Homer wants to stress Odysseus' despair at being so far from his family for so long, without having us dwell on the fact that he has shared Kalypso's bed for perhaps longer than he has ever shared Penelope's. After Hermes makes his announcement to her, Kalypso searches

> after great-hearted Odysseus,
> and found him sitting on the seashore, and his eyes
>   were never
> wiped dry of tears, and the sweet lifetime was draining
>   out of him,
> as he wept for a way home, since the nymph was no
>   longer pleasing
> to him. By nights he would lie beside her, of necessity,
> in the hollow caverns, against his will, by one who
>   was willing,
> but all the days he would sit upon the rocks, at the
>   seaside,
> breaking his heart in tears and lamentation and sorrow
> as weeping tears he looked out over the barren
>   water.   (5.150–158)

This is how Homer has Odysseus at last enter the poem: the beloved of a goddess who can surround him with comfort, love, and even the promise of immortality, he nevertheless weeps inconsolably every day for home.

Not knowing of the gods' decree, and wary of her motives, Odysseus secures Kalypso's oath that she will not thwart him. He constructs and launches a raft, still unaware of divine sanction, let alone support, for his homecoming. Poseidon, his one implacable foe among the gods, returns from his Aithiopian sojourn, spies Odysseus on his raft, and unleashes a storm that sweeps him into the bitter sea. Aware of Poseidon's enmity, Odysseus fears imminent destruction, but finds some relief from a local goddess, Ino, who urges him to abandon the raft he has now regained and swim for shore. Dreading deceit, he clings to the raft, only for Poseidon to smash it asunder and leave Odysseus bobbing in the angry waves. As Poseidon resumes his homeward course, Athene secretly intervenes, calming the seas. When breakers threaten to dash Odysseus against the rocky strand, she prompts him, again in secret, to an escape. He scrambles ashore, naked, chilled, exhausted, fearing death from cold or whatever beasts prowl these parts. Within 150 lines, we have seen him buffeted by one life-threatening challenge after another, in a miniature of his long ordeals on sea and land. Homer makes us feel Odysseus' dread and despair, his sense that he is alone, without support in a hostile world, even as he also allows us to see that the gods will ensure his return to Penelope, however unlikely life, let alone happiness, has just seemed to him at sea and on this lonely shore.

Nausikaa

From the start, we expect Odysseus' return. Not only does tradition decree it, but we have heard it freshly guaranteed by Zeus, not once, but twice. Yet Odysseus has no such reassurance. And suddenly we, too, face a surprise.

Odysseus has not seen another mortal for seven years.[17] Now, sleeping exhausted under a blanket of leaves, still naked and caked with salt, he has reached his lowest ebb. But he wakes the next day to an idyll, a vision of purity: the cries of young maidens at play, amid their gleaming, freshly washed linen spread out to dry in the sun. The night before, Athene visited Nausikaa, the Phaiakian princess, in her sleep, and reminded her that she is nearing the age of marriage, and that part of a wife's duties will be to keep the household linen clean. Early next morning, with the permission of her father, King Alkinoös, and his delicate recognition that she is readying herself for marriage, she rides with her young handmaidens to wash the palace linen in the river Odysseus staggered from the night before.

When Odysseus wakes with a start, her friends scatter in fear, but the princess remains. He hails her with "mortal or goddess?" If she is mortal, her parents and brothers must be blessed "with happiness at the thought of you, seeing such a slip of beauty . . . but blessed at the heart, even beyond these others, is that one who, after loading you down with gifts, leads you as his bride home. I have never with these eyes seen anything like you." He asks her for some rag to wrap himself in, "and then may the gods give you everything that your heart longs for; may they grant you a husband and a house, and sweet agreement in all things." As he bathes and clothes himself, Athene silently beautifies him. Nausikaa confides to her handmaids that although the unkempt stranger had seemed unpromising, he now looks like a god: "If only the man to be called my husband could be like this one, a man living here, if only this one were pleased to stay here."

The hints become still stronger. Nausikaa asks him not to accompany her into town lest people say "Who is this large and handsome stranger whom Nausikaa has with her, and where did she find him? Surely, he is to be her husband." Her father, not yet aware of Odysseus' identity, hears him recount his escape from Ogygia, his landing on Scheria, his meeting with Nausikaa. Alkinoös exclaims: "how I wish that, being the man you are and thinking the way that I do, you could have my daughter and be called my son-in-law, staying here with me. I would dower you with a house and properties, if you stayed by your own good will. Against that, no Phaiakian shall detain you."

The *Odyssey* opens with a proem announcing Odysseus as hero; the first four books surprise us by keeping him offstage. The next book seems to promise a more predictable course. True, it starts with a different kind of surprise as we find Odysseus loved by a goddess ready to offer him immortality, yet still weeping every day for home. But we expect him to be released, and he is. We expect him to face life-threatening ordeals at sea, and one after another he does.

We also expect him to be reunited with Penelope and to face down the suitors. Yet now he meets the lovely Nausikaa, and suddenly we wonder, are all bets off? We both know what is coming in the story, which is a pleasant reassurance, and don't quite know what's coming, which is a pleasant surprise.

So far Athene has been the prime planner and instigator of the action, and here she marshals both Odysseus and Nausikaa onto the beach, and there transforms and revitalizes her depleted champion for Nausikaa's wonderment. Why?

Homer goes to some trouble to keep alive for as long as he can the possibility that Odysseus could become a husband for Nausikaa. In Greek etiquette, a stranger should be fed before being asked to identify himself, but then he should say who he is. Notoriously, Odysseus does not.[18] If he announced himself as Odysseus, a figure familiar even to these remote Phaiakians, he would be known at once as the husband of Penelope. He makes no secret of his anxiety to return, yet he not only avoids identifying himself but even avoids disclosing that he has a wife and son: "and let life leave me when I have once more seen my property, my serving people, and my great-roofed house" (7.224–225).

The possibility that Odysseus might become a husband for Nausikaa lingers. There is a fairy-tale pattern in the air—the frog who turns into a handsome prince, or the forlorn stranger, the anonymous ragtag knight who unexpectedly wins the tournament and the hand of the princess.[19] Odysseus, provoked into proving himself at the Phaiakian games, easily wins the one contest he enters.

But despite the admiration Odysseus and Nausikaa express for each other, despite Alkinoös' encouragement to the stranger to ask for her hand, nothing comes of it. Much earlier in his travels Odysseus has faced temptations to linger and forget home. Some of his crew succumbed to the narcotic dreams of the Lotus-Eaters, although he himself was never tempted. But he does enjoy a whole year with Circe, before his companions remind him they should all move on. Forewarned, he enjoys but protects himself against the fatal enchantment of the Sirens' song. Washed up alone on Ogygia, he finds a refuge in Kalypso's arms, but soon realizes she is no surrogate for home. And now among the Phaiakians he enters a wish-fulfillment world, a kind of paradise, where the trees fruit all year round, yet a human paradise, where he would not be alone as on Kalypso's island, but surrounded by a civilized, courteous, almost godlike people, among whom he could have a young princess for wife.

Whereas Kalypso held him fast until ordered to release him, Alkinoös both invites Odysseus to ask for his daughter if he chooses and in his next breath promises not to detain him. Odysseus never wavers from his resolve to return. For Odysseus and for Nausikaa, the new face each encounters on the beach is a sudden bright possibility, but for Odysseus no mere possibility can compare with the actuality of Penelope. As he says to Nausikaa at the end of his first speech to her, while still covering his nakedness: "may the gods . . . grant you a husband and a house and sweet agreement in all things, for nothing is better than this, more steadfast than when two people, a man and his wife, keep a harmonious household" (6.180–184).

Nausikaa is no Circe, or Siren, or Kalypso, no deflector of Odysseus' course, but a measure of his focus and determination.

Throughout the *Odyssey,* Homer offers us the pleasures of anticipation and foreknowledge, by making us privy to the decisions of the gods and the decrees of fate, the pleasures of sympathizing with characters coping with their plight *without* the luxury of foreknowledge, and the pleasures of the unexpected. I can say "us," because Homer's appeal to his audience works both for his original listeners, steeped in the Greek narrative tradition, and for much later audiences who may have to rediscover the tradition as they read.

Homer sustains the narrative drive of his long story through the power and clarity of Odysseus' goal of reaching home. He complicates that goal, raising both our apprehension and eagerness, by showing from the first the threat the suitors pose to Ithaka and to Odysseus should he return.

Despite its unity of action, Homer cannot tell his story in one sitting, and so divides it into four-book blocks each of which can hold our attention over one sustained session. He not only changes the scene and cast from block to block but thereby also realigns the ironies that saturate his story. *Within* each block he prepares marked changes of subject and mood. In the second block, our first encounter with Odysseus, we watch him isolated and weeping his life away on Kalypso's island, fighting for dear life on the seas, or surrounded by admiration and ceremony on Scheria. Even in the Phaiakian world, the mood changes from the idyll with Nausikaa, to the pride and pathos of reliving his Trojan exploits in the songs of Demodokos, to challenge and victory at the Phaiakian games.

Not only does Homer invite, sustain, and refresh the attention of a first-time audience, in his own time or later; he also prepares for a repeat audience. A second time around, we can enjoy the ironies of a first encounter in a new way, reliving the old pleasures of confident but still uncertain anticipation in light of the new satisfactions of exact foreknowledge. Something similar can happen in any rereading or retelling, but Homer sharpens the tang of a repeat encounter by appealing to our ability to assess the precise proportion of truth in his endless partial anticipations, divine decisions, mortal guesses or wishes or hopes, portents or prophesies or omens heeded or ignored.

Homer has also shaped his story so tightly that on each new reading we can discover patterns, new ironies or symmetries or tensions, within the whole Trojan saga, or in the *Odyssey*'s relation to the *Iliad,* or purely within the *Odyssey.* To take just some that we can notice in the details already discussed in Odysseus' three days with the Phaiakians:

1. The *Iliad* is at one level a story of wife capture: Helen, Briseis, Chriseis. We can recognize the *Odyssey* as a contrasting story of husband capture. Circe, Kalypso, and Nausikaa all want Odysseus for their husband.[20] The first two, being goddesses, hold him for a year and seven years, respectively, the third for no more than a momentary pang of mortal enchantment.

2. Or we can see that among the Phaiakians Odysseus has enjoyed peaceful human company and lavish hospitality for the first time in over twenty years, and for the only time in the *Odyssey.* His gracious reception on Scheria echoes and expands on Telemachos' entertainment at Pylos and Sparta, and will contrast in different ways with his warm, intimate, but humble welcome at the swineherd's hut on Ithaka and the heartless abuse he receives from the suitors treating his home as theirs.

3. On Scheria, far from home, among a people he has never heard of, and still unknown to them, Odysseus triumphs, swiftly capturing the heart of Nausikaa, winning a lordly welcome and the option of a permanent place of honor from Alkinoös, hearing his deeds sung by the divine singer Demodokos, overmastering all at the discus, and enthralling his audience as he recounts the story of his wanderings. On Ithaka, his own home, by contrast, he will have to remain anonymous and in disguise as a beggar, to withstand abuse and hostility, to tell false versions of his recent past, and to overcome Penelope's persistent doubts, even after his conquest of the suitors, that he really is her husband.

We may feel much in these contrasts on our first time through the story. But as we reread we can appreciate them anew when we return to Phaiakia and savor Odysseus' reception there in contrast to what we now know in detail he faces in Ithaka.

A biocultural approach to literature does not invalidate the way we have customarily read literature. After all, it stresses both common evolved understandings and our human ability to refine understanding through our evolved capacity to share so much via culture. But it will place new emphases on literature: new, at least, to modern criticism, if not to the way people have actually told or listened to tales. It will see authors as appealing to the attention and emotion of their audiences, or in biological cost-and-benefit terms, as individuals seeking strategically to maximize the benefits they can earn against the compositional costs they are prepared to pay. In the case of an author like Homer, aware of his rare storytelling skills, that challenge can involve a high cost in imaginative effort, but in return for an extraordinary benefit: others' attention to his story in his own day and far into the future, whether they are surprised newcomers or satisfied oldtimers.

If we take due note of the centrality of shared attention to all storytelling, we will also recognize the astonishing inventive accomplishment of classic authors, able to appeal to audiences of many kinds over many years, and the accomplishment of *any* storyteller who can secure an audience, wide or select, brief or enduring. There is nothing "mere" about audience appeal.

A story's appeal begins in character and plot. Art, like anything else, catches our attention against a background of expectations: in the case of story, especially expectations of human nature (character) and experience (plot). But we also bring to stories specifically artistic or narrative expectations: expectations about the story form, already developed or emerging as we proceed, if the form is unfamiliar, as hexameter epics will often be to modern readers; expectations about the storyteller, if we have encountered other work by the same person; and expectations generated by the work so far. These generic, author-specific, and work-specific expectations may drastically modify the way we attend.

I have suggested that we should see art as cognitive play, as play with pattern, and that great art offers high-intensity pattern at multiple levels. In the case of fiction the most immediate patterns are those of character and event (goal, action, obstacle, outcome) and their combination in plot. Vivid patterns of character, such as we find in Odysseus and Penelope, Athene and Poseidon, the lead suitor Antinoös and the loyal swineherd Eumaios, allow for rich expectations, and their combination at the level of plot allows for still more compelling patterns, like the tension between Odysseus' return and the suitors' attempt to establish themselves in his place. But if, as our acquaintance with a story deepens, new patterns emerge into view —character contrasts, structural ironies, narrative designs—we will have all the more reason to attend with renewed interest. Such patterns will be less prominent than those of character and event, which we have evolved to notice and process efficiently, but will modulate our sense of relevance even without our conscious awareness and will deepen our sense of artistry when we do consciously recognize them. In most cases our recognition of structural patterns will swell and fade as the more humanly insistent patterns of character and event compete for our attention; but that constant complex interplay of the immediate and the remote, of shape and shimmer, of clear tone and half-caught undertone, can make a familiar classic forever new.

An evolutionary perspective on storytelling will not make us ignore meaning. But we will not rush to reduce complex strategies for holding attention and evoking emotion, and the rich psychological intuition and invention they require, to portable, take-home meanings. Meaning will deserve discussion, but as one effect among many.[21] And many of the possible meanings of a work will mean still more when we consider them in the long perspective of human evolution: the evolution, for instance, of intelligence or cooperation. To these we now turn.

16

THE EVOLUTION OF INTELLIGENCE (1):
IN THE HERE AND NOW

IN GREEK LEGEND ODYSSEUS is the most intelligent of mortals, "beyond all other men in mind" (*Odyssey* 1.66), in the words of Zeus himself. Intelligence matters far more in the *Odyssey* than in earlier or later epic. Whereas Achilleus embodies pure strength, Odysseus has not only physical strength but also strength of spirit and mind: resilience and resourcefulness. Athene, goddess of wisdom, plays a larger role in the *Odyssey* than all other gods combined, aiding and adoring Odysseus precisely *because* of his intelligence, "because you are fluent, and reason closely, and keep your head always" (13.332). Penelope, too, is "so dowered with the wisdom bestowed by Athene ... to have ... cleverness, such as we are not told of, even of the ancient queens" (2.116–119).[1]

Despite the *Odyssey*'s promotion of intelligence, twentieth-century scholars, notably Bruno Snell (1946) and Julian Jaynes (1976), have advanced bold claims about the lowly evolution of mind in Homer. Snell argued that the notion of "mind" was absent in Homer and discovered only later, and that "Homeric man does not yet regard himself as the source of his own decisions."[2] Jaynes claimed that there is not even consciousness in the *Iliad*: its characters are "not at all like us" but "noble automatons who knew not what they did," with no awareness of their "awareness of the world, no internal mind-space to introspect upon." Like schizophrenics they acted on dictates they heard from the other (right) side of their brain, which they attributed to the gods but which we would call hallucinations. In the *Od-*

yssey, which he regards as at least a century later, he sees "a new and different world inhabited by new and different beings" and a marked growth toward subjective consciousness.[3]

Not only is it implausible that a writer gifted enough to last three millennia—and to inspire authors as major as Virgil, Shakespeare, Goethe, and Joyce—should have lacked a conscious mind or a notion of mind; it is also wrong. Others have ably critiqued the arguments of Snell, Jaynes, and others—Bernard Williams most wittily, Jeffrey Barnouw most comprehensively.[4] But to appreciate what Homer *does* see, show, and appeal to in the mind, we need an evolutionary perspective, which can reveal that the *Odyssey* not only encapsulates but even advances the growth of intelligence.

There are three main reasons for claims such as those of Snell and Jaynes: (1) Homer does not have a precise or clear set of mental terms. But as philosophers love to show, our own nontechnical terms for mental phenomena are far from clear, even at the level of *mind, brain, consciousness, soul, spirit, will, reason,* let alone in everyday terms like "heart" or "guts." Adding terms like "Oedipal complex," "anima," or "mirror stage" only thickens the fog. (2) Homer has fewer ways of rendering mental states than Horace or Montaigne, let alone Tolstoy or Proust. He tends to express thought in external fashion, through actions like weeping, or through speech, to others or to oneself, or through dreams involving a god's visitation to the dreamer. (3) He shows many human actions as prompted by the gods.

Common sense can make us suspect the claims of Snell and Jaynes, but common sense can be a poor guide, mixing natural intuition and cultural elaborations we mistakenly assume to be natural. Homer has no words or concepts equivalent to the faculties of emotion, reason, and will, established in Western thought for so long that we think they demarcate definite features of mind. In fact they remain mere folk terms, approximating aspects of mind but imprecise enough to need the thorough revision that only scientific psychology can make possible.* To assert that Homer's not having *these* notions

---

* Neuroscientist Antonio Damasio (1996) provides good arguments for distinguishing emotions from feelings, and for not making a sharp distinction between reason and emotion. Without emotions, he shows, we cannot reason, since we cannot decide among multiple options.

is proof he has no notion of mind is provincial, as if our own local habits contained the only valid ways of thinking about minds.

If we examine Homer in terms not of current Western folk notions of mind, but of what he shows in his characters and expects in his audience, and of what science is currently disclosing about the mind, we will see not only that it is wrong to suppose he has no notion of mind but that he has rich gifts as an intuitive psychologist and a considerable contribution to make to our understanding of the evolution of human intelligence.[5]

Bernard Williams rightly argued that a belief-desire-intention psychology links Homer's world and ours.[6] An evolutionary perspective would have made his case swifter and stronger: animals other than humans understand desire and intention, human children understand intention in their first year and desire by age two, and this much (and more) is part of human intuitive psychology across cultures.[7] Understanding others in terms of desires and intentions, our evolved *intuitive* psychology, is very different from understanding minds as having distinct faculties such as reason and emotion, a much more culturally specific theory—even if now widespread enough to be part of Western *folk* psychology—and, as it turns out, a much less tenable one.

An understanding of others in terms of desires and intentions has been a major force, perhaps *the* major force, in evolving intelligence in all social species. Without it stories would be impossible and the *Odyssey* incomprehensible. Odysseus desires to return to Ithaka and Penelope, and intends to attain these goals, and that aim drives the whole epic. But although we share with other animals an understanding of desire and intention, *human* intuitive psychology has moved a major step further. We understand one another in terms not only of desires and intentions but also of beliefs (see Chapter 10).

The *Odyssey* depends on belief as much as on desire and intention. The suitors believe that Odysseus is dead and that even if he were alive he would be no match for them all, and therefore think they can carry on with impunity. Penelope believes her husband might well be dead, but is not sure; hence her dilemma. We, and Homer and his original audience, understand even the gods in terms of a belief-desire-intention psychology. Zeus and Athene believe Odysseus to be of exceptional worth and deserts, and therefore want to act quickly, while the absent Poseidon still believes Odysseus to be safely stranded on Kalypso's island.

The impact of the *Odyssey*'s story, like that of almost all stories, depends so much on belief-desire psychology that we do not even register it. But belief-desire psychology is a human minimum. What do we find at the maximum, in a storyteller as rare as Homer and a character as proverbially clever as Odysseus?

Homer has a formulaic way of representing characters coming to major decisions that scholars call the deliberation type-scene. A character wants to act on plan A, but then considers plan B, and usually, deciding that the latter is the better way, acts on B. Sometimes the character may act on neither, but simply perform C.

As we saw in Chapter 3, we can explain the initial emergence of mind in terms of more or less automatic, heuristic-based responses to recurrent environmental conditions. Only much later did there evolve a flexible intelligence, especially in humans, that can with effort arrive at novel solutions to novel problems. Heuristic responses are fast, often parallel-processed in the mind, through multichanneled, specialized, automatic subsystems that need little conscious attention. In order for novel solutions to be found, these rapid-fire reactions have to be inhibited (in the orbitofrontal cortex) so that there is time to formulate and assess new options (in the dorsolateral cortex) before acting on them. Formulating these options is comparatively slow, because the ideas must be processed not in parallel, with many neural channels operating at once, but serially, with one input manipulated under the spotlight of attention until it activates the next.

Situations that cannot be coped with by routine responses activate working memory's central executive, which allocates attention, inhibits information irrelevant to the problem, and amplifies relevant information into conscious, explicit representations within working memory's slave systems, especially the most vivid, the *phonological loop* (our mind's ear, our inner speech) and the *visuospatial sketchpad* (our mind's eye).[8] Moreover, "The ability to mentally generate, maintain, and manipulate these abstract representations is limited by . . . attentional and working memory resources . . . and is thus effortful."[9] Our capacity to focus on information relevant to a problem and to inhibit irrelevant stimuli and responses expanded greatly with language, which allows the compression of information for comparatively efficient manipulation within working memory.

Generating new responses to difficult situations, then, involves stopping automatic responses and thinking with effort, in a highly conscious way, to solve problems. Homer may not have had ready-made terms for aspects of the mind, but he describes exactly this process. Formulaic yet flexible, he shows characters deliberating thoughtfully in the new or challenging situations for which flexible intelligence offers such an advantage: the inhibition of the initial response, the conscious, explicit attention to the problem, and the generation and assessment of a new response before further action.[10]

The most memorable deliberation scene occurs in the cave of Polyphemos. After he has picked up and dashed out the brains of two of Odysseus' companions, the Cyclops eats them, "entrails, flesh and the marrowy bones alike," then, comfortably replete, lies down to sleep. Odysseus tells the story (9.298–306):

> Then I
> took counsel with myself in my great-hearted spirit
> to go up close, drawing from beside my thigh the sharp
>> sword,
> and stab him in the chest, where the midriff joins
>> the liver,
> feeling for the place with my hand; but the second
>> thought stayed me;
> for there too we would have perished away in sheer
>> destruction,
> seeing that our hands could never have pushed from
>> the lofty
> gate of the cave the ponderous boulder he had
>> propped there.
> So mourning we waited, just as we were, for the
>> divine Dawn.

Notice how Odysseus rehearses the swordthrust in his mind, only to be stopped by a glimpse of unintended consequences. For the moment, he has no other plan; he merely knows that this one is fatal. He has to wait until Polyphemos takes his flocks out to pasture, leaving his "guests" trapped in the cave behind him, with Odysseus

> mumbling my black thoughts
> of how I might punish him, how Athene might give me
>   that glory.
> And as I thought, this was the plan that seemed best
>   to me.
> The Cyclops had lying there beside the pen a great
>   bludgeon
> of olive wood, still green . . .   (9.316–320)

Here plan B requires time to formulate, involves several stages of foreplanning, and, rather than being summarily described and then executed, unfurls suspensefully before our eyes as Odysseus prepares each new step.

His plan, to blind Polyphemos and then have his men steal out the next day beneath the Cyclops' flock, has been recorded in 200 folk variants.[11] Whether or not these versions derive from or feed into the *Odyssey,* Homer has made the incident his own and promoted it to pride of place, the first detailed adventure, after our and the Phaiakian audience's long wait, in Odysseus' account of his wanderings. He has made it pivotal, since the blinding of Poseidon's son arouses the sea god's wrath and leads to the smashing of the raft Odysseus builds to flee from Kalypso. Twice later Odysseus will invoke the ordeal of the cave as a morale-booster (12.209–212, 20.19–20), a reassuring example of his resourceful resolution in direst circumstances.

The blinding of a one-eyed giant is memorable enough, but this incident especially haunts the mind because Homer makes it such a paradigmatic example of flexible intelligence at work. Killing a deadly sleeping enemy who intends to eat you and your men, one by one or two by two: how could *that* be a wrong move? But in this case it is. Odysseus stops, thinks, and after hours of brooding devises a multipart plan based on what lies at hand in the cave. If he only *blinds* Polyphemos, the Cyclops will still be able to move the boulder blocking the cave and to offer them all a chance of escape. But when a stake pokes his eye out, Polyphemos will naturally howl with pain, and the noise could bring other Cyclopes to his aid. To counter that danger, Odysseus tells Polyphemos that his name is "Nobody," so that when the other giants do indeed rush to his cave and ask who is hurting him, Polyphemos answers, "Nobody"—and they promptly withdraw. When he lets his flock out to pasture, the blinded Cyclops will monitor their passage by touch, to ensure that none but his sheep leave the cave. Odysseus solves *that* problem by tying each of his men under the middle of a trio of sheep, and by himself hanging underneath the fleece of the largest ram. The Wooden Horse is the classic instance of Odysseus' invaluable ingenuity at Troy, not mentioned in the *Iliad* but retold three times from three angles in the *Odyssey*. The brilliant ruse of *entering* the enemy stronghold at Troy, inside an artificial horse, seems to be restaged and reworked in the *exit* from the giant's cave, hidden under live sheep.[12]

Odysseus will later present the encounter with Polyphemos as a triumph, but Homer makes the incident more problematic. As he takes his leave of Polyphemos after the escape from the cave, Odysseus undoes much of his success. When he thinks himself out of range, he calls to Polyphemos from his ship, vaunting like the heroes of the *Iliad*. Polyphemos replies by snapping off the peak of a mountain and hurling it into the sea toward Odysseus' voice, almost swamping his vessel. Sailing farther out, Odysseus calls again, although his companions urge silence (9.502–505):

> Cyclops, if any mortal man ever asks you who it was
> that inflicted upon your eye this shameful blinding,
> tell him that you were blinded by Odysseus, sacker
>   of cities.
> Laertes is his father, and he makes his home in Ithaka.

Even after Polyphemos replies that he will invoke the aid of his father, Poseidon, Odysseus taunts him again, and elicits in reply Polyphemos' prayer to his father (9.530–535):

> ... grant that Odysseus, sacker of cities, son of Laertes,
> who makes his home in Ithaka, may never reach
>> that home,
> but if it is decided that he shall see his own people,
> and come to his strong-founded house and to his
>> own country,
> let him come late, in bad case, with the loss of all his
>> companions,
> in someone else's ship, and find troubles in his
>> household.

At this point, almost ten years before Odysseus' return to Ithaka, most of the companions and all the ships returning from Troy are still with him, and the suitors have yet to begin their assault on Penelope. Polyphemos' curse establishes the terms of Odysseus' fate over the next ten years. Its echo of Odysseus' taunt and its accurate prefiguration of his future suggest how much Odysseus is to blame for the woes on his way home, and all because for once he does not restrain himself at this moment of intense anger and relief.

The Cyclops episode resonates because of its complex mix of Odyssean forethought and un-Odyssean impulse. The inhibition theme stretches backward and forward from here throughout the story. The first mention of Odysseus' great triumph at Troy, the Wooden Horse, comes from Menelaos, who reports that Helen, standing with her second Trojan husband, Deiphobos, called to the heroes inside the horse in the voices of their wives, whom they had not seen or heard for ten years. Odysseus saved the day and the chance of a Greek victory at Troy, already created by his own ingenuity, by persuading Menelaos and Diomedes not to respond, and even forcibly muffling Antiklos as he was about to answer. Odysseus' failure to be silenced by his companions below the cave of Polyphemos pointedly reverses this earlier strategic success.

Odysseus' failure of judgment in taunting Polyphemos is central, but strikingly exceptional. Critics have often wondered why the advance summary of the story in the *Odyssey*'s proem singles out only one much less celebrated adventure, the oxen of Helios (1.4–9):

many the pains he suffered in his spirit on the wide sea,
struggling for his own life and the homecoming of his
    companions.
Even so he could not save his companions, hard though
he strove to; they were destroyed by their own wild
    recklessness,
fools, who devoured the oxen of Helios, the Sun God,
and he took away the day of their homecoming.

At the start of the poem, to ensure sympathy for his hero, Homer contrasts Odysseus favorably with his companions. He did all he could for them, but their own lack of self-control in eating the oxen sacred to the sun god, even after he had sternly warned them, led to their deaths.

Just as the Cyclops episode earns structural prominence in various ways, so, too, does the Oxen of the Sun episode. Circe instructs Odysseus that he must travel to Hades, into the dread land of the dead, to learn from the shade of the seer Teiresias how to ensure his return. Teiresias tells him that Poseidon will not make his return easy, but that he might still come back, "if you can contain your own desire, and contain your companions," when they reach Thrinakia and the herds of Helios:

Then, if you keep your mind on homecoming, and leave
    these unharmed,
you might all make your way to Ithaka, after much suffering;
but if you do harm them, then I testify to the destruction
of your ship and your companions, but if you yourself
    get clear,
you will come home in bad case, with the loss of all your
    companions,
in someone else's ship, and find troubles in your household,
insolent men . . . (11.110–116)

Teiresias' series of conditional outcomes (*if . . . but if . . . but if*) stresses the companions' own responsibility for their final losses, then slides into the terms of Polyphemos' curse before elaborating further. When Odysseus returns from Hades to Circe, she repeats the warning (including the first five lines above). As his ship approaches Thrinakia, Odysseus thinks of the warnings from Teiresias and Circe: "Both had told me many times over"—incompatible, in fact, with the evidence so far, since there has been only one, hurried, occasion for Odysseus in Hades to listen to Teiresias; but this recollection makes all the plainer Homer's wish to weight these warnings. Then Odysseus repeats to his men the double repetition of the admonitions. Nevertheless they rebel against his order to sail straight past Thrinakia. Although they at first heed his warnings not to touch the sacred cattle, contrary winds trap them on the island, and when they run out of the food they have brought and are forced to catch fish and birds, they slaughter the cattle while Odysseus sleeps. Henceforth they are doomed, and are struck by Zeus's lightning as soon as they head to sea.

The inhibition of automatic responses is essential to higher intelligence. It is also essential to morality, to overcoming instinctive but unwise responses to, for instance, anger (Odysseus' challenge after escaping from the Cyclops who has eaten his companions) or hunger (the companions who slaughter the sun god's cattle). Odysseus is reckless that one time, and that lapse has grave consequences for him and his men. But the companions are reckless throughout the wanderings, even despite warnings from the gods. Homer repeatedly compares the recklessness of the companions with the recklessness (Zeus uses that very word, in the first speech in the poem) of Aigisthos in ignoring the gods' warnings against killing Agamemnon, and with the recklessness of the suitors, repeatedly warned by gods, mortals, omens, prophets, but unready to curtail their pleasures.[13]

In his first reference to Odysseus, Zeus responds to Athene's plea for her favorite (1.65–67):

> How could I forget Odysseus the godlike, he who
> is beyond all other men in mind, and who beyond others
> has given sacrifice to the gods, who hold wide heaven?

Odysseus is the darling of the gods, and the scourge of the suitors, almost the hand of divine justice, because—with that one costly exception against Polyphemos—he takes the larger view: he overmasters the impulses of the moment.[14]

In order to *become* the scourge of the suitors and the instrument of divine wrath, Odysseus on his return to Ithaka has to inhibit himself for longer, and under more trying circumstances, than ever before. Almost the entire second half of the *Odyssey* focuses on him back in Ithaka, hiding his identity in the guise of an aged beggar. He has to maintain the disguise despite two opposite provocations: the repeated insults to his heroic pride by the suitors and their acolytes, and the impulse to melt into tender tears on his reunion with his family and loyal followers.

Athene has warned him in advance: "you must endure much grief in silence, standing and facing men in their violence" (13.309–310). Odysseus alerts Eumaios at the start of book 16 to someone's approach: "His whole word had not been spoken when his beloved son stood in the forecourt" (16.11–12). He has not seen Telemachos since he was an infant at Penelope's breast, but he has yearned for him ever since and recognizes him instantly. Eumaios has become a kind of surrogate father during Odysseus' long absence, and he and Telemachos greet each other as warmly as actual father and son. Odysseus longs to tell his son who he is and clasp him to his breast, but he has to watch these two act out the reunion that he himself has craved all these years, allowing no sign of his identity or his emotions.[15] Not that Homer says so explicitly, here, but he does expect us to infer his hero's feelings.[16] Those emotions are made manifest later, when Eumaios has left and Odysseus can disclose to Telemachos who he is: "So he spoke, and kissed his son, and the tears running down his cheeks splashed on the ground. Until now, he was always unyielding" (16.190–191). Later Homer marks the irony immediately when Penelope confides in this beggar-stranger, whom she does not recognize as her husband and who claims to have entertained Odysseus (19.209–212):

> her beautiful cheeks were streaming tears, as Penelope
> wept for her man, who was sitting there by her side. But Odysseus
> in his heart had pity for his wife as she mourned him,

> but his eyes stayed, as if they were made of horn or iron,
> steady under his lids. He hid his tears and deceived her.

Homeric heroes usually have a high consciousness of status and an aggressive reaction to any perceived slight. But Odysseus has to prepare his surprise assault on the suitors while enduring the lowly role of a beggar and the taunts of the suitors and their sycophants. Note the deliberation-scene pattern when the goatherd Melanthios, an acolyte of the suitors, derides Odysseus-as-beggar and tries to kick him off a path. Odysseus withstands the blow

> while he pondered within him
> whether to go for him with his cudgel, and take the life
> from him,
> or pick him up like a jug and break his head on the
> ground. Yet
> still he stood it, and kept it all inside him.   (17.235–238)

In this case Odysseus entertains both a plan A and a plan B but rejects both, however immediately satisfying.

Inhibiting an instinctive response to seek a more adequate solution is central to Odysseus' character and his special intelligence in the *Odyssey*. But one person matches him. Penelope's most common epithet is "circumspect." She, like Odysseus, sees the big picture.

Before her husband's return Penelope fobs off the suitors with her famous web ruse, knowing she does not want to marry any of them yet unwilling to rule out the refuge of remarriage should Odysseus never come back. After the suitors discover her ploy, she still cannot decide: "so my mind is divided and starts one way, then another. Shall I stay here by my son and keep all in order . . . or go away at last with the best of all those Achaians who court me here in the palace, with endless gifts to win me?" (19.524–529). Even after Odysseus returns, kills the suitors, and stands before her undisguised, Penelope will not acknowledge him. She has heard false reports of Odysseus for years, and she knows that Zeus impersonated Amphitryon, while he was off at war, in order to possess Alcmena and sire Herakles. No longer expecting Odysseus' return, she fears that she, too, may be deceived by the gods—who have indeed flawlessly impersonated mortals throughout the *Odyssey*. Much to her son's vexation, Penelope restrains her desire to acknowledge this man who looks like her husband until she has put him to a test that only Odysseus himself could pass. By instructing Eurykleia to move their bed out of her bedroom for him to sleep on, she tricks him into exploding into anger—at this he *cannot* inhibit his response—and only then does she give way to her feelings (23.205–208):

> So he spoke, and her knees and heart within her
> > went slack
> as she recognized the clear proofs that Odysseus
> > had given;
> but then she burst into tears and ran straight to him,
> > throwing
> her arms around the neck of Odysseus, and kissed his
> > head . . .

Both Athene and Homer are themselves exemplars, at different levels, of the role of inhibition within intelligence. Both stop, reject the obvious course, and search several moves ahead for a better solution. Instead of rushing straight to Odysseus when she has been given Zeus's permission to release him from Kalypso, Athene chooses to transport herself to Ithaka to develop Telemachos as the ally Odysseus will need when he confronts the suitors. Instead of introducing Odysseus at the start, or following his fate since the end of the Trojan War in chronological order, or beginning with the thrills of his most famous adventures, Homer searches for a less obvious and far richer solution.

At every level, Homer shows the power of inhibition, the top-down control of thought. We can now understand this in neurological terms, as the prefrontal cortex—the part of the human brain that has expanded most dramatically over the last three million years, and the part that makes abstract reasoning possible—suppressing immediate natural responses for the sake of long-term strategic advantage. Homer had no notion of the mind as modern neuroscience can explain it, and the terms he had at his disposition were vague.

But from his own intelligence and his experience in shaping stories he knew enough about how thought worked to portray the power of the mind at its best in Odysseus, Penelope, Telemachos, and Athene: intelligence resisting immediate impulse in order to seek more satisfactory solutions.

17

THE EVOLUTION OF INTELLIGENCE (2):
BEYOND THE HERE AND NOW

The Metarepresentational Mind

Children between one and two start to entertain multiple models of reality, to recall the past and recognize it as *no-longer*, to anticipate the future and recognize it as *not-yet*, and to enjoy pretense as *not-really*. But although they can reason within these frames, they cannot yet clearly understand the relationships between one model and another. Only during their fourth year do they begin to develop fully human metarepresentational minds, which allow them to understand readily past, present, and future; real, pretend, supposed, or counterfactual; and the perspectives of others and even their own in the past.[1]

In the most famous twentieth-century discussion of the *Odyssey*, and perhaps of any literary text, Erich Auerbach wrote that subjective or perspectival effects are "entirely foreign to the Homeric style; the Homeric style knows only a foreground, only a uniformly illuminated, uniformly objective present . . . despite much going back and forth, it yet causes what is momentarily being narrated to give the impression that it is the only present, pure and without perspective."[2] Like the readings of Snell and Jaynes in the previous chapter, Auerbach's claim would seem prima facie highly unlikely: that a major masterpiece operates below the level that modern children reach before age five. And in fact the *Odyssey* not only multiplies perspectives but makes the most of their complex strategic interaction.

We learn at the start that Odysseus is alive, "detained by the queenly nymph Kalypso" (1.14) and that the gods have decided to release him. This knowledge makes all the difference to our sense of the first scenes in Ithaka, where Penelope and Telemachos pine for an Odysseus whom they imagine to be dead at sea or a pile of bare bones on land. Our knowledge also colors our first glimpse of Odysseus, who believes himself forever trapped on Ogygia, and yearns for an Ithaka that he has not seen for twenty years and that we have learned has changed in ways he cannot suspect. Not only are the characters' notions and the audience's emotions saturated with the knowledge of other perspectives, but characters easily take other characters' perspectives into consideration. Athene as Mentes tries to reassure Telemachos, when she first meets him, that his father is not dead but "held captive on a sea-washed island, and savage men have him in their keeping" (1.197–198). She invents these "savage men" so as not to disturb him with the truth: that his father is held, however unhappily, in the amorous arms of a goddess.[3] Much of the poem's pleasure derives from Homer's allowing us to see things simultaneously from the perspective of one character, or of many, or in a larger view altogether.

Auerbach's Exhibit A (he saw no need for a B) is the digression on Odysseus' scar in book 19. Penelope has ordered her serving-women to bathe the stranger's feet: unlike the suitors, she will give him due hospitality despite his beggarly status. Odysseus, angry that he might be handled by the abusive young serving-women who betray the household by sleeping with the suitors, asks for an old serving-maid instead. As Eurykleia steps forward, he realizes that she might recognize the scar on his leg and turns away from the firelight to hide the telltale mark (19.392–395):

> She came up close and washed her lord, and at once she recognized
> that scar, which once the boar with his white tusk had inflicted
> on him, when he went to Parnassos, to Autolykos and his children.
> This was his mother's noble father . . .

The digression explaining how he received his scar runs another seventy lines. It ends by noting that on his return to Ithaka his parents

> were glad in his homecoming, and asked about all that
>   had happened,
> and how he came by his wound, and he told well his
>   story,
> how in the hunt the boar with his white tusk had
>   wounded him
> as he went up to Parnassos with the sons of Autolykos.
>   The old woman, holding him in the palms of her
>   hands, recognized
> this scar as she handled it. She let his foot go . . .
>   (19.463–468)

Auerbach sees this passage as typical of Homer, and as proof that he "knows no background. What he narrates is for the time being the only present, and fills both the stage and the reader's mind completely. So it is with the passage before us. When the young Euryclea (vv. 401 ff.) sets the infant Odysseus on his grandfather Autolykos' lap after the banquet, the aged Euryclea, who a few lines earlier had touched the wanderer's foot, has entirely vanished from the stage and from the reader's mind."[4]

Not from *this* reader's mind, or the minds of many others.[5] Auerbach adds that the scar digression could have been presented subjectively, rather than authorially,

> as a recollection which awakens in Odysseus' mind at this particular moment. It would have been perfectly easy to do; the story of the scar had only to be inserted two verses earlier, at the first mention of the word scar, where the motifs "Odysseus" and "recollection" were already at hand. But any such subjectivistic-perspectivistic procedure, creating a foreground and a background, resulting in the present lying open to the depths of the past, is entirely foreign to the Homeric style; the Homeric style knows only a foreground . . . And so the excursus does not begin until two lines later,

> when Euryclea has discovered the scar—the possibility for a
> perspectivistic connection no longer exists, and the story of
> the wound becomes an independent and exclusive present.[6]

Auerbach could hardly be more wrong. The story of the scar is not presented as *Odysseus'* memory, true, but it *is* presented as Eurykleia's, as Irene de Jong convincingly shows.[7] "She recognized that scar" begins the digression, and "the old woman . . . recognized this scar" ends it. The digression begins with a description of Autolykos, his visit to his infant grandson, and Eurykleia's laying the babe on Autolykos' knees and suggesting he should find a name for the child. There and then he names him Odysseus, and looks forward to seeing the boy grown up and visiting him. This account provides the transition to Odysseus' visit as a youth, his being gored by a wild boar, and his vividly retelling the tale on his return to Ithaka. The vividness of that version remains fresh in Eurykleia's mind.

Homer achieves through the scar digression effects far richer and subtler than Auerbach notices: (1) When she sees the scar, Eurykleia immediately recognizes her master. Her mind floods with the knowledge that she has cared for this "stranger" since infancy and has watched him grow into a hero. The long digression stresses how long she has known him, how vividly she recalls him, how happily she will greet his return. She recognized him at once; she caused him to be named as an infant; will she now name him again, and destroy his strategy of secrecy?[8] (2) In name and in much else, Eurykleia is a doublet of Odysseus' mother, Antikleia, whom Odysseus was troubled to find already in Hades, dead of grief at his long absence.[9] Laertes had bought Eurykleia "when she was still in her first youth . . . and he favored her in his house as much as his own devoted wife, but never slept with her, for fear of his wife's anger" (1.431–433). Eurykleia has suckled Odysseus, has suggested that he be named, has watched him grow up. She has been another mother to him—and in her agitation at recognizing him, and wanting to announce his name to Penelope, she drops his leg back into the basin, causing it to spill. (3) Within the digression, we see Odysseus traveling away, proving himself a brave young hero, and returning to a joyful welcome from his parents. We have just seen Telemachos traveling away from home and proving *himself* his father's son, a hero in readiness, welcomed back by a relieved mother—and also by the father he has consciously seen for the first time. Now a second mother wants to welcome Odysseus back, after a much longer and more harrowing absence. (4) When Telemachos left on his secret journey, he confided in only one person, Eurykleia. He swore her to secrecy for at least twelve days. She kept the secret then—but can she do so now? (5) Odysseus sees the tears of delight in Eurykleia's eyes, and hears her name him. Though not privy to her memories of him, he knows how far they reach and how vividly she will recall his earning the scar. He takes her by the throat and swears the astonished old nurse to silence on pain of death—just as once before, in the Wooden Horse at Troy, he had been able to stop Antiklos[10]* from responding to Helen's impersonation of his wife's voice only by "brutally squeezing his mouth in the clutch of his powerful hands" (4.287–288). (6) The instantaneousness of Odysseus' recognition by this second "mother" will stand in contrast to—and increase the sheer surprise of—the slowness of his recognition by his wife.

Far from having no background or perspective, Homer creates multiple perspectives, present, past, future possible, future foreglimpsed or future preordained, mortal, divine, postmortal, divinely objective or humanly subjective, observed, dreamed, or remembered. He portrays sophisticated multilevel metarepresentational minds in his characters, and he expects such minds in his audience, easily able to imagine and distinguish memories, projections, perspectives, guesses, mistakes, and lies, and effortlessly understand their status. He assumes we will understand individual perspectives and strategies, even when he does not comment on them, and compare them with the larger picture the poem lets us see. True, he does all this with a comparatively poor toolbox for fine mental detail, one lacking many terms and narrative tricks that others would invent over the coming millennia, and therefore he often represents thought via external action. But throughout the *Odyssey* he both creates and ap-

---

* His name, like Antikleia's, pointedly echoes Eurykleia's.

peals to metarepresentational minds that can and must negotiate multiple perspectives and multiple levels.

Deception

A metarepresentational mind allows us, among much else, to understand what others might know or infer from what they see or hear, and how that knowledge may affect what they do. We can use this understanding for the purposes of cooperation—for example, in explicitly teaching others—or of competition.

In the world of the *Odyssey,* as we shall see in the next two chapters, trust is scarce. Many habitual locutions suggest low expectations of honesty. One asks not who these strangers are, but who they "announce themselves as being" (4.139). When Hermes visits Kalypso to inform her of the decision of the gods' council, he answers her inquiry: "You, a goddess, ask me, a god, why I came, and therefore I will tell you the whole truth of the tale" (5.97), as if, were either or both of them mortal, there would be little reason to expect a true answer. Alkinoös says to Odysseus, amid the long tale of his wanderings, "Odysseus, we as we look upon you do not imagine that you are a deceptive or thievish man, the sort that the black earth breeds in great numbers, people who wander widely, making up lying stories, from which no one could learn anything" (11.363–366)—a deeply ironic comment, given Odysseus' own talent for deceit. Deceit is rife and Odysseus the master deceiver.

Nature itself employs deception in mimicry, camouflage, and bluff. Deceptive *behavior* is less widespread across species, and its frequency in a given species appears proportionate to the expansion of its neocortex or equivalent.[11] Once thought to be uniquely human, deceit has been found in animals from scrub jays to wolves and all families of monkeys and apes.[12] In most of these species it involves simply suppressing information, such as one's knowledge of a food source, so that one can feed without having to share. In the *Odyssey,* too, Odysseus deceives most often by withholding information, especially the one piece of information he knows better than anyone, his identity. He withholds his name from the Phaiakians until he can stall no longer if he wants their aid in returning to Ithaka.

Much earlier in his travels, Polyphemos asks him where his ship is, and he pretends that it broke up on the shore, since he senses that the Cyclops might want to destroy it to cut off his escape.

Many researchers suggest that deception in other species does not usually involve intentions to create false beliefs.[13] Such intentions are certainly a human specialty and Odysseus a supreme specialist. After he plies Polyphemos with strong drink, the Cyclops demands to know his name. Punning on his name, Odysseus answers *ou tis,* "nobody." A little later, blinded by the stake, Polyphemos bellows out in pain and brings his fellow Cyclopes running. They call into his cave: "Surely none [*mē tis*] can be killing you by force or treachery?" He answers: "Good friends, Nobody is killing me by force or treachery." The other Cyclopes leave, assuming their help is not needed. Odysseus gleefully tells the tale: "the heart within me laughed over how my name and my perfect planning [*metis*] had fooled him" (9.406–414). Not for nothing is he *polymetis* Odysseus, Odysseus of many devices.

At last back on Ithaka, though he does not yet know it, Odysseus sees a youth—Athene in disguise—and asks her where he has been put ashore. She answers, playing the part of an untraveled Ithakan: "the name of Ithaka has gone even to Troy, though they say that is very far from Achaian country." Odysseus responds,

> but he did not tell her the truth, but checked that word
>   from the outset,
> forever using to every advantage the mind that was in
>   him:*
> "I heard the name of Ithaka when I was in wide
>   Crete."   (13.254–256)

He goes on to regale the "youth" with an elaborate lie about being in exile for killing a man who "tried to deprive me of all my share of the plunder from Troy" (13.262–263)—a warning to the youth not to try to steal the guest-gifts he has brought ashore from Phaiakia. Athene

---

* Notice inhibition here explicitly linked to intelligence, and both to deception.

now appears before him in more or less her own shape, expressing her delight at his fluent falsehoods (13.291–299):

> It would be a sharp one, and a stealthy one, who would
>   ever get past you
> in any contriving; even if it were a god against you.
> You wretch, so devious, never weary of tricks, then you
>   would not
> even in your own country give over your ways of
>   deceiving
> and your thievish tales. They are near to you in your very
>   nature.
> But come, let us talk no more of this, for you and I
>   both know
> sharp practice, since you are far the best of all mortal
> men for counsel and stories, and I among all the
>   divinities
> am famous for wit and sharpness.

Throughout the second half of the poem, indeed, Odysseus will not only remain almost always in disguise as a beggar, but will tell one slick lie after another, to Eumaios, to Penelope, to his father, inventing elaborate versions of his past cunningly shaped to his immediate audience and purpose.

Most of us live in societies with higher levels of trust than Odysseus' world, where raiding for cattle, women, and other movable goods was seen as normal and even enterprising behavior. The majority of those researching the evolution of intelligence consider that the prime impetus for the acceleration of intelligence has been sociality, the need to understand and perhaps outwit others. At first dubbed the Machiavellian intelligence hypothesis, as we have seen, others rightly objected to this term, since social intelligence can strengthen cooperation as well as sharpen competition. In the human case our present ability to cooperate as well as we do in such large groups of unrelated individuals is unique, and a tribute to our intelligence, and to the cultural means, the values and institutions, we have elaborated to amplify trust.[14] But the virtuosic deceptiveness with which Odysseus deploys his intelligence and the delight it arouses in the goddess of wisdom suggest that in times and places of low trust, like the world of the *Odyssey*, a Machiavellian intelligence may have been a necessary defensive and offensive weapon.

Just as Odysseus is rated "beyond all other men in mind," Penelope knows herself to "surpass the rest of women, in mind and thoughtful good sense" (19.326). Despite being antiquity's paragon of female fidelity, she, too, is an archdeceiver. Her ruse of the web misleads the suitors for years. She outdoes even Athene. While Athene's disguise fools Odysseus on his first return to Ithaka, she cannot make him give himself away.[15] That is precisely what Penelope accomplishes— and indeed *has* to do, to be sure that this man really is her Odysseus, almost back from the dead—when she pretends that the bed he built for them, out of the base of a tree still rooted to the ground, has become movable, as if her affections were themselves now mobile. His anger at the idea—his inability, for once, to inhibit his outrage—discloses that he is no impostor, while her ruse testifies to the cunning they share.[16]

FALSE BELIEF

Children under the age of four find it almost impossible to understand false belief (see Chapter 10), but over the next year they develop the capacity to comprehend that ideas can be wrong.[17] Minds have evolved for many millions of years to take in information from the reality around them, but to pass the Sally-Anne test, to understand that false beliefs can exist, children have to learn to inhibit what they know of the *actual* state of things to grasp Sally's or their own wrong idea. Inhibiting the immediate and known to reason out a false reading of the known is a highly unusual thing to ask of a brain, and there seems no firm evidence that any nonhuman animals manage the task.[18] But for us it soon becomes automatic and effortless to track what others might know or not know.

Understanding that belief underpins desire and intention, and that belief can be wrong, allows us much finer accuracy in predicting others' desires, intentions, and behavior. It makes possible the complexity and sophistication of human interaction. We effortlessly assess others' beliefs about social situations and the consequences of any errors. That skill, central to social life, explains the force of *dramatic irony,* which pervades all storytelling and saturates the *Odyssey.*

Dramatic irony occurs in any story in which the audience or some characters realize they know more than others. From the start of the *Odyssey,* we know more than any of the mortal characters. We know that Odysseus still survives and the gods have decreed his return. We derive pleasure from seeing the whole picture and from foreseeing the good fortune ahead for the woebegone Odysseus, Penelope, and Telemachos, and the come-uppance looming for the smug suitors.[19] We keep track of those who know more than others, although we enjoy knowing still more. In book 2, Telemachos, as the only person in Ithaka who knows he has been visited by a god, finds himself in a uniquely powerful position. But he does not know which god he encountered. We do, and we know it is the divinity with the most intense devotion to Odysseus, and we know just why she has prompted Telemachos on his quest, and what answer he will find to his query.

Although dramatic ironies swarm through the poem, they keep on changing. In the first four books, many characters vividly imagine Odysseus' death or wish for his return, not thinking it at all likely, while we know him to be alive and about to return. In the next four-book block, Odysseus manages a dangerous escape but does not know as we do that Athene is keeping a close eye on him. In the next block, comprising Odysseus' tale of his wanderings, *he* discloses much that is new to *us,* yet fresh dramatic ironies still arise. When he ventures into Hades, he is surprised to find Agamemnon there. He does not know of his old leader's murder, on Zeus's mind when the council of the gods opens in book 1, or of Phemios' song later in that first book, and the subject of a long report by Nestor in book 3 and a shorter reminder from Menelaos in book 4. *Everyone* knows this story by now—everybody except Odysseus.

Upon Odysseus' return to Ithaka, the ironies deepen. Athene disguises him as a gray-haired old beggar or restores him to his glorious prime according to strategic need. Disguise, like dramatic irony, pervades story, because our understanding of false belief shapes so much of what we think about behavior. For many decades critics and teachers have proposed "appearance and reality" as "the theme" of endless numbers of literary works. That it is ever *the* theme—or that any work of much interest can have only one theme—is unlikely, but that it is *a* theme is almost inevitable, since it is a central part of human intelligence to understand that we can make mistakes, whether about appearance and reality (like the Smarties tube, or Odysseus in disguise) or merely through not knowing some crucial piece of information (such as that Anne moved the marble, or that Odysseus is still alive).

Dramatic irony, though a narrative staple, is unusually ubiquitous in the *Odyssey,* and for good reason. Like Athene, Homer is fascinated by Odysseus because of his intelligence (and by Athene because of hers). He builds into his story the audience's pleasure at knowing more than even the most intelligent of mortals, and then at siding with this mortal when he, too, has been apprised of some of the gods' secrets.

Many animals, especially omnivores like crows, parrots, rats, and chimpanzees, are curious, but human curiosity extends far further. False belief makes all the difference. If we realize we can make mistakes through not knowing the whole situation, our realization adds a new spur to curiosity. We know we will often need to know *more* in order to choose what to do.

Athene pricks Telemachos' natural curiosity about his father, rouses it to new urgency, and starts him on a voyage of discovery. But Odysseus himself, the archetypal voyager, has an even deeper curiosity, emphasized from the start. He explores the island and then the cave of Polyphemos out of sheer thirst for the unknown. He ventures into Hades, despite his dread of entering the land of the dead; even after he has heard what he needs to learn from Teiresias he lingers to ask questions of the familiar or the famous. He wants to hear the sound of the Sirens, despite the risk, so has himself tied to the mast with unstopped ears in order to enjoy the spellbinding new sensation. He always wants to take the long view, to see actions in their widest context, a quality in him that the Cynic and the Stoic philosophers later singled out admiringly.[20] Unlike them, Odysseus himself is no intellectual, for his world still *has* no such individuals; but it is no accident that Dante, Joyce, and Kazantzakis follow Homer in characterizing Odysseus or Ulysses as defined by curiosity.

But before and behind Odysseus there stands Homer. If Odysseus stops to consider the whole picture, Homer does so compulsively—and did it in a different way in the *Iliad*. As we have seen, he focuses intently in each epic on a particular phase within the story of Troy or the return, but he does so in ways that are also magisterially sweeping. Even within the phase of action on which he focuses, he encompasses the largest possible panorama and invites us to enjoy our privileged perspective. From the start, he shows us the decisions of the gods and confirms the pleasure of knowing the full story with each new irony, each wrong guess about the fate of Odysseus. Without spoiling the story's surprises, which continue into the last two books, he allows us to see as much as or even more than Zeus.

Throughout the ironies that pervade the *Odyssey*, Homer plays constantly on our awareness of false belief, on our recognition that there may be hidden aspects of a situation that invalidate what we—or the characters—know. Again and again he prepares for us the satisfaction of knowing the whole story and anticipating the pleasures awaiting the harried first household of Ithaka and the punishments awaiting the suitors.

Auerbach thought that Homer knew no background. Not so, as we have seen. But his famous discussion *was* right about the deep dissimilarity between Homer and Genesis. In the Hebrew text, God's will is inscrutable, permanently unknowable except when he chooses to reveal it, and even then remains of an order fundamentally beyond human fathoming.[21] In Homer, the comprehensive vision *is* in principle humanly obtainable, not unimaginably incomprehensible, even if we do not usually have access to all the relevant mortal facts, let alone to divine plans. But in the story of the *Odyssey*, we do have such access, and it places us in a position well beyond that of even the "wisest of mortals."

The *Odyssey* derives much of its power from the gap between the characters' limited perspectives and our all-encompassing vision. Homer reinforces our recognition that in life our grasp of situations may often be partial, and that full understanding involves seeing beyond the immediate. He incorporates the broadest possible perspective, he exudes confidence that it is in principle accessible to human minds, he easily incorporates us into his comprehensive vision and the pleasures and power it allows. These qualities in Homer may well have hothoused the flowering of Greek thought in the centuries that followed, when his work was the center of Greek education, thought, pride, and confidence. Homer has not only focused on a hero who embodies human intelligence, but he may have advanced intelligence itself, by increasing the confidence of his heirs within Greek culture that the whole world can in principle be understood by the human mind.[22]

RELIGION

Both Snell and Jaynes claim that Homer lacks a notion of mind—or that he and his characters altogether lack consciousness—especially on the basis of the gods' role in motivating human action. To both scholars, the presence of the Olympian gods confirms in Homer the absence of a modern mind or at least a modern notion of mind. I claim the converse. Religion plays such a prominent role for Homer and his characters precisely because they have a fully human sense of mind. They read others in terms of belief as well as desire and intention, and they understand—and find fascinating, to judge by all the dramatic irony in the *Odyssey*—the risks of false belief, of not knowing enough about a particular situation.

Our human awareness of false belief evolved out of a need to predict the behavior of others, but it has had much wider consequences for the growth of human knowledge. Once we realize that we often lack the information needed to explain a situation fully, we naturally wish, if the situation matters enough, to find a deeper explanation. Of course there may be none readily available, and the realization that there may be strategically essential information that we cannot access can disturb.

Because other agents—prey, predators, and especially human friends and foes—make the most dramatic difference to our chances and choices day by day, our understanding of other minds has evolved into our richest natural mental capacity. We explain the visible behavior of agents of all kinds, especially human agents, in terms of things we cannot see—beliefs, desires, and intentions—and we find these explanations powerful. We see psychological cause as a paradigm of *all* cause. So when we realize the implications of false belief, we naturally try to supplement what we do know and to counter our sense of the vulnerability of not knowing by resorting to the most powerful form of explanation at our disposal. If visible agents can be moved by invisible forces, perhaps there exist other invisible forces with still more explanatory value, unseen causes of movement and change? And though *we* often lack information, perhaps these others can always monitor what matters?

Agents with exceptional powers especially command our attention and haunt our memories. Agents who violate the intuitive ontologies by which we understand other living things—agents like the Greek gods, who are immortal, can know things at a distance, can assume invisibility or whatever form they wish, or prompt humans in sleep or awake—seize on our imaginations precisely because they overstep natural boundaries. But although these supernatural beings transgress the natural order, we can still understand them in terms of a belief-desire-intention psychology. In the Greek case, if the gods know we have ignored their decrees or failed to offer due sacrifice they will feel affronted and react accordingly.

Precisely because we understand false belief, because we realize we may err in action if we err in knowledge, we try to explain events more deeply. And in all human cultures, that has meant trying to explain events through supernatural agents. We do not set aside the natural, but we add the supernatural as another, apparently more powerful explanation. Storms at sea can sink a ship through waves, wind, or lightning, and knowing that, sailors can avoid setting sail in hazardous conditions. But to explain why *this* ship encounters a sudden storm, or lightning strikes *that* one, the Greeks invoke what they consider a more profound and pointed explanation, like the wrath of Poseidon or Zeus. Classicists call this "double determination": explanation in terms of the natural, including ordinary physical and psychological causes, *and* in terms of the supernatural. The gods add to but do not eliminate human responsibility. Athene helps Odysseus and Telemachos in their slaughter of the suitors, but *they* fire the arrows and hurl the spears and in most cases deflect their opponents' blows.

There are inconsistencies in Greek religious thinking, as in any other. "The gods know everything," we hear repeatedly. Sometimes they know immediately and from afar (as when Zeus sends eagles to confirm a mortal's chance remark), and sometimes they do not know unless they happen to encounter an event directly (as when Poseidon, returning from Aithiopia, happens to espy Odysseus at sea) or learn indirectly (as when Polyphemos' curse informs Poseidon just who has blinded his son). Homer makes the most of these uncertainties for narrative impact, at times allowing Athene confident foreknowledge, at times making her seem genuinely apprehensive that Odysseus' case has been forgotten, at times having to act herself in the moment, uncertain quite what to do, at others having to incite her favorites to act in ways she has not anticipated.

To us Olympian gods explain nothing. But for Homer and his world, humanly aware of the possible incompleteness of their knowledge, these gods offered the deeper explanation they sought and confirmed their confidence in the intelligibility of their world. Homer had a singular assurance in his power to explain the immediate through the remote, the seen through the unseen, the local through the all-encompassing. Within a couple of centuries his work was already at the center of Greek tradition. He evokes with incomparable force the incompleteness of human knowledge in the here and now and yet the possibility of attaining comprehensive knowledge. Surely this combination helped to inspire the first philosophers to seek out still deeper explanations, even if they no longer depended on the gods, or even if, as for Xenophanes, human intelligence could recognize the gods themselves as parochial.

Religion offers not only the appearance and reassurance of deeper explanation, but also an apparent solution to problems of control.[23] Brains are "anticipation machines" whose task is to guess what will happen next in order to stay one step ahead.[24] Most brains focus only on the here and now and try to anticipate only the immediate future. Human brains by contrast have a broader perspective, and can search for clues to the future in the past, or test sequences of imminent possible actions and outcomes, or imagine still longer-term futures. In the *Odyssey*, as in our world, people can vividly imagine elsewheres and elsewhens. But that capacity also puts us at risk of conjuring up futures filled with uncertainties and possible dangers—especially if our augmented ability to imagine the future fuses with our awareness that we often lack key information.

Religion can allay this special human anxiety about the future. The gods and their signs in Homer offer predictions as well as explanations. The terms of Zeus's decision in favor of Odysseus' return explain the hero's ordeal, and can themselves be "explained" still more deeply in terms of Zeus's need to heed the animosity toward Odysseus of a god as powerful as Poseidon. But we in the audience can see these explanations only because we are outside the story's world and have untrammeled access to it through the might of Homer's muse. Inside the story, much remains uncertain to the characters. Yet much can also be seen by those who care to look.

The signs Zeus sends—birds flying or fighting on the right or left of the characters, a lightning flash, a sudden sound after someone has prayed for an omen—confirm his control of events. Despite the portents' ambiguities, they can be validly interpreted by the discerning, like the seers Halitherses or Theoklymenos. Or the deserving can directly receive supernatural instruction about ways to cope with future dangers. The less deserving receive divine admonitions but ignore them, like Odysseus' companions ignoring his firm instructions from Circe and Teiresias not to harm the oxen of Helios, or the suitors' heedlessly laughing off the warnings of Athene, Halitherses, Theoklymenos, and Zeus himself. Homer rarely comments in his own voice about his characters, but he repeatedly calls "fools" those who ignore the signs of the gods' wills, and he presents Odysseus as someone whose intelligence stands out as much as anything else by his always seeking to know and abide by the will of the gods.

Our conceptions have changed since Homer—although people all over the world still seek the reassurance of control through religion, superstition, and astrology. But at least in educated secular societies we think we have found in science a better form of explanation, prediction, and control. We still look for present signs and patterns as a way of predicting future risks—climate change, pandemics, earthquakes, or asteroids—and of gaining a sense of control over a future whose anxieties still trouble us. Our methods have changed, but we still have a uniquely human drive to seek more comprehensive explanations and more powerful predictions. In that sense Odysseus descending even into Hades to learn the will of the gods stands for us all.

Religion offers another reassurance to creatures with both the blessings and the curse of imagination. Our metarepresentational minds allow us to entertain images of our own death or our absence from the world after death. That such imaginings have haunted our species since long before the first civilizations is a matter of archeological record.

Odysseus' long absence from Ithaka amounts to a kind of death.[25] Many in Ithaka imagine him dead, and we first see his homeland suffering under the effects of life without him. Early in his wanderings he almost reaches Ithaka with his men, only for them to untie the winds of Aiolos and drive their ship back where they came from, causing Odysseus to contemplate suicide. Later he steps into Hades and talks with the dead. He comes near to death before being washed up on Ogygia, where Kalypso offers him immortality. He declines, because he wants to return to Ithaka, preferring mortality's real rewards even at the cost of its real risks. He almost dies again on his way to Scheria. As the Phaiakians take him from Scheria back to Ithaka, upon his eyes "there fell a sleep, gentle, the sweetest kind of sleep with no awakening, most like death" (13.79–81). He returns to a land where many envision him already dead, and where, unrecognized by them, like a returning ghost, he can appraise others in terms of their attachment or hostility toward him and apportion reward or punishment accordingly.

Homer does not promise us immortality. As much as any poet, and more than most, he knows death close up. Death, in his depictions of Hades, is not an end, but it is not life, nor even a half-life. It does not inspire terror, like the Christian Hell, but it does deserve dread, for all the challenge and joy of life have drained away. Life is what matters, and the only immortality worth wanting is in the memory of the living. But he does suggest that through intelligence, through thinking beyond the moment and taking thought for the gods, we may have our best chance for a long mortal life and a lasting memorial in the minds of others—as his hero and heroine and he himself have found.

18

THE EVOLUTION OF COOPERATION (1):
EXPANDING THE CIRCLE

WHAT COULD THE *Odyssey* possibly tell us about human relations? We first hear of Odysseus trapped on an island as the
paramour-cum-prisoner of a goddess. At the end he arms to resist a
revenge attack by the male relatives of the hundred-plus suitors he
has killed the previous day, only for Zeus to intervene and impose
peace from above.

We may see the beginning, the bloody climax, and the ending of
the *Odyssey* as highly unrealistic, although they were not to Homer
and his audience. Does the gap between Homer's beliefs and ours
not make it unlikely that we can learn anything about our common
humanity from the *Odyssey*?

The extreme situations certainly catch our attention. But they do
much more. They allow Homer to explore solitude and society, competition and cooperation in fundamental ways. We can see the whole
story as a thought experiment. Unlike Ursula Le Guin, in the social-
thought experiments of her science fiction, Homer has no idea of
experiment or evolution. But an evolutionary approach to his story
can suggest how searchingly he explores social life, even if there is inevitably much he does *not* question; how and why he uses the moral
emotions to engage his original audience, in ways that still engage us,
despite our very different social assumptions; and even how and why
these emotions arose, and what part stories have played in shaping
them.

Homer builds on the traditional outline: Odysseus returns, the last of the Greeks fighting at Troy, to find his wife still faithful after twenty years, despite being pursued by suitors whom she has managed to keep at bay through her ruse at the loom.[1] Most of the rest the poet seems to have added, like the major role Telemachos plays, and the key character and role of Eumaios. These characters, and also the unusual scale of the poem, the tight chronological focus, and the complex chronological rearrangement that throw so much weight onto the Ithaka scenes, suggest that the lengthy buildup to a decisive confrontation with an army of suitors was also new with Homer.

Homer says some profound and deeply moving things about human nature through balancing the Greek and Trojan forces in the *Iliad*, and through balancing the gods who back each side's warriors. In the *Odyssey* he manages to say some very different, and some very similar, things, through the opposite strategy. He divides the characters sharply into good and bad, and places the ruler of the gods and the most active of goddesses on Odysseus' side, and no gods at all behind the suitors.

Stories regularly engage the moral emotions, because these emotions matter so much to social life. Before a year and a half, infants "look longer at visual displays depicting violations of arbitrary social rules" than at similar displays that do not.[2] Children from the age of three, like adults, look out for and detect violations of social rules, especially for "permissions, obligations, prohibitions, promises, and warnings," but they do not notice and detect violations in other types of reasoning.[3] We attend to social violations, and they stir our emotions. And across cultures, whether Apache, Dogon, or Sami, stories and storytellers are valued for their ability to inculcate the values of the community by depicting social violations and their consequences.

An evolutionary approach to storytelling can consider morality in terms of the conflicting demands of social life as they affect any species. Far from being solely human, the social emotions are present in rudimentary form even in creatures like the worm *C. elegans,* with a brain containing only 302 neurons, compared with the billions in the human brain.[4] Complex social emotions, like empathy and a sense of fairness, or behaviors like punishment and reconciliation, have been observed and tested in many species, especially in social primates.[5] Many social features often explained only in terms of a particular human culture actually reflect factors common to all or many human societies, and even to other species, so that attempts to explain them solely through local conditions are doomed to incompleteness. Not that the local should be ignored, but to explain it adequately we need to consider the problems that a particular cultural practice attempts to solve, the evolved dispositions it deploys, and the problems it *fails* to solve or the new problems it generates. And to understand these problems in their local situation we first need to understand the problems of social life in general.

As we saw in Chapter 4, cooperation exists in many species yet is extremely difficult to evolve. It begins in mutualism, animals benefiting one another merely as they go about their ordinary self-interested activities like feeding and keeping watch. It seems probable that sociality in multicellular animals began with defense, with safety in numbers. This first impulse to sociality depends for its effectiveness on emotions of comfort and pleasure in the company of conspecifics, and anxiety in being too far away from them. Blind human babies smile at the sound of a human voice and focus their sightless eyes on the place where the sound issues from.[6] Universally humans withdraw attention from others to punish them, and one of our severest forms of punishment is imprisonment in solitary confinement.

Odysseus begins the *Odyssey* deeply miserable. Although he is safe on an island with a comely goddess who supplies his every physical need, he has lost all his companions, he has not seen another of his kind for seven years, and he feels utterly bereft. He does not hesitate to risk death in crossing the sea for the chance of rejoining the human world. In opposition to him and his solitude, literally and structurally, and indeed currently occupying the place in Ithaka to which he yearns to return, are the suitors, a whole "swarm" of them. Although they nominally compete for Penelope's favors, these are currently off-limits, so meanwhile they enjoy Odysseus' resources, his livestock, his wine, his servants. Their numbers give them confidence in their safety even should the lord of the estate return in righteous anger (2.246–251).

Except for identical twins, any two individuals of the same sexually reproducing species will be genetically different and will to that degree have competing interests wherever resources are limited. This competition of interests produces dominance disputes over access to resources. A stable hierarchy reduces the need for repeated squabbles over precedence and for that reason tends to establish itself in many animal societies.

Whatever the precise political structure on Ithaka and the neighboring islands, Odysseus is undisputedly at the top of the hierarchy among his companions when they leave for Troy and as they head back for Ithaka. He ends up alone, naked, on the island of the Phaiakians. At their athletic contests, the Phaiakian youths jeer at Odysseus as a mere merchant sailor, since he declines to participate. Angered, he proves his noble status, even before disclosing his identity, through a displaced dominance contest, by throwing a discus far beyond any of theirs. Back in Ithaka, he must adopt the lowly position of a beggar before he can win back the dominance the suitors have challenged. The confirmation that he can do so comes in another ritualized physical contest, when he alone can string and draw the bow—at which point he unleashes his full power, and the bow's, on those who have dared to challenge his position.

Hierarchy, so long as it is accepted, can reduce the competitive friction in sociality. But what establishes the benefits of cooperation? As William Hamilton's theory of inclusive fitness notes, our genes exist not only in ourselves but also in our kin. We therefore recognize kin and are motivated to help them through feelings of attachment, usually mediated in mammals by the neurochemical oxytocin.

In the *Odyssey* Odysseus instantly recognizes Telemachos, though he last saw his son as an infant at Penelope's breast. Despite his strategy of secrecy, Odysseus discloses his identity to Telemachos as soon as they are alone, but keeps it still concealed from his swineherd Eumaios, although he has known and loved the "godlike swineherd" for decades, and although Eumaios has freshly and fully demonstrated his love and staunch loyalty. Later, just before the battle, Odysseus quizzes Eumaios about his preparedness to fight, but he does not need to ask for Telemachos' help. From the first he assumes that his son will aid him even in the potentially deadly showdown with the suitors.

At the moment when Telemachos realizes that the man before him is indeed his father, son and father embrace and weep for all their long years apart. Homer amplifies the poignancy via the universality of parent-child love in an unexpected simile (16.213–219):

> now Telemachos
> folded his great father in his arms and lamented,
> shedding tears, and desire for mourning rose in both
> of them;
> and they cried shrill in a pulsing voice, even more than
> the outcry
> of birds, ospreys or vultures with hooked claws, whose
> children
> were stolen away by the men of the fields, before their
> wings grew
> strong; such was their pitiful cry and the tears their
> eyes wept.

Since I won't discuss Homer's bold and brilliant similes again, let me linger over this. Here he resorts to an animal simile to express not so much the clamor of the cry issuing from Telemachos and "his great father"—although he does that, too—as the depth of their loss. In order to stress the father-son bond, he chooses not warm and cuddly mammals but birds, since avian fathers and mothers usually both take close responsibility for their chicks, whereas mammalian fathers rarely help rear their offspring. The species Homer names are rapacious and hard-edged, yet they, too, suffer when their young are taken from them. In this characteristically complex Homeric reverse simile, human father and son have just been reunited but are compared to parent birds just parted from their children for life. And that is the recognition that hits human father and son: they have been parted virtually all of Telemachos' life.

Homer expects his audience to feel the force of the comparison. The universality of the emotion, the fact that it can rend the heart even of a vulture, does not make it seem in the least commonplace. On the contrary it signals that the love of parent and child is so central to life that we can recognize it even in an almost alien species— let alone in our two heroes. Homer is closer than chronology would suggest to the Darwin of *The Expression of the Emotions in Man and Animals*. Although these are not Homer's or Darwin's terms, this emotion has deep evolutionary roots and deep neural routes. It goes to the heart of the mind.

Cooperation spreads wider, beyond kin, through reciprocal altruism, the "you-scratch-my-back-and-I'll-scratch-yours" principle, helping now in return for similar help later. This trait requires the ability to remember individuals (the person to repay) and what they have done for us (the reason for repayment), the probability of repeat encounters (a reason not to fail to repay) in a nondispersing population (an occasion to repay), and a relatively long life (time to repay).

We take the ability to identify individuals for granted, and easily keep track of them. Yet noncooperators will have strong incentives to mislead others about their identity, a problem that manifests itself repeatedly in the far-flung and low-trust world of the *Odyssey,* where characters name themselves by declaring "I claim to be So-and-So, the son of So-and-So." Odysseus himself hides his identity in the *Odyssey* more than he reveals it, and readily invents spurious identities that others believe, even though they have all had past experience of travelers as liars.

Besides an ability to track individuals, reciprocal altruism requires an awareness of who does not repay, and a readiness to punish, or withhold cooperation from, noncooperators, through emotions such as trust, gratitude, guilt, suspicion, and indignation.

From the moment Odysseus walks into the swineherd's hut in the guise of an old beggar, Eumaios reveals his devotion and gratitude to the master he still believes is absent and will never return. No one in the *Odyssey* has suffered more than Odysseus, kept from his homeland for twenty years, but in one respect at least Eumaios suffers more. A king's son, he was abducted as a boy from his homeland and sold off to Odysseus' father: in this story of homecoming, *he* will never see his homeland again. Yet though he has plummeted in status from prince to slave, he happily serves Odysseus, who has appointed him chief swineherd, a lowly position but one that still allows him the paradoxical dignity of being "the godlike swineherd," "the swineherd, leader of men" (14.121):

> never again now
> will I find again a lord as kind as he, wherever
> I go; even if I could come back to my father and mother's
> house, where first I was born, and they raised me when I
> was little.   (14.138–142)

Just before pulling the bow and slaughtering the suitors, and still disguised as an old beggar, Odysseus tests Eumaios and the oxherd Philoitios, asking them whether they would fight for Odysseus against the suitors if he came home. Eumaios prays to all the divinities that this might happen, Odysseus announces himself and, even before showing his scar in proof of his identity, pledges that in return for their aid "I shall get wives for you both, and grant you possessions and houses built next to mine, and think of you in the future always as companions of Telemachos, and his brothers" (21.214–216).

Eumaios' relationship with Odysseus includes elements of reciprocal altruism, but it runs much deeper. Odysseus' kindness as a master has earned him enough credits that the swineherd does not hesitate at his request to join a potentially deadly mission, although the solicitous Odysseus offers him and Philoitios still ampler repayment should they survive. But Eumaios and Odysseus feel much more than a sense that their debts to each other should be repaid in full. Tooby and Cosmides point to the banker's paradox: when we seek to borrow money, we are in our least secure financial position and therefore offer the greatest risk to the bank. In reciprocal altruism, too, I am likely to call on your aid when I am in my most desperate plight, and when aid may therefore often be dangerous to lend, and when you are least likely to want to venture your support. But friendship overcomes that problem, since it is based not on the expected exchange of services over time, but on the attachment of individuals, as something akin to, if less strong than, the emotional bonds in real kinship.[7] Despite the master-slave relationship between them, that is certainly the case for Odysseus and Eumaios. As Eumaios pointedly confesses to Odysseus, not knowing who this "stranger" is or how much it will mean to him (14.144–147):

> the longing is on me for Odysseus, and he is gone
> > from me;
> and even when he is not here, my friend, I feel some
> > modesty
> about naming him, for in his heart he cared for
> > me greatly
> and loved me.

When much later Odysseus discloses who he is, they weep, embrace, and kiss before heading into battle.

Friendship can prove an even stronger cooperative cement than reciprocal altruism, and reciprocal altruism can make cooperation possible in a group whose members regularly and repeatedly interact. But what of those in larger populations or wider spaces, whom we might never encounter again? How can trust expand even there?

Groups larger than a few hundred individuals are "'unnatural' as far as genetic evolution is concerned because to the best of our knowledge they never existed prior to the invention of agriculture. This means that culturally evolved mechanisms are absolutely required for human society to hang together above the level of face-to-face groups."[8] There has not been time in the 10,000 years or so since the agricultural revolution for the slow workings of evolution to reconfigure human emotions to the point where we could automatically trust even strangers. We often *do* now trust strangers, but usually we do so because we know that if things go wrong we have recourse to a whole system of legal safeguards.

The society of the Greeks in the *Odyssey* is not large by modern standards, but travel around the eastern Mediterranean, though often fraught with danger, is common enough, if sporadic. Many in Odysseus' world come into contact with others they may never see again, or cannot easily oblige to repay a debt. In this situation, cooperation is even less assured, since reciprocal altruism poses too great a risk. Why put yourself out for someone whom you may never see again?

Here the solution has to be cultural, not purely evolutionary. The main solution in the world of the *Odyssey* is the institution of *xenia*, hospitality.[9] The word *xenos*, stranger-guest-host-friend, tells a whole exemplary tale in a single word. When a *stranger* arrives at my doorstep, I am obligated to welcome and feed him (and it will be *him*: women did not travel alone), even before asking him who he is, and then perhaps to offer him a bath and a place to sleep. *Anybody* from outside my social unit should be received, simply because of his mere common humanity, as a person away from home, in need of food and shelter and company, in this world with no inns, restaurants, or even money. The stranger becomes my *guest*, and to signify that he has therefore also become my *friend*, I should bestow on him a valuable gift at his departure, and help him on his onward journey. He is then obliged, should I arrive on *his* threshold, to become my *host*: to repay the hospitality, a gift on parting, and onward assistance. But more than that: the bond of *xenía* created by the initial act of welcome and cemented by the gift should endure between us for life and between our descendants. A Greek and a Trojan who meet in battle in the *Iliad* put down their weapons when they discover they are "hereditary friends" because their forebears were *xenoi*.

In light of the problem of extending cooperation simply by evolved human impulses, especially at the fringes of one's community, *xenia* offers an excellent solution, provided that people abide by it. And by and large Homer's people do abide by a code so ingrained into their beliefs.

Xenía is a central value and a central issue in both the *Iliad* and the *Odyssey*. In the back story to the *Iliad*, Paris has been lavishly treated as a guest by Menelaos, but he breaks not only the rules of marriage but also the tenets of *xenía* by leaving with Menelaos' wife and a good deal of his property: he is explicitly decried as "doing evil to a kindly host, who has given him friendship" (*Iliad* 3.354), and Menelaos denounces the Trojans in general: "You who in vanity went away taking with you my wedded wife, and many possessions, when she had received you in kindness" (13.626–627).

In the *Odyssey*, from the moment we first reach the mortal world, the suitors systematically and flagrantly violate *xenia*. Athene arrives in the form of Mentes as Telemachos broods on his father (1.118–124):

> With such thoughts, sitting among the suitors, he saw
>> Athene
> and went straight to the forecourt, the heart within him
>> scandalized
> that a guest should still be standing at the doors. He
>> stood beside her
> and took her by the right hand, and relieved her of the
>> bronze spear,
> and spoke to her and addressed her in winged words:
>> "Welcome, stranger.
> You shall be entertained as a guest among us. Afterward,
> when you have tasted dinner, you shall tell us what your
>> need is."

Although the suitors have all but taken over the palace, they ignore guests or, as we discover when Odysseus arrives as a beggar, actively abuse them. In contrast to the chaos of Ithaka, we see unstinting *xenia* when Telemachos is welcomed to Pylos by Nestor and his close and open family and to Sparta by the brilliant but brittle Menelaos and Helen. Odysseus, arriving with nothing but his naked, salt-caked, battered body, is entertained in even grander style by the Phaiakians, who show their ideal hospitality not only by not pressing him for his name for almost two days, but also by spiriting him onward—after first offering him to stay as long as he likes—all the way to Ithaka.

In partial repayment, and in the successful hope of winning more gifts, Odysseus tells the Phaiakians who he is and enthralls them with his account of his travels. The first episode that he describes in any detail is the encounter with Polyphemos, who violates *xenia* as grossly as can be imagined. On seeing Odysseus and his men, Polyphemos immediately asks, "Who are you?" Aware that this question has already breached decorum, Odysseus explains that they are Greeks, and tries to prime Polyphemos to his duties as host (9.267–271):

> we come to you and are suppliants
> at your knees, if you might give us a guest present or
>   otherwise
> some gift of grace, for such is the right of strangers.
>   Therefore
> respect the gods, O best of men. We are your suppliants,
> and Zeus the guest god, who stands behind all strangers
>   with honors
> due them, avenges any wrong toward strangers and
>   suppliants.

Ominously, Polyphemos laughs away Odysseus' guidelines, gloating that the Cyclopes "do not concern themselves over Zeus of the aegis, nor any of the rest of the blessed gods, since we are far better than they." Instead of feeding his guests, he begins to feed *on* them, after bashing their brains out.[10] Later the Cyclops asks Odysseus his name, "so I can give you a guest present to make you happy," and when he hears Odysseus' (false) name, promises to eat him last, "and that shall be my guest present to you." When Odysseus manages to blind Polyphemos and lead the rest of the survivors to freedom, he vaunts at the Cyclops, "you, hard one, who dared to eat your own guests in your own house, so Zeus and the rest of the gods have punished you" (9.478–479). Confident he is in the right, he announces who he really is, upon which the giant offers him another mocking "guest gift" (9.517), the curse of Poseidon's wrath.

Odysseus and his men cannot escape Polyphemos' "hospitality" soon enough. But another central facet of *xenia* is to aid guests to move on when ready. Just as Odysseus and his crew encounter horrific violations of the principle of feeding guests in Polyphemos, the cannibalistic Lestrygonians, and the man-devouring Scylla, so they face repeated risks of not being allowed to leave their hosts: the Lotus-Eaters, whose narcotics can trap visitors forever; Circe, who comes close to enthralling Odysseus' crew permanently as pigs; and Kalypso, who keeps Odysseus as her lover until the gods decree she must release him.[11]

When Odysseus completes his tale to the Phaiakians, *these* exemplary hosts swiftly transport him to Ithaka, where the first mortal he sees, after Athene has transformed him outwardly into an aged beggar, is his swineherd Eumaios. The swineherd's love for the master he thinks absent wins us over instantly, and confirms Odysseus in his old affection for him. Although his resources are modest, Eumaios welcomes the beggar unreservedly and with unfeigned solicitude. When Odysseus thanks him, praying that Zeus and the other gods will reward him, "for you have received me heartily," Eumaios answers (14.56–58):

> Stranger, I have no right to deny the stranger, not even
> if one came to me who was meaner than you. All
>   vagabonds
> and strangers are under Zeus . . .

Eumaios' respect for the principle of *xenía* matches his perfect practice, but his welcome is more than mere duty. It arises from a compassion so imbued as to have become utterly spontaneous—founded on innate human empathy, but reaching this level only through the salience of *xenía* as a local cultural value.

One line before Eumaios' reply to Odysseus' thanks, Homer thematizes *xenía*, I suggest, for immediately after Odysseus' "you have received me heartily," he introduces Eumaios' response in these words: "Then, O swineherd Eumaios, you said to him in answer . . ." It as if Homer as narrator picks up on Odysseus' direct thanks to Eumaios and, almost in his own rush of gratitude, himself addresses his character directly. From this point on—but not before—Homer repeatedly reports the swineherd's actions in the second person: "Eumaios, you [did this or that]." No one else in the *Odyssey* is singled out in this striking way, but the precise location of Homer's own first "you," echoing Odysseus' first "you" to Eumaios, as he thanks him for his hospitality, and preceding the swineherd's straightforward declaration that he simply has no right to deny hospitality, seems no accident.

The heartwarming reception that Eumaios gives Odysseus contrasts starkly with the abuse that Odysseus receives when he turns up, still disguised as a beggar, at his own palace—a move that Eumaios has cautioned him against. The suitors jeer at him, and Antinoös, the foremost and most brazen, hurls a footstool. Penelope, when she hears of this episode, reacts with outrage and compassion (17.500–502). When she meets the stranger, she offers him food, a bath, bedding, and all her attention as he regales her with his—false—accounts of his travels, his meeting with Odysseus, his assurances of Odysseus' imminent return.

As a value entrenched in Greek life, as a touchstone of decency, *xenía* extends reciprocal altruism beyond its natural limits. Another amplifier of reciprocal altruism is *indirect reciprocity*, our readiness to deal with people because we have seen or heard of their dealings with others: reputation, in other words—an inevitable feature of social life in a species whose members monitor and report on one another. And the force of reputation can itself be culturally amplified, as it is by the high value set on *kleos* (report, reputation, glory) in Homer's world. Not the least component of *kleos* is a reputation for *xenía*. As Penelope, ironically, tells Odysseus-as-stranger in response to his assurance that her husband will very soon return, no one could match Odysseus in this respect (19.313–316):

> Odysseus will never come home again, nor will you
>   be given
> conveyance, for there are none to give orders left in the
>   household
> such as Odysseus was among men—if he ever existed—
> for receiving respected strangers and sending them off on
>   their journeys.

Someone with a reputation for *xenía*, for dealing honorably even with strangers, should be trustworthy in dealing with familiars, and will therefore stand to benefit from the opportunities opened by the trust of others. But *xenía* is upheld by something regarded as even more powerful than social vigilance: the vigilance of the gods. As Odysseus tries to remind Polyphemos, as Eumaios states to Odysseus as a mere fact of life, the gods, and especially Zeus, monitor and protect vagabonds and strangers. In this role the most powerful of the gods is known as Zeus *Xeinios*, the protector of strangers. In the social world Homer depicts, strangers exist on or beyond the fringes of the familiar community that could protect them. They are at their most vulnerable, and for that reason Zeus protects them by punishing those who refuse to help them. Zeus, the supporter of kings, also takes an interest in those in the positions of least power, beggars and suppliants, often mentioned in the same breath as strangers. Belief in Zeus's protection of those without power in their own community, those who have ventured beyond the safety of home, and those who otherwise find themselves in dire need mitigates the potential harshness of hierarchy in Homeric society and helps solve the problem of extending the range of cooperation.

The suitors abuse Odysseus because he is a beggar and a stranger, because he has no backing or standing—and, as transformed by Athene, no apparent strength of his own. But we know that the house they lord it over is his, like the food they feast on and begrudge sharing with him.

Even if we do not share the Greek code of *xenia,* we respond to the suitors' treatment of Odysseus in multiple ways. Humans have a hierarchical streak, an urge to dominance, an awe for the dominant, and a dislike of domination, that we share with many of our closest primate relatives. Chimpanzee politics are unstable because an ambitious male entering maturity can form alliances with one or two others to topple the current alpha male, in return for a share in the rewards of power (reciprocal altruism, in this case in pursuit of power). But humans can cooperate far more flexibly than chimpanzees, and can therefore more successfully resist domination. Dominance is particularly hard to prevent in societies in which resources can be stored, as they could after the invention of agriculture. But in all known hunter-gatherer societies, where no one can accumulate much in the way of resources, a tight egalitarianism operates: all cooperate to resist any attempts by individuals to dominate or even display special abilities.

We all share a desire to dominate, a resistance to being dominated, and an awe before dominance. We, like Homer's original audience, cannot help feeling admiration at the might of an Odysseus. But we also empathize with those who are dominated, we respect those who use their dominance to help those lower in status, and we despise those who dominate without compassion. Odysseus is the natural and hereditary leader of his community, at the top of his local hierarchy, and in his role as beggar and stranger he also experiences life at the bottom. Even without the Greek sense of *xenía*—although Homer has made the principle explicit enough for us to understand it—we admire someone who gives generously of what he has to someone in need, as Eumaios does to Odysseus as stranger. Odysseus himself as ruler in Ithaka has an impeccable record: "he did no act and spoke no word in his own country that was unfair . . . Odysseus was never outrageous at all to any man" (4.690–693). Both the wise Ithakan, Mentor, and Athene, wisdom itself, rate Odysseus as "sceptered king" as "gentle and kind . . . schooled in justice . . . kind, like a father" (2.230–234, 5.8–12). We feel a compounded outrage at the suitors for denying someone in need what they freely help themselves to, despite not being themselves needy, and for humiliating someone already apparently low in status, when the stranger not only needs a share of the food they happily waste, but is actually the person whose food they are squandering and whose status they are aspiring to, without his strengths or his consideration for others' weaknesses.

Xenía is axiomatic for the altruists in Odysseus' world, but even for them cooperation can be deeply unstable. Although one must welcome a guest in the proper manner, one also has a perfect right when abroad, and in a position to do so, to raid another town or homestead.[12] Odysseus observes *xenía* incomparably, yet one of his formulaic honorifics is "Odysseus, sacker of cities," which he has earned especially but not only for his role in the sack of Troy. In his account to the Phaiakians, the first "adventure" he records is the raid he and his men undertook against the Kikonians, allies of Troy (9.40–43):

> I sacked their city and killed their people,
> and out of their city taking their wives and many
> possessions

> we shared them out, so none might go cheated of
> > his proper
> portion.

Proper portion, indeed! And at the end of the *Odyssey* the returned hero announces that he will restock the flocks culled to feed the suitors partly "by raiding" (23.357), as if this were merely a matter of prudent management. This behavior appears intended to characterize him not as rapacious, but as capable and enterprising, like Achilleus in the *Iliad,* or Menelaos on *his* return from Troy, or Odysseus' father, Laertes, before him. Even Odysseus' protective deity, Athene, goddess of wisdom, is also known as "Athene, the giver of plunder" (16.207).

In a world of low trust such as this, where raids for women, slaves, livestock, and valuables were commonplace, it is no wonder that *xenia* developed as a way of offering a guest what he immediately needed, and more, before he simply took whatever he could. The code of *xenia* was especially a means of appeasing the leaders of visiting forces and forging alliances. Leaders could benefit not only from the interchange of goods (as Menelaos says, "there is both honor and brilliance in it, and there also is profit"; 15.78) but also from the increased capacity of a broader alliance to resist the raids of others or to initiate raids against them. For this reason, democratic Athens would see *xenia* as a reactionary legacy of aristocratic privilege.[13]

Extending the boundaries within which cooperation works does not remove the temptation to use it for the short-term advantages of competition at a still higher level: "We are motivated to preach and impose morality—to exhort and force others to do their duties—and to punish transgressions, but we are also tempted to cheat when we believe we can get away with it."[14] Out on the fringes of the familiar, we often believe that we *can* get away with it, and in the world of Odysseus, doing so is not even seen as cheating. You must look after strangers who arrive on your doorstep. But if you arrive on a stranger's, let them take you in—or just take what you want. After all, they are outside the circle of cooperation.

19

THE EVOLUTION OF COOPERATION (2):
PUNISHMENT

IN HOMER'S TIME *xenía,* hospitality to strangers, enlarged the
circle of trust in a world where strangers were widely mistrusted. Yet
xenía's instability on the fringes of the familiar glaringly illustrates
the problem of expanding cooperation. Even nearer home, though,
cooperation can be abused.

Free-riding, taking advantage of the benefits of cooperation without paying a fair share of the costs, is "*the* fundamental problem
of social life."[1] It always offers a short-term advantage and hence a
temptation, unless punishment reduces benefits and increases costs.

Penelope's suitors are free-riders par excellence. As I suggested
in the previous chapter, Homer largely invents the suitors, at least as
a virtual army confronting Odysseus. They form an extraordinary
group: a bunch of noncooperators in competition over a resource,
Penelope, that only one of them can win, cooperating with each
other just enough to last for years as noncooperators with the rest of
their world.

At one level, they can be seen as an alliance of young upstarts
threatening to replace the alpha male. In *Chimpanzee Politics* primatologist Frans de Waal reports in depth on successive alliances and
challenges to current leaders in the Arnhem chimpanzee colony. Anthropologist Robin Fox notes that a younger human generation's
challenge to the older generation's power recurs in epic cross-
culturally, from the *Mahabharata* to Nordic saga.[2] Throughout the

Iliad and the *Odyssey,* Homer depicts young males—especially when they congregate, like the suitors or the competitors at the Phaiakian games—as heedless and reckless, a challenge to order.[3]

At another level, the suitors betray the hope that human social intelligence can compensate for the effects of hierarchy. We *do* share the widespread primate urge to dominate, and we feel awe before dominance: we are naturally hierarchical. But we also dislike *being* dominated and can form alliances to challenge existing power combinations and thereby mitigate hierarchy.

Yet the suitors, although they challenge or usurp the absent Odysseus' position at the top of the Ithakan hierarchy, sharply raise the cost of anyone else's lower status, abusing Eumaios as a slave or Odysseus or Itos as beggars. To the unrecognized Odysseus, Eumaios laments that "never again now will I find again a lord as kind as he" (14.138–139). The suitors, by contrast, especially their leaders, bolster their sense of superiority by stressing the inferiority of those below them. When Antinoös refuses bread to Odysseus as beggar, impugning *him* as a free-rider, Odysseus rebukes him: "You would not give a bit of salt to a servant in your own house, since now, sitting at another's, you could not take a bit of bread and give it to me" (17.455–457). In reply Antinoös hurls a footstool at him, sparking a protest even from the other suitors and from Penelope herself.

From the moment we first see the suitors, as they ignore Athene waiting as the visitor Mentes, through their abuse of the stranger, to their final, fatal, feast, the suitors explicitly violate *xenia.* The essence of *xenia* is reciprocation, but the suitors think only of *taking* what they want—Odysseus' livestock and wine, and ultimately his wife and position—and *giving* others only insult and injury.

Like Aigisthos before he murders Agamemnon, the suitors have been warned repeatedly—by Penelope and Telemachos, by their far-sighted townsmen Mentor and Halitherses, by the prophet Theoklymenes, and by the omens of the gods—but have heeded nothing. Odysseus therefore punishes them for himself, his family, his community, and the gods.

As we saw in Chapter 4, punishment proves essential in establishing cooperation beyond immediate kin. To punish others involves costs in time, energy, and the risk of resistance or retaliation, but some punishments are more costly to impose than others. Standing up firmly to 108 armed suitors, the prime of the local youth, would exact too great a price for Telemachos acting alone, even though at the bow-contest he proves mightier than any one of them and has powerful but distant sympathizers in Nestor and Menelaos. But Odysseus never hesitates to pay the price.

Odysseus of course is fired by anger, an emotion that evolved to ensure that we resist others' taking what we consider ours. His world does not yet have institutions that can depersonalize punishment by sharing across the whole community the burden of detecting infringements and imposing penalties. It therefore operates, like many small-scale societies even today, on the basis of personal revenge. If we do not look out for our own, if we do not signal that we will make others pay if they infringe our interests, then we can easily be exploited: "retaliation is absolutely essential to keep the wolves of selfishness at bay."[4]

We can see the social emotions as solutions to the problems of commitment in social life.[5] If we do not feel anger and sometimes act upon it, others may well encroach on our rights. Once an offense occurs, of course, full redress may be impossible, but if we lack a sense of outrage sufficient to make us retaliate, others may feel they can take piecemeal what is ours—as the suitors do to Odysseus before he reappears. Once Odysseus has killed Antinoös and told the rest he will deal with them the same way, Eurymachos, next among the suitors, offers Odysseus compensation from them all for whatever he has lost. Odysseus remains implacable: whatever the compensation, "I would not stay my hands from the slaughter, until I had taken revenge for all the suitors' transgressions" (22.63–64).

For Odysseus routing the suitors is partly personal retaliation for the losses he and his family have suffered through the suitors' three years of feasting. It also reestablishes his honor: he should not be messed with—crucial to show in pastoral economies such as Ithaka's, where low population densities make policing difficult and the wealth concentrated in herd animals is easy to steal and difficult to guard. In such ecological circumstances, across the world, the best form of defense has been a reputation for aggressively retaliating to any challenge.[6]

But Odysseus also punishes the suitors as a public duty, as part of a commitment to restore order—and his own position as its upholder. Even in primate societies perceived violations of the social code are the single most common cause of aggression. Observers have noted that "animal societies often resemble human feudal societies in that high-status individuals typically take on the role of enforcing these implicit social norms, aggressing against those who violate them,"[7] partly because they have most to gain or lose. In Ithakan society, the norms are explicit rather than implicit, but otherwise the pattern seems identical. Odysseus has more to gain than others from reestablishing the social order. By removing the threat of the suitors he reconfirms his sole access to his wife, his sole command of his estate, and his position as first among the Ithakan lords, and he scotches the suitors' plan to murder his son. Homer, indeed, shapes his whole epic around Ithaka's inability to police the suitors' flagrant, protracted violations of the community's norms until their leader returns.

But Homer stresses another side to Odysseus' confrontation with the suitors. They are "sinners" (20.121) who "had first begun the wrongdoing" (20.394). Odysseus punishes them on behalf of the gods, and by the gods' design, almost at their prompting and certainly with their backing.

As soon as she hears Telemachos describe the suitors' conduct, Athene declares: "How great your need is now of the absent Odysseus, who would lay his hands on these shameless suitors. I wish he could come now to stand in the outer doorway of his house, and wearing a helmet and carrying shield and two spears" (1.253–256). Penelope, Mentor, Halitherses, Nestor, and Menelaos all confirm that mortals as well as gods feel outraged by the suitors, and Zeus in the second council of the gods impatiently confirms Athene's hope: of course "Odysseus shall make his way back, and punish those others" (5.23–24). In Hades the seer Teiresias prophesies, even before the suitors have begun besieging Penelope, that if Odysseus returns he will find "insolent men . . . eating away your livelihood . . . You may punish the violences of these men, when you come home," killing them "either by treachery, or openly with the sharp bronze" (11.116–120). When at last Odysseus reaches Ithaka, Athene for the first time since the start of the story discloses herself to him. Before he has even sighted them himself, she tells him to "consider how you can lay your hands on these shameless suitors" (13.376). When he declares himself ready to fight with her help against even 300 men, she affirms (13.393–395):

> I will indeed be at your side, you will not be forgotten
> at the time when we two go to this work, and I look
> > for endless
> ground to be spattered by the blood and brains of the
> > suitors.

Throughout the poem, Homer does his utmost to show the suitors undermining cooperation, taking all for themselves without thought of repayment or fear of reprisal. For three years, despite openly offending against reciprocity, *xenia,* and the gods, they have multiplied their misdeeds with impunity, thanks to the safety of numbers and the absence of opposing power. Their success amplifies their arrogance to the point where it almost calls into question the justice of the gods—just as, in the opposite sense, Odysseus' suffering, despite his deserts and devotion to the gods, explicitly raises the question of divine justice, in the doubts Athene herself voices in the two Olympian councils. Homer has worked at the problem from both ends: the misery of the good, the prosperity of the evil. By having the deserving Odysseus enforce the severest penalties against the undeserving suitors, Homer and Zeus affirm cooperation and the power of punishment to uphold it, even when it seems to have been flouted as insolently as the suitors have done.

The gods or other supernatural agencies have been invoked to reinforce cooperative norms in almost all human societies. Whatever the advantages of cooperation, we, like other social animals, "behave selfishly in some contexts," as indeed we "must to survive any immediate competition for resources."[8] Simple consent rarely sustains cooperation among many for long,[9] especially when we think no one will notice our selfish behavior. A community's common belief in the tireless, ubiquitous, invisible vigilance of supernatural agents acts as a strong deterrent to uncooperative behavior.

From the start of the *Odyssey* Homer stresses the role of the gods in monitoring human behavior, in the prominence he gives in the proem to Helios' punishment of Odysseus' companions for devouring his sacred cattle, in the central role this incident has in at last leaving Odysseus totally isolated on his homeward route, and in repeatedly noting and showing the power of the gods to know all.

Yet in the *Odyssey* the most egregious sinners do not hide their sin. Indeed the very brazenness of the suitors' unchecked behavior makes them a key challenge to Greek norms and their divine enforcers. Eumaios calls them "that swarm of suitors whose outrageous violence goes up into the iron sky" (15.328–330). For three years, nevertheless, the skies have left them unpunished. But when the suitors do fall to Odysseus, he insists that they have been "destroyed by the doom of the gods and their own hard actions" (22.413). And Homer structures his poem so that from the first we have no doubt of the gods' vigilance, and their preparations for the punishment of the suitors, however remote either may seem to the troubled citizens of Ithaka.

Not only can the gods see all; they will severely punish mortals whose conduct offends. Since they are irascible, mortals need to be wary about what they do and whom they mix with. Many Greeks returning from Troy pay for Athene's wrath against them because of the rape of a woman in her temple during the sacking of Troy, whether or not they had anything to do with it. All Odysseus' crew die for their part in killing and eating the oxen of the sun god despite stern warnings not to. Amphinomos proves himself the least selfish of the suitors, and Odysseus issues him a warning that Odysseus is about to return, and that he should leave before the "reckoning, not without blood, between that man and the suitors" (18.149–150). But Amphinomos walks back to his seat, not, like the rest of the suitors, dismissing the vengeance ahead, yet unable to escape. He

> went back across the room, heart saddened within him,
> shaking his head, for in his spirit he saw the evil,
> but still could not escape his doom, for Athene had
>   bound him
> fast, to be strongly killed by the hands and spear of
>   Telemachos.   (18.153–156)

Since the gods are mighty and irascible, they must be appeased. Appeasement involves both abiding by their precepts and offering sacrifices in tribute. Yet Odysseus has suffered for ten years since the end of the Trojan War, despite always honoring the gods (19.363–367):

> Surely Zeus hated you
> beyond all other men, though you had a godly spirit;
> for no man among mortals ever has burned so many
> thigh pieces to Zeus who delights in the thunder, nor
>   given so many
> choice and grand sacrifices.

What does ritual sacrifice have to do with the evolution of cooperation? A good deal, in fact. Evolutionary biology has discovered the importance of costly signaling. Senders will often exaggerate signals to others—for instance, of their formidableness as rivals or their attractiveness as mates. Costly signals, however, cannot readily be faked, since the signals themselves cost senders and indicate that they have resources and the readiness to commit them. Rituals are a human form of costly signaling. Anthropological studies have confirmed that religious groups, such as Israeli *kibbutzim,* are consistently more cohesive and profitable than their secular equivalents, apparently thanks to ritual commitments that increase internal levels of trust.[10]

Being prepared to expend time and valuable meat sacrificing to the gods indicates that you take their powers and purposes seriously, and are therefore more likely to abide by the values they endorse—like Zeus's *xenia*—than are others who skimp on ritual observance. Odysseus, Nestor, and Menelaos all signal by their elaborate sacrifice their respect for the gods, and therefore their trustworthy place within the nexus of social cooperation. The suitors, by contrast, though they do not mock the gods as overtly as the brutish Polyphemos, repeatedly pay no heed to the warnings of those who *do* see the signs of the gods. That heedlessness emboldens them to ignore even the values that "Zeus who delights in thunder" supports.

If we wish to assess religion's social benefits, its truth matters less than its power to motivate.[11] So long as they are widely believed, false ideas can powerfully spur social cooperation: that the gods can watch every human move, that sooner or later they will punish infringements, that ritual can signal our respect for them and the precepts they safeguard. The contrary notion, that the gods can be ignored with impunity, would undermine cooperation among the Greeks. Homer structures his whole epic around the contrast between the view from Ithaka, where the suitors appear immune from punishment, and our own view, which lets us see from the start the plans of the gods to unleash Odysseus on the suitors.

Odysseus himself has a larger view than any other mortal within the *Odyssey,* a fact that reflects both his intelligence and his role as exemplar and upholder of his society's values. Whenever about to make an important move, he almost invariably stops and reflects: is this the right choice, practically and "morally," in terms of the gods' approval? High-level inhibition within the prefrontal cortex is essential to the human capacity to live by moral codes subtler than other animals' social emotions allow, even though these emotions also underpin human morality.[12] But people vary in their impulsivity and self-control, and poor self-inhibitors often find themselves eventually inhibited by others, as the suitors are at last by Odysseus: "Weak motivational inhibition is thought to underlie the aggressive problems of conduct disorder . . . Conduct-disordered children and psychopaths perform poorly on passive avoidance tasks (that require response inhibition) as a result of their diminished sensitivity to and reflection on cues for adverse consequences."[13] The suitors pay no heed to likely adverse consequences, even when sent by Zeus, and therefore Odysseus, by executing his plan, by inhibiting his own outrage at their fresh insults, finally inhibits *them*. They will be "reckless" no more.

Over the years audiences have responded differently to the slaughter of the suitors, but most of us agree with the moral pattern of the poem as I have summarized it. The suitors act with impunity, as if they can escape the usual social give-and-take. They take the goods they want and give bad in return. They cooperate against the very principle of cooperation and despite repeated warnings assume that their numbers place them above retribution. They have it coming.

Nevertheless we may also ask ourselves what exactly the often-noted "violence" of the suitors actually amounts to. Antinoös himself admits that they are violent men (21.289–290), but what evidence do we see? Antinoös' throwing a footstool at Odysseus, Ktesippos' throwing a cow's hoof, the collective plan to kill Telemachos, and years of conspicuous freeloading. Should 120 die for this?

No, we are likely to agree—although, as the *Oxford Classical Dictionary* notes, under "reciprocity," the Greek idea of reciprocal justice was not an eye for an eye but a head for an eye. The gods themselves assume a norm of massive retaliation. Outraged at the affront to herself in the Greek forces' desecration of her temple when they sack Troy, Athene ensures that all the Greeks encounter hardship on their return, and many ships are lost; on behalf of Helios, Zeus strikes a thunderbolt at Odysseus' ship in return for his men's killing one of the sun god's sacred oxen, and kills the entire crew, sparing only Odysseus himself. Like mortals, the gods, too, applaud mortal retribution, like Orestes' revenge killing not only of Aigisthos but of his own mother.

Part of us accepts the killing of the suitors because we appreciate both the challenge to social cooperation that the shameless free-riding of the suitors poses, and the need for punishment to eradicate it. Part of us simply conforms to the unanimous judgment of Ithaka, its neighbors, and the gods against the suitors—and social conformity, the sharing of common norms, is itself a powerful contributor to the evolution of cooperation.[14] Part of us actively thirsts for revenge on behalf of Odysseus, Penelope, Telemachos, and Eumaios—even if part of us also recognizes that the suitors are not as emphatically or uniformly violent as they are made out to be, and deserving of punishment but hardly eradication. Armed with an evolutionary understanding of human nature, we can see both why evolution has built into us an impulse for revenge and why nevertheless it is a crude, costly emotion that modern systems of impersonal justice have harnessed and tamed.[15]

But those systems were not yet invented in Homer's world. There, justice still depended on individuals' readiness to stand up for their own rights, despite the risks.

Game-theory models show how cooperation can evolve through the simple rule of Tit for Tat: cooperate on the first move, and if the other partner does not cooperate in reply, withhold cooperation (or actively punish) on the next move. Yet Tit-for-Tat has one serious weakness: when one side makes a noncooperative move, the other side will do the same in response, and both will find themselves locked into a vicious circle of retaliation. Computer simulations suggest lifelike refinements to the rules to avoid this impasse. "Generous tit-for-tat" allows two defections before it punishes, so that cycles of retribution can be avoided—though, in this particular tweak, at the cost of the other side's realizing just how much it can get away with before cooperating again to avoid punishment. In a sense, even the harsh world of Odysseus offers a slightly generous tit-for-tat system, since the gods often issue warnings before exacting penalties. Penelope's suitors receive repeated warnings, if they would only heed them, from the soundest citizens in Ithaka, from Telemachos, and via the omens of the gods, and Odysseus even personally warns Amphinomos, the most reasonable of the suitors. But when none acts on the admonitions, even the apprehensive Amphinomos, no room remains for generosity, and retaliation is absolute and final.

For the suitors, at least. They are dead—but their relatives survive. And *they* want to exact revenge on Odysseus and Telemachos for slaughtering their sons and brothers. When justice operates at a personal level, retributive feuding presents an ever-present risk, as Homer shows himself very aware. The epic ends with Antinoös' father leading the male relatives of the slaughtered suitors in revenge against Odysseus and his father and son and their supporters. Eupithes, Antinoös' father, falls to Laertes, Odysseus' father. Despite being again outnumbered, Odysseus' side would have won the day, but Athene orders the fighting to cease, Zeus throws down a warning thunderbolt, and Athene instructs Odysseus to "stop this quarrel in closing combat, for fear Zeus of the wide brows . . . may be angry with you":

> So spoke Athene, and with happy heart he obeyed her.
> And pledges for the days to come, sworn to by both sides,
> were settled by Pallas Athene, daughter of Zeus of the
>   aegis,
> who had likened herself in appearance and voice to
>   Mentor.

Thus ends the *Odyssey*, with Homer's tacit admission that he can see no *mortal* solution to breaking the cycle of revenge. The gods intervene because they choose that Odysseus deserves a peaceful old age as compensation for his trials and reward for his achievements. But only their uncharacteristic intervention has imposed an improbable calm.

Unlike the *Iliad*, the *Odyssey* has from the first signaled a "happy ending." There is no doubt that the "good guy" will defeat the "bad guys," and he does. But although Odysseus triumphs against the suitors, in pitched battle against overwhelming odds, the tone never waxes triumphal. When Eurykleia sees the fallen suitors and begins "to raise the cry of triumph," Odysseus silences her: "Keep your joy in your heart, old dame; stop, do not raise up the cry. It is not piety to glory so over slain men" (22.408–412). Even as he discloses his identity to Penelope and awaits her final recognition, Odysseus broods on the looming retaliation by the suitors' families and next day braces himself reluctantly for another battle. Combat ceases after the first death, thanks to the gods' intervention, but just as in the *Iliad*, Homer closes with a provisional peace: not as brief as the lull at Troy, which after all occurs in the middle of a war, but with no lasting confidence that calm can long prevail in human affairs. The *Odyssey* ends in relief and resignation, and a long way from exultation.

The suitors have subverted cooperation and the Greek attempt to extend it through *xenía*. They have also perverted hierarchy through their arrogance toward both inferiors and superiors. Although it seems they might get away with their outrageous conduct, they pay the ultimate price. The prudent heads in Ithaka had counseled the families of the suitors to curb their sons, as the aged Halitherses reminds them when he tries to dissuade them from marching against Odysseus (24.451–462).

We can see the earlier attempts of Halitherses and Mentor to shame the suitors' families into restraining their unruly sons in the light of recent research into cooperation. Evolutionary biologists and economists have shown not only that monitoring and punishing noncooperators are essential to maintaining reciprocal altruistic cooperation, but also that we may face temptation to leave punishment and its costs to others—in itself a selfish ploy, rendering punishment perhaps too frail to ensure that cooperation survives. But cooperation becomes comparatively robust if nonpunishers are themselves punished, since there will remain fewer nonpunishers, and therefore fewer who profit from noncooperativeness, fewer who need to punish, and, ultimately, fewer who need to punish nonpunishers.

Halitherses and Mentor speak against the suitors in the assembly convoked by Telemachos in book 2. Warning of "a great disaster . . . wheeling down" on the suitors, Halitherses urges: "let us think how we can make them stop, or better let them stop themselves" (2.163–169). Mentor declares to the Ithakans: "I hold it against you other people, how you all sit there in silence, and never with an assault of words try to check the suitors, though they are so few, and you so many" (2.239–241). But Halitherses and Mentor cannot punish the nonpunishers in Ithaka who outnumber *them*, even if those nonpunishers themselves outnumber the offending suitors, any more than Telemachos himself can punish the suitors directly, though he does reprove them. Nestor and Menelaos also agree on the need to punish the suitors, but they live too far away to take responsibility (or action), and can only look forward, like so many others, to Odysseus' return. Only his return and his resolve to take on all the suitors, despite the odds, will at last restore the principle of reciprocity.

Homer ends the poem with three climaxes: the confrontation with the suitors, anticipated from the start; Penelope's recognition of Odysseus, also expected from the start, but taking us by surprise in its difficulty and details; and the aftermath of the slaughter. In this most unexpected ending of all, Homer shows that he is too aware of the cost of death—as of course he is throughout the *Iliad*—to make the killing of the suitors the triumph of right over wrong that we almost expect. At the close of the poem he stresses the difficulty of ensuring cooperation: the costliness of the failure to punish the suitors earlier, before their offenses warranted death; the costliness of the revenge the suitors' families intend; and the necessity and the difficulty of reconciliation, when earlier corrective punishments have been lacking and tensions have been allowed to run too high.

Odysseus has strongly personal reasons for anger at the suitors, but he also shoulders the burden of the community's outrage when he leads his son and his two loyal herdsmen against them. But because the punishment is personal, even if also on behalf of all, it precipitates personal revenge and a cycle of bloodletting that Homer can stanch only through the gods' intervention.

In highlighting Odysseus' central role in meting out punishment and restoring order in Ithaka, the *Odyssey* implicitly supports a firm hierarchy headed by one individual, although it softens the effects of his dominance by stressing his sympathy for beggars, strangers, and suppliants and others below him—or at least those who honor the values of sympathy and hierarchy. Hierarchies converging on single individuals have remained the norm in larger-scale societies for most of history ever since, in monarchies and autocracies, and although we have invented new ways of distributing power in democracies, power still concentrates in the hands of leaders.

Although we may not agree with the arrangements entrenched by autocracy or with justice in the form of massive, bloody revenge, we can see the problems that cooperation entails, the attempted solutions Greek society has invented, and the challenges the suitors pose. The social emotions evolved to help humans and other animals solve problems of cooperation. Building on such dispositions, local cultural institutions allow us to extend cooperation still further, in Homer's society through norms like *xenía* and a belief in the gods as monitors and admonishers of mortal doings. Since Homer we have invented new ways of distributing power and responsibility, of detecting crime and administering punishment so as to break the cycle of transgression and personal retaliation. Yet we still accept the need for hierarchy, for concentrating power and responsibility, and we still feel a visceral desire to punish those who unfairly exploit the cooperation that makes social life possible and potentially profitable to us all.

Our own solutions remain imperfect, and we still need to extend cooperation further, and to fine-tune the detection and punishment of noncooperation. But the underlying emotions are the same, even if inflected by local norms. Our awareness of the sheer difficulty of establishing cooperation, of the distance we have come and the distance we need to go, becomes all the sharper when we look at the similarities and differences between Homer's world and ours.

The *Odyssey* galvanizes our moral emotions, and thereby holds our attention through our desire, with Athene, to see Odysseus reap his just rewards, and our desire, with Athene and most of the mortals in the story, to see the suitors get their just deserts. The poem relies on and rouses our moral emotions, and it defines and vividly demonstrates, chiefly by contrast, the principles of *xenía* and divine justice that build on these emotions: Polyphemos eating his guests and mocking the idea of guest-gifts; the suitors hurling abuse and injury at Odysseus and Eumaios; Eumaios offering the beggarly Odysseus everything at his disposal; Odysseus, Telemachos, and Eumaios, aided by Athene, punishing the suitors. For the original audience a public performance of the *Odyssey* would have helped to clarify and consolidate Greek social principles and, simply by reiterating the norms so memorably and publicly, to confirm for the whole audience that all share these emotions and recognize these norms.

Homer reinforces *xenía,* the gods' role as arbiters of justice, and, ultimately, the principle of cooperation. Through his characters he repudiates and exorcizes free-riders, exalts punishers, and rebukes nonpunishers. Mentor and Halitherses reproach the people of Ithaka for not having punished the suitors earlier, and the leader of those whom they reproach, the father of the foremost and most offensive of the suitors, Antinoös, is killed, the last person to die in this tale of the deaths of hundreds upon hundreds. Some have concluded that the last book-and-a bit of the *Odyssey,* after the reunion of Odysseus and Penelope, could not have been composed by the author of the rest of the poem. But an awareness of the problems of the evolution of cooperation explains why the story ends where it does, and why the ending is both right and unsatisfying. Odysseus' actions solve the problem of cooperation as well as possible, given the emotions and the institutions prevailing in his world; yet Homer lets us sense that the problem is not really solved—as, of course, it is still not solved for us, even if we have found means to establish cooperation on a much larger scale and a sometimes surer basis than in Homer's world.

PART 5

ONTOGENY: *HORTON HEARS A WHO!*

"I know it is wet
And the sun is not sunny.
But we can have
Lots of good fun that is funny!"

Dr. Seuss, *The Cat in the Hat* (1957)

20

PROBLEMS AND SOLUTIONS: WORKING AT PLAY

IF BRIGHT THREE-YEAR-OLDS can understand and enjoy Dr. Seuss's picture-story book *Horton Hears a Who!* (1954), why does it need adult analysis? Theodor Geisel, the man behind the Seuss name, claimed he dropped out of Oxford and the Sorbonne—not that he had clambered far in—because "they were taking life too damn seriously."[1] But a serious investigation of his funny little children's book need not be *too* serious, and it can be serious enough only if it adds to the fun. And although even the very young can appreciate Dr. Seuss, we can understand literary works, like anything else, at different levels. I want to show that a biocultural approach makes it possible to explain stories both more comprehensively and more precisely than through other current approaches.

Like Homer, Dr. Seuss appeals to his audience over multiple generations and multiple readings. In Homer's case the lack of documentary evidence makes it difficult to tell where his own contributions start and stop, and where the tradition or other individuals have shaped what we attribute to him. In Dr. Seuss's case we have rich evidence for his work and its contexts. His classic children's books allow us to explore another origin of stories: not their historical or phylogenic but their individual or ontogenic origin. As we saw in Chapter 12, the earliest stories that children "compose" are rudimentary, evanescent, improvised, as much play as story, dependent on their situation, and not easily reproducible. Dr. Seuss's stories by contrast prove ideal specimens—as "replicable" as scientific experiments—for investigating the origin of stories.

The brevity and modernity of Dr. Seuss's children's stories allow us not only the maximum contrast with Homer but also a conveniently compact example that leaves us space to introduce some general principles of an evolutionary approach to literature and criticism. Apart from standard concerns—author, work, audience, context; intention, effect, meaning; interpretation, evaluation, explanation—we will follow three main lines: (1) a problem-solution model that links the long term of evolution to the short term of an author making choices about this or that detail; (2) earning attention as prior to generating meaning in the problems an author faces; and (3) a multileveled system of explanation.

I propose four interconnected levels of explanation appropriate to Dr. Seuss's story or any other work of literature: a *universal* level, which considers aspects of human nature in general; a *local* level, which focuses on particular cultural, historical, social, economic, technological, intellectual, or artistic contexts; an *individual* level, which assesses the dispositions and experience of an author (or, alternatively, of a reader or critic); and a *particular* level, which examines the specific problem situation of the author composing *this* story, or of a reader reading it in a certain situation (for the first time or the *n*th time) or for a given purpose.

The Origins of the Story

Bright young readers and their parents almost never realize what prompted Dr. Seuss to write *Horton Hears a Who!* Nor do they need to. While we normally understand what others do and say partly through inferring intentions from contexts, we often know nothing about works of art other than their appeal for attention. The more surely something rates as a work of art, the more it can appeal to us independently of its origin. *Horton Hears a Who!* succeeds on its own terms: as the story of a kindly elephant who unhesitatingly offers to protect the tiny Whos who call for help from a dust speck. The other jungle animals, who cannot hear the Whos, not only deride him for his eccentricity but threaten to cage him and boil the clover on which the dust speck settled. Only when Horton exhorts the Whos to make themselves heard, and all shout together, even a last young boy who had shirked the initial call to add his voice, do the other animals hear the Whos, realize Horton was right all along, and vow to help him protect his minuscule charges.

Nevertheless we do know the origin of *Horton Hears a Who!*—and it proves most unexpected. Dr. Seuss traveled to Japan in 1953, a year after the end of the Allied occupation. As he later recalled: "I conceived the idea of *Horton Hears a Who!* from my experiences there ... Japan was just emerging, the people were voting for the first time, running their own lives—and the theme was obvious: 'A person's a person, no matter how small,' though I don't know how I ended up using elephants. And of course when the little boy stands up and yells 'Yopp!' and saves the whole place, that's my statement about voting—*everyone* counts."[2]

Academics have acted on this knowledge. Ruth MacDonald writes that Horton "represents postwar United States in the international community of nations. The Whos of Who-ville are Dr. Seuss's characterization of the Japanese after Hiroshima, a people whom he found optimistic, hardworking, and particularly eager to vote in their elections."[3] Historian Richard Minear concurs: "If Who-ville is Japan, Horton must stand for the United States."[4] But there seem to be many steps missing between such an origin and such a meaning. How did Dr. Seuss's desire to foster democracy in Japan lead, as he wonders himself, to Horton the elephant and the microscopic Whos? And why should we read the story in terms of 1953 politics? The story's text does not direct us there: Horton is an elephant, and about to be caged by a family of jungle monkeys, the Wickershams, and a mother-child pair of kangaroos; if he is the United States, who would they be? A political allegory would be a strange key to a story aimed at children too young to follow current politics and at future readers.

I suggest two related problems here. First, the academic impulse to search for meaning results from not really understanding the key problem facing storytellers, the need to secure and maximize their audiences' attention. Second, the origin of Dr. Seuss's story lies much deeper than 1953, both in himself as an individual and in the human nature he shares and appeals to.

PROBLEMS AND SOLUTIONS IN LIFE AND ART

Let me offer a model of artistic explanation that focuses on artists' problems and solutions and can lead us to what we genuinely enjoy in Dr. Seuss. This model builds from the ground up, links science and art, and takes us from the long term of evolution to the moment of the individual artist trying to solve *this* problem in *this* work.

Biologically we can see all organisms as problem-solvers, each action or process as an attempt to solve a problem, however minor or routine. Evolution generates *problems and solutions* as it generates life. Rocks may crack and erode, but they do not have problems. They do not lose anything if they do not maintain or pass on their form. Amoebas and apes do. Natural selection creates complex new possibilities, and therefore new problems, as it assembles self-sustaining organisms piecemeal, cycle after cycle, by generating partial solutions, testing them, and regenerating from the basis of the best solutions available in the current cycle. In time, it can create richer solutions to richer problems.

Species are integrated systems of specific solutions—developmental patterns, organs, and behaviors—to recurrent problems in multifarious modes of life, especially to the persistent unpredictability of the environment. *Sexual reproduction* partially solves the problem of unpredictability by maximizing variations in form—new sets of possible genetic solutions—while still reproducing overall design. *Flexible behavior* solves the problem of unpredictability at the level of the organism by opening up new options for action. *Minds* solve the problem of directing agents to better choices for future actions, especially by predicting the near future from the present via patterns relevant to a species in its past. *Sociality* solves the problem of gaining the most from pooling individual capacities. *Culture* solves the problem of making available to the whole group behavioral solutions more responsive to novel circumstances than genetic evolution can offer.

Any solution can create new problems. *Play* solves the new problem that if organisms need to *learn* flexible behaviors, they cannot come fully preloaded. It encourages animals especially while young to enjoy repeating actions in intense or extreme forms in situations in which they will not face life-or-death risks, until responses become both rapid and automatic yet finely sensitive to conditions. *Art,* I have suggested, solves a similar problem for humans. We derive most of our advantages from intelligence, and our minds solve the new problem of maximizing their capacity to process information patterns in flexible ways, in low-risk situations, through the cognitive play of art, our compulsive engagement in high-intensity doses of the patterns that matter most to us.

A desire for *status* solves the new problem that the solution of sociality causes: the need to compete for the resources that social living makes available. Because of sexual reproduction, humans other than identical twins are genetically unique, and our differences are compounded by our natural and cultural behavioral flexibility. Our ultrasocial species gains still more from sociality than other species, partly through our capacity and desire to imitate others; but we solve the corresponding problem of maximizing our status and our access to the rich resources our ultrasociality earns by making the most of our considerable differences. We assess our individual capacities in comparison with those of others, and develop those abilities that offer us the greatest chance of success in competition with our peers. Through the rigorous techniques of behavioral genetics, Judith Rich Harris eliminates the usual explanations for the bulk of human personality difference (55 percent) unaccountable for by genetic variation. Comparing identical, nonidentical, and adopted and non-adopted twins and nontwins, she shows that most nongenetic personality difference results not from environment (which accounts for about 9 percent) but from what she calls the "status system" within each of us: "the status system searches for self-knowledge in the social cues provided by others. It then uses this information to plot a long-term strategy that will involve direct competition only in those areas of endeavor in which the individual has a hope of succeeding and, if possible, avoid competition in other arenas. The result is that individuals seek unoccupied niches; they specialize in different things."[5]

One way in which humans can specialize is through art, through providing occasions for cognitive play that engage the attention of others and thereby earn status and access to resources. As artistic modes and traditions develop, they accumulate extensive repertoires of problems and solutions for earning attention and so modify audience expectations. As individual artists specialize within an artistic tradition, they will also accumulate over years of creative effort personal repertoires of artistic problems and solutions that make the most of their individual difference and allow their work to earn attention by their novel recombinations. In each new work they will seek to raise the benefit—the attention-earning power—of their compositional efforts and lower their composition costs, through recombining existing solutions in new ways, while also raising the benefits and lowering their audience's costs in time and effort.

All this may sound abstract, but with Dr. Seuss as subject I promised fun. Ted Geisel singled out as his niche making the most of the spirit of play that underlies art. It earned him high status—acclaim as a classic and a genius—and it earned him rich resources as he sold, among much else, hundreds of millions of Dr. Seuss books.

DR. SEUSS: WORKING AT PLAY

From his youth Geisel made the most of his singular ability and inclination to make others laugh. He turned these into his specialty: he worked and worked and worked at play. In high school he was voted class wit and artist. At Dartmouth College he channeled almost all his energies into the college humor magazine, *Jack o' Lantern,* as its editor and most frequent contributor;[6] a college contemporary later recalled: "That he is a rare and loopy genius has been common knowledge from an early epoch of his undergrad troubles."[7] When he was just twenty-four his extravagantly Seussian cartoons for an insect spray, Flit, caused such a sensation across America that an advertising magazine could proclaim: "The most momentous theme of the summer of 1928 was not Prohibition, presidential election, aviation, or world peace. It was mosquitoes."[8] From 1929 his work was famous enough to be published in collected form, year after year, and to earn the homage of parody. In the 1930s he became a one-man industry producing laugh lines like the unprecedented, idiosyncratically comic Seuss Navy advertising campaign he created for Essomarine.

Dr. Seuss's individuality leaps out from almost everything he produced after settling on his pseudonym at twenty-four: from every illustration, from every story in his children's books, and if not from every phrase he wrote, then from all the verse of his mature children's stories. He confessed, mock-apologetically: "I have a Seuss astigmatism in both eyes so that I see things as if they've been put through a Mixmaster or viewed through the wrong end of a telescope. It's not intentional. That's just the way I see things."[9] In 1952, the year before he wrote *Horton Hears a Who!*, he lamented that most adults lose their capacity to laugh as freely as they did as children,[10] and he set out to remedy the situation.

Play and Thought

Dr. Seuss developed his intense sense of nonsense by rethinking play from the roots of thought up. In Chapter 10 we saw how our capacity to understand events begins with our intuitive understanding of things, kinds, bond and minds, with our folk physics, biology, sociology, and psychology. Dr. Seuss learned how to turn each systematically into play.

Developmental psychologists test infants' first understandings of their world by noting which categories they see as the same (and therefore not worth fresh notice) or different (and therefore worth new attention, as they look longer at an image or suck harder on a teat in concentration). Similarity and difference, identity and repetition, underlie logic and mathematics and all understanding, and Dr. Seuss plays on them with aplomb. Throughout his artistic life he loved to confuse same and different in images like deer whose horns—or men whose beards—grow together. In an enchanting cover of *Judge* magazine his humor derives almost entirely from the comical repetition of parents' profiles and poses in their offspring. A Seussian mother has twenty-three sons and names them all Dave. From the book *The 500 Hats of Bartholomew Cubbins* Dr. Seuss leaped to the film *The 5000 Fingers of Dr. T.* On one page of *Scrambled Eggs Super* the Single-File Zummzian Zuks advance in a soft Seussian curve from infinity, identical except as they dwindle sinuously into the distance. On another page, fifteen solo representatives of fifteen preposterous species strut below two families of identical birds, one family all equal in size, the other all steadily diminishing.

Dr. Seuss: cover of *Judge* magazine, June 1933.

Intuitive physics unfolds so early in young minds that by eight months infants have a basic understanding of folk physics, a "theory of things." Dr. Seuss never lost a childlike fascination with the fun to be had with the rudiments of physics, with gravity, height, depth, size, in unstable piles or arrays (the Cat in the Hat's fish in a bowl on a rake on a cat on a ball, or turtles precariously stacked below Yertle) or heavyweights hoisted aloft (a whale in a tree or on a volcano, a cow milked as she flies) or animals endlessly out of kilter.

After theory of things children develop a theory of kinds, a capacity to distinguish between natural kinds and artifacts, and between kinds of natural things. Children and adults alike naturally notice difference in animal forms and readily recall animal names, but no one creates critters crazier than Dr. Seuss in nature, name, number, and form, in his *If I Ran the Zoo* and *If I Ran the Circus,* in his Schnopps and Sneetches and tizzle-topped tufted Mazurkas galore.

Before age three, children understand the difference between natural kinds and artifacts, and know to look for the *identity* of natural kinds but for the *function* of artifacts. Dr. Seuss turns function to play in his Fix-it-Up Chappie's Star-On and Star-Off Machines *(The Sneetches)* or the nameless but handy and many-handed cleaner-upper machine that allows the Cat in the Hat to restore order after making mayhem while the sun is not sunny.

During the "terrible twos," children begin to understand the rudiments of social relationships. Dr. Seuss exploits these, too, in his play with attachment (Matilda the Elephant, who yearns to rear a chickadee chick), with cooperation (Horton helping the Whos), competition (the North-Going and the South-Going Zax refusing to yield to or bypass one another), status (the Sneetches priding themselves on having or not having stars), domination (Yertle standing on his fellow turtles), exploitation (the parasites infesting Thidwick's mighty antlers).

The last, most sophisticated, and computationally most challenging of these intuitive understandings to emerge is theory of mind. As we have noted, children do not attain an assuredly human theory of mind until age five. Accordingly Dr. Seuss plays with theory of mind sparingly but, as we shall see in *Horton Hears a Who!,* in a form he makes telling and accessible even for the young.

Play as Boundary-Crossing

Recent research into religion shows that we particularly notice and recall whatever crosses intuitive ontological boundaries.[11] The same surprise factors work in play. Even before he became Dr. Seuss, Ted Geisel tirelessly undermined ontological barriers, between natural and artificial (an animal with an umbrella for a tail, a bird with a strawberry for a body and metal nuts for knees, a kangaroo whose pouch is a flower-vase) or between one species and another (fish with antlers, deep-sea chamois). Dr. Seuss subverts the human-animal boundary through an extravagantly overt anthropomorphizing that makes animal features both easy to read and distinctively and outrageously Seussian: "None of my animals have joints and none of them balance . . . none of them are animals. They're all people."[12]

Play and Line

Even when he draws a single image of an ordinary animal or artifact Dr. Seuss remains unmistakably Seussian, unmistakably playful in his uniquely recognizable way. In *The Expression of the Emotions* Darwin, describing what he called "the principle of antithesis," showed how all aspects of a dog's stance, expression, and movement are contrasted in moods of anger and moods of joy. It would be hard to better his characterization of the joyful or playful state as "flexuous movements."[13] This looseness of position, movement, and expression has evolved in pointed contrast to serious behavior precisely in order to send an unambiguous signal of playfulness. Although we and other animals easily "read" playfulness in others, it took a Seuss to "write" it in images. By the time he was twenty-five he had developed a graphic style marked by an exuberant lolloping loopiness that he never abandoned. His soft curves signal the relaxed and harmless.[14] Even ugly, fearsome, smug, or tragic figures seem almost beguiling in his world of supple sinuosity. And he made his animals not only playful but comically human, from the long flat floppy feet of his quadrupeds, to their wide-mouthed smiles, with ancillary smiles even in the creases at the corners of their mouths, and their eyes wide open with whites showing or softly shut in serene beatitude, with each closed eyelid itself a smile, beneath sweet-surprised humanoid eyebrows.

PLAYING WITH EXPECTATION

Minds are bundles of expectations. The better we anticipate what happens next, the better moves we can make. The better we predict others' ideas, intentions, and actions, the more we have over them—although they will also try to predict *our* next moves. As we saw in Chapter 3, the capacity to anticipate others and how they will anticipate us has been a main driver of intelligence. And not only in competition: we can also anticipate others for cooperative ends, like dolphins or humans hunting or playing together.

Predators and enemies prefer to take their victims by surprise. But surprise, like expectation, can be used not just competitively but also cooperatively, not just for antagonism but also for friendship. Play and laughter test expectations, pushing them to or beyond usual limits. In social play or humor we catch others off guard not to overcome them but only to unbalance them just long enough to invite their gleeful rebound. Playful humor outruns others' expectations not to defeat but to delight them.[15]

In humans and other animals that enjoy tickles, play-fights, and peekaboo, the most unexpected movements cause the most intense squeals or pants of laughter—provided a friendly atmosphere has been established, especially through the relaxed expressions, postures, and movements of play. Play and humor work best not when we are caught completely unawares but when we are primed for surprise yet still find our expectations outstripped, as when an infant expects a tickle or boo but cannot know just where, when, or how—or when we turn a new page in Dr. Seuss.

When humor at its best transforms the outstripping of expectations from threat to treat, it affirms how much our minds share even as we push into the unlikely or unknown. No one captures that affirmation of what we share, even in wild surprise, better than Dr. Seuss.

From "Classic!" to Classic

In advertisements, cartoons, magazine covers, and stories, Dr. Seuss's images and swoopy humor caught the eyes of readers across America. From the late 1920s his work brought him wealth and fame, but he wanted more. Books would allow him to hold an audience and invite them to return and reread. Books could pay more, they could highlight his name, they could keep his achievement in permanent rather than ephemeral form. His advertising contract with Standard Oil allowed him children's books as his only sideline, but this accident coincided with his natural penchant for play.

Although little of Dr. Seuss's early work had consisted of sustained stories, much of it had a strong narrative streak. From the first he had turned character and event, the core of story, into play, from his invented creatures of every stripe and species to his fantastic situations—like the Flit advertisements, under the catchcry "Quick, Henry, the Flit," featuring, say, a tightrope walker on the verge of a backward topple, or a bullfighter inches from a snorting bull, but each panicked instead by diving mosquitoes.[16]

Dr. Seuss's visual language had become very much his own by 1929, although he still had more to learn. His *verbal* playfulness had developed even earlier, and had been inventive from the start, but was still lax, prolix, and lame in much of his output of the 1920s and 1930s. Just as he worked to turn logic, physics, biology, technology, and sociology into play, just as he turned line and color, expression, posture, gesture, movement, and form into an inviting game, so he gradually learned to make his verbal play match the punch of his visual game.

In traditional verse around the world the need to focus and refocus attention has led to rhythm and to line-lengths of about three seconds, perhaps in instinctive reflection of the three-second span of the human conscious present.[17] But in most of his children's books Dr. Seuss returns through this adult norm to the childhood play behind it. He selects an anapestic rhythm that like nursery rhymes and unlike the iambic norm of English poetry skips around natural English intonations: dit-dit-DA dit-dit-DA dit-dit-DA dit-dit-DA. (As if to confirm Ellen Dissanayake's claim for the importance of mother-infant protoconversations as a start for art, Geisel recalled that it was his mother more than anyone else who was responsible "for the rhythms in which I write and the urgency with which I do it.")[18] He twists and plays with words in sense, sound, syntax, word formation, names, rhythm, and rhyme. He had always been meticulous, but he became even more of a perfectionist, taking care with every word, "rewriting tediously, draft after draft," so that his readers could read without a care except to read on: "He insisted on momentum in his work and demanded excitement. Ted wanted to write a book whose young readers would 'turn page after page' until there were no more."[19]

Even after his first picture books, Dr. Seuss tried other media: political cartoons, films for the U.S. forces in World War II, sculpture, painting, and more. Movies could earn the largest audience of all, but despite a string of Oscar successes, he preferred the creative independence and control of writing and illustrating his books. He succeeded so well as a children's author that by 1959, at the end of his most creative decade, Rudolf Flesch could write: "A hundred years from now . . . children and their parents will still eagerly read the books of a fellow called Ted Geisel . . . I predict that Dr. Seuss will emerge as one of the great classics of this era."[20]

Homer composed in the high art form of his time. Dr. Seuss knew that *his* world, unlike Homer's, was saturated with multiple appeals to attention and that within that cacophony his children's books were considered "a second-class citizen of the arts,"[21] but he painstakingly marshaled his stories through every stage of the production process. His care paid off, selling over 400 million Dr. Seuss books and earning the affection of generations of readers.

From as early as Dr. Seuss's first children's book, *And to Think That I Saw It on Mulberry Street* (1938), Clifton Fadiman praised him for writing for adults as well as for children, despite the simplicity of his stories and the breeziness of his draftsmanship.[22] Seuss took pride in composing for both children and the adults who might read to them—or merely for their own pleasure: adults learning to read in English prisons spurned other children's books but eagerly devoured Dr. Seuss.[23] He wrote so that his stories would captivate imaginations young and old, impel them through to the end, and provide new pleasures upon rereading. He wrote for immediate success, and achieved it, but also for future audiences. He could be topical, as in his Prohibition protests as a student cartoonist (his family had been brewers) or in his World War II political cartoons, but over the course of his career he aimed his highly individual sense of play more squarely at the universal, toward the pleasure of toppling our shared expectations. He wrote for audiences young and old, first-time and repeat, present and future, native and foreign. Like Homer, he *made* himself a classic.

21

LEVELS OF EXPLANATION:
UNIVERSAL, LOCAL, AND INDIVIDUAL

DR. SEUSS RECALLED that he wrote *Horton Hears a Who!* because he wanted to encourage democracy in Japan after his visit there. No one else, not even another children's writer of genius, would have started from that problem and reached anything remotely like his solution. Looking back, even he wondered how he "ended up using elephants."[1]

I have begun to suggest, and will go on to show in more detail, how we can explain Dr. Seuss's special stamp and *Horton Hears a Who!* in particular by seeing artists as individual problem-solvers aiming above all to earn the attention of their audiences. Especially enduring artists arrive at solutions that appeal richly to human cognitive universals, even as the most confident of them intensify their own idiosyncrasies in order to catch attention by their difference from others. Dr. Seuss made the most of his uniqueness by refining his individual ways of appealing to universals like our predilection for cognitive play, for story, for the pleasures of amused surprise.

Explanations of art in terms of individual artists' shaping their work to appeal to human nature may seem not only natural but even obvious. Yet they are out of line with much recent academic criticism, which tends to underplay both the individual and the universal. Let me compare the current norm of academic criticism with an evolutionary model based on problems and solutions and on universal, individual, and particular as well as local levels of explanation.

CULTURAL CRITIQUE: THE LOCAL AS STANDARD

Much criticism, especially recent academic criticism, has overstressed the local, the limited perspectives of an era, and understressed other levels of explanation, especially the commonalities of human nature as the subject and object of stories, and storytellers as individuals and as solvers of particular artistic problems in each work. Scholarly insistence on the differences between one time or place or culture and others makes it hard to account for the success of art across generations and frontiers. I concur with the Ghanaian-American philosopher Kwame Anthony Appiah that we approach art "not through identity but *despite* difference. We can respond to art that is not ours; indeed, we can only fully respond to 'our' art if we move beyond thinking of it as ours and start to respond to it as art . . . My people— human beings—made the Great Wall of China, the Sistine Chapel, the Chrysler Building: these things were made by creatures like me, through the exercise of skill and imagination."[2]

Historian Richard Minear criticizes Dr. Seuss for his "willful amnesia" about the devastation of Japan that American bombing caused in World War II. Despite his disclaimers, he also reproves Dr. Seuss for not being "qualitatively different from his contemporaries . . . racism was an ingredient in much if not all American wartime thinking about Japan." Minear's attitude typifies the currently dominant critical mode, Cultural Critique, which critiques past perspectives from the standpoint of the present: "In visceral fashion, Dr. Seuss's cartoons, films, and books from that era take us back into a mind-set that reminds us painfully of the pitfalls of racism, of the distance we have traveled."[3] There is indeed a whiff of hysteria in Dr. Seuss's depiction of the Japanese in his political cartoons of 1941–42, although we can hardly call it racism, since the prime target of his exasperation was America's false sense of security and refusal to join the Allies in resisting the aggressive imperialism of Germany, Italy, and Japan. And Minear's historicist indictment of Dr. Seuss is both too narrow and too blunt, since it underplays his particular artistic problems and his individual way of deploying universals to solve them.

By the time he wrote *Horton Hears a Who!*, Geisel's attitude to Japan had long been his own, unlike his unreflecting acceptance of crude racial and ethnic stereotypes as a source for humor in the 1920s, in his early twenties. In the political cartoons he began producing in 1941 Dr. Seuss had stressed the dangers of American isolationism in the face of both German and Japanese aggression, and after Pearl Harbor, he had briefly warned about the dangers of Japanese immigrants in the American West, even as he also began to satirize racial prejudice, against blacks, Jews, or anyone else. General Douglas MacArthur considered his film for the American army headed for Japan, *Your Job in Japan* (1945), too sympathetic to the Japanese to be screened.[4] Geisel and his wife Helen then cowrote the film *Design for Death* (1947), which "portrayed the Japanese people as victims of seven centuries of class dictatorship."[5] Geisel described it to a friend as "an attempt to show the means used on one particular nation to whip its people into war. But the point we try to make is that this particular nation is no different, actually, than any other."[6]

Geisel's personal response to postwar Japan after his 1953 trip, his desire to promote democracy there, seems a concern remote from his usual audience, whether in the United States or in Japan. But in his desire to interest children and adults he transformed the initial impulse to advocate the value of even the smallest voice from a message of contemporary political relevance into a timeless tale. In search of an appeal to all he overcomes the limitations of his time, as he had *not* transcended those limitations when he used racial and ethnic stereotyping as a source of humor in his youthful work. That work of the mid-1920s holds no interest now except as a sad testimony to the pervasiveness of racial prejudice early in the twentieth century. His mature work by contrast appeals because it does *not* merely reflect his time. Just as in his brilliant satire of racism, *The Sneetches,* also written in 1953, Dr. Seuss tries to engage the imaginations of children in *Horton Hears a Who!* by appealing to human ethical universals, to values of individual and community, of independence and interdependence, of support for the weak and for the strong, common to Japan and the United States and to human beings everywhere.

HISTORICISM: THE LOCAL AS CAUSE

Minear inappropriately impugns Dr. Seuss for accepting the stereotypes of his time, but at least he uses some personal detail—Dr.
Seuss's political cartoons of 1941–42 and his trip to Japan—to do so.
But sometimes in the past, and often in recent Cultural Critique,
criticism has allowed even less room for the individual as it elevates
the supposed distinctiveness of an era into a causal trump card. Recent literary theory has tended to treat cultures or societies as almost
autonomous agents, and literary works as impersonal texts, even "social texts," and to underplay both the universal ("Man does not exist
prior to language, either as a species or an individual": Roland Barthes)[7] and the individual ("Since the 1960s, criticism . . . has denied,
on theoretical grounds, the relevance of the single historically definable author": A. R. Braunmuller).[8]

Literary critic Louis Menand, for instance, claims that "*The Cat in
the Hat* was a Cold War invention. His [Dr. Seuss's] value as an analyst of the psychology of his time, the late nineteen-fifties, is readily
appreciated: transgression and hypocrisy are the principal themes of
his little story."[9] In one sense, the first claim is irrefutable: *The Cat in
the Hat* was written during the Cold War. But Menand's phrasing
seems to imply that the book was invented *as a consequence of,* or
even somehow *by,* the Cold War, with the surprise value that either
of these claims would entail, but without committing himself to evidence that he would need in order to substantiate either. This vague-
ness about causation, and the tendency to transfer agency from the
individual to the local, pervades Cultural Critique.[10]

Menand links the concern to improve children's reading to the
Cold War (and to the spread of comics and television), although, as
he acknowledges, *The Cat in the Hat* was published some months
before Sputnik was launched and America suddenly feared it might
be lagging behind Russia in education, science, and technology. But
Dr. Seuss felt a personal urgency about education that long predated
the Cold War. He began to compose a child's first ABC in 1931–32[11]
and wrote his first published children's book in 1936–37. During
World War II he was involved in army education films. Late in the

war, stationed in Europe, he identified childhood education as one of four major problems facing postwar Germany.[12] By 1949 he was articulating his dissatisfaction with the stolidity of Dick and Jane readers and realized by chance, when he heard a three-year-old recite the whole of *Thidwick the Big-Hearted Moose,* that his own books could reach children much younger than he had supposed, and as a result rethought his tactics.[13] John Hersey, in a widely publicized 1954 *Life* magazine lament about the readers used in public schools, had already singled out Dr. Seuss as one of the rare "imaginative geniuses" who could offer an alternative to dull Dick and joyless Jane.[14] Thus in fact the Cold War seems to have had little to do with Dr. Seuss's interest in awakening children's capacity to enjoy reading.

Menand's characterization of Dr. Seuss or his Cat as an analyst "of the psychology of his time, the late nineteen-fifties" seems to assume a substantial shift in human psychology every few years, and to praise Dr. Seuss for having his finger on the mental pulse of the moment. But in most of his children's books Dr. Seuss avoids topical allusions and local appeals. He aims at the enduring rather than the ephemeral—a policy reflected in the sales of his books generation after generation. Are the themes that Menand identifies, transgression and hypocrisy, unique to the late 1950s, or were they just especially salient then?

Addressing the annual conference of the Modern Language Association, the bastion of North American academic literary criticism, in 2004, Menand voiced the widespread recognition that something has gone wrong with literary studies and that university literature departments "could use some younger people who think that the grownups got it all wrong."[15] He could not suggest what they should say his generation had got wrong, but he deplored the absence of a challenge to the reigning ideas in the discipline. He lamented its "culture of conformity" in professors and graduate students alike, and noted with regret that the profession "is not reproducing itself so much as cloning itself."[16] Curiously, however, he then insisted that what humanities departments should definitely *not* seek was "consilience, which is a bargain with the devil."[17] Consilience, in biologist E. O. Wilson's 1998 book of that name, is the idea that the sciences, the humanities, and the arts should connect with one another, so that science, especially the life sciences, can inform the humanities and the arts, and vice versa.[18] Menand claims that he wants someone to say, "You got it all wrong,"[19] but he rules out anyone's challenging the position in which he and his generation have entrenched themselves.

In his address Menand stressed the importance of what he calls "difference," which literary studies learned from "the greatest generation" of Barthes, Jacques Derrida, and Michel Foucault: that there are no universal truths and no universal human nature, but only local cultural and historical differences. But this stance leads to multiple problems. First is the simple logical one. The idea that all is difference, merely local and situated, must apply, if true, to itself; and if this disqualifies its claim to truth, as the implication seems to be, then it contradicts itself. The only way out of the muddle of such paradoxes is to assume that the propositions are false: then no self-contradictions arise.

A second problem arises from the attempt to define difference as uniquely human. "Culture . . . is constitutive of species identity," writes Menand, meaning human species identity.[20] The implied corollary is that culture is always local, always marked by difference. Actually, culture by itself is not uniquely constitutive of human identity, for, as we have seen, many other species have culture. But Menand's declaration also contradicts his claim of difference, since it presupposes a distinctive, species-typical trait, a common feature, as he thinks, uniting all humans and only humans. Yet this is exactly what the doctrine that all is difference purports to deny: that there are some features common to all human natures. In fact, not everything in human lives is difference. Commonalities also exist, and without commonalities among people, human culture would be impossible, since it could not pass from one person to another or from one tradition to another. Cultural Critique wants to stress the "situatedness" of all that is human, but wants to define that situation only in terms of particular cultures. But why not also include the unique situation of being human, with the special powers evolution has made possible in us?

A third problem looms when Menand draws implications for criticism: "A nineteenth-century novel is a report on the nineteenth century; it is not an advice manual for life out here on the twenty-first-century street."[21] Does this mean that all those who in our own times have read *Pride and Prejudice* and felt that it reveals something about the dangers of first impressions, and the error of equating social ease with merit and social stiffness with coldness or disdain, have been wrong? Or, to return to Dr. Seuss, and Menand's claim that *The Cat in the Hat* analyzes "the psychology of his time, the late nineteen-fifties," by focusing on transgression and hypocrisy, an evolutionary approach to *The Cat in the Hat* will take seriously the particulars of the text, the individual who wrote it, and the universals he appeals to. It can therefore suggest other themes like the importance of play in human lives, and the tension between play and seriousness, play as a way of testing real-world risks without real-world consequences, the importance of imagination, the naturalness of parent-offspring conflict, and the value and risks of deception, that link "Sally and me" of *The Cat in the Hat* not just to other American children of the late 1950s but to children and others anywhere. An evolutionary approach can highlight more of the book and its author without obscuring its particulars.

Local versus Universal

Over the last two centuries many have proposed explanations of art in terms of the spirit or ideas or conditions of the times: the rise of Puritanism, the Enlightenment, late capitalism, and so on. If one set of conditions applies to and explains the actions of many, then an individual-level explanation may indeed be insufficient or superfluous. But common factors cannot explain uncommon choices. What allows a work to outlast its time when almost everything else of the time has been forgotten cannot by definition be common.

There are other fundamental problems with such explanations tied to the ethos of an era. They tend to posit a fundamental spiritual or intellectual difference between now and then. But our default evolutionary, commonsense, and scholarly assumption should be that human nature remains everywhere roughly the same. The similarity between ourselves and others is a precondition for our interpreting others at all, even for anthropologists or historians who wish to amplify their discoveries by stressing the exoticism of difference.[22]

Unless we find strong contrary evidence, we should expect ordinary human concerns to persist. Art critic Ernst Gombrich reports how, during his years as a graduate student, concrete instances of human similarity across time called into question for him

> the current interpretation of Mannerism as an expression of
> a great spiritual crisis of the Renaissance. If you sit down in
> an archive and read one letter after another by the family of
> the Gonzaga, the children and the hangers-on and so on, you
> become gradually much more aware that these were human
> beings and not "ages" or "periods" or anything of that kind. I
> wondered about these people undergoing such a tremendous
> spiritual crisis. Federigo Gonzaga, the patron of Giulio Romano, was in fact a very sensuous prince, particularly interested in his horses, his mistresses, and his falcons. He was
> certainly not a great spiritual leader. Yet, Mannerism was the
> style in which he had built his castle.[23]

A recent strategy has been to claim that notions we may take for granted were unthinkable in this or that past era. Some have argued, for instance, that there was "no concept of childhood in the past . . . parents were, at best, indifferent to their offspring and, at worst, cruel to them."[24] Such ideas are "still adhered to and used by many literary critics in spite of powerful criticism from other historians"[25] and despite clear evidence of special care for the young demonstrated in many species of birds and mammals. Similar arguments have been advanced for romantic love, supposed by some to have been invented in "the twelfth century, first in southern and then in northern France,"[26] although cross-cultural, neurological, and cross-species studies have demonstrated the workings of romantic love across societies and even species.[27] Such cultural history has made "difference the absolute in human history,"[28] but an evolutionary perspective, with its cross-cultural and cross-species comparisons, makes it hard to sustain the notion of only radical difference in human experiences as central as childhood and love, however variable the local circumstances.

Explanations in terms of cultural difference tend to lack many links in their proposed causal chains. "Reflectionist" explanations of art, which assume that art immediately *reflects* its time or place, presuppose either that ages have a unitary spirit or that different pursuits within a period are inevitably contesting representations of the age. Film critic David Bordwell notes that top-down explanations in terms of an era repeatedly begin from preconceived notions and are very selective in their presentation of supporting evidence, first in the historical data and then in the artistic works they choose and the details they choose from them.[29] He also observes that scholars who commit themselves "to a search for a single overarching pattern tend not to treat historical actions as shaped by a multitude of factors."[30] Such sweeping explanations turn people into passive conduits of the impulse of the age or participants in an unavoidable common debate, rather than treating individuals as different in susceptibility to influence, according to their capacities, positions, roles, aims, and interests.

Local Possibilities, Problems, and Solutions

Evolutionary criticism will have a special focus on the species, but it will not ignore local contexts. Neither artists' activity nor the artistic forms they employ would be possible without preexisting institutions, technologies, norms, and examples.

An evolutionary approach to the human needs to acknowledge the special importance of culture in human life. In the world of problems and solutions that emerges with life, culture "can act as a potent problem-solving device."[31] But *a* culture or *an* era is not a consistent whole or a solution to a particular problem. The institutions, ideas, and practices of a given place and time will contain a patchwork of solutions to many different kinds of often unrelated problems. An explanation of a work of art will need to consider problems and solutions offered at the local level, but it will need to demonstrate the relevance and salience of such local problems and solutions to the problems of individual artists creating particular works.

After Bordwell shows the weakness of top-down schemes seeking to explain stylistic difference via cultural or historical difference, he proposes his own far more subtle and sensitive model of stylistic history. He happens to be writing about film style, but his arguments apply to any medium. Explanations in terms of the spirit or conditions of an era, he notes, impose a priori notions on the complexity of events, presuppose biologically implausible whole-scale but short-term shifts of psychology, and deny an active role to individuals. He proposes instead seeing style not as a necessary reflection of this or that aspect of a passing phase of wider history, but as a network of problems and solutions, variously related to one another, in which individuals make choices within the institutions that provide the opportunities and the norms of the period. These institutions therefore make certain kinds of options more likely and certain kinds of solutions more salient:

> By granting a role to the artist's grasp of the task and of her own talents, the problem/solution framework acknowledges various reasons for the agent to act . . . The artist's choices are informed and constrained by the rules and roles of artmaking. The artistic institution formulates tasks, puts problems on the agenda, and rewards effective solutions. Gombrich points out that even that precious resource individuality can be achieved only when the artist asks "What is there for me to do?" within the artistic institution and the larger culture.

Bordwell concludes by noting that "individual initiative matters" but that

> group norms matter too . . . So the history of a technique is not likely to consist of one problem and one solution; often, a problem links to a solution and thence to a new problem. For the same reason, the problem/solution model does not commit itself to a neat outline of overarching change. There is no guarantee of a rise and fall, a birth or maturity or decline . . . Similarly, the dynamic of problem and solution can lead to quite diverse, competing outcomes, all coexisting at

the same moment, none of them emerging as the preferred solution.[32]

DR. SEUSS'S PROBLEMS AND SOLUTIONS

How would a problem-solution model apply to Dr. Seuss? As Bordwell's model suggests, much of what any artist chooses to do is made possible by traditions, institutions, and norms. Dr. Seuss draws on the immemorial tradition of fictional storytelling; on local traditions of written and especially printed storytelling that have developed ways of compensating for the absence of the storyteller in person, like narrative economy, scenic focus, clarity of character, and so on; and on the established genre of the illustrated children's story.

He lives in an urbanized society in which the division of labor is normal, artistic specialization possible, and earning status and resources through writing stories a professional option. He could not have had the career he had without the institution of book publishing, and its specialized branch of illustrated children's fiction, and without traditions of authorship, royalties, attribution (unlike Dr. Seuss's books, Shakespeare's plays were often published without his name), reviews, and a wide distribution system. Had these not been available, he would never have left advertising for children's books, or would have left it for film, and would not have been able to take many months to rework a single story into a masterpiece. Local cultural conditions both offered opportunities and imposed constraints. The cost of early color photolithography explains the limited color range and flat tones of Dr. Seuss's early books. He had an acute eye for color and drafted his first children's book with a whole range of subtle hues—until he learned it would cost buyers "about $150" at a time when a wool sweater cost one dollar.[33]

Dr. Seuss makes the most of the traditions of English, including the patterns of internal and end rhyme and anapestic rhythm that poets have developed over the centuries. He writes mostly in a modern standard English, with the concreteness and directness generally favored in twentieth-century literary English, but will eclectically

From Palmer Cox's *Brownies*, 1887.

adopt an archaism like "'Twas" for the sake of rhythm—in Bordwell's terms, an old solution to a recurrent problem—or an Americanism or Briticism ("It sure was a . . ." "What rot!") for rhythm or rhyme. He absorbs the norms of mid-twentieth-century verse in English: he eschews the poetic inversions that he encountered in the comic children's verse of his childhood, like that of Hilaire Belloc, or which he parodied and added to in a schoolboy imitation of Walt Whitman,[34] but which were by now seen as poor taste, a clumsy twisting of English idiom into the straitjacket of rhythm and rhyme.

From the world of graphic art he took many ideas. From George Herriman's long-running and groundbreaking comic *Krazy Kat* he borrowed its surreal role-reversals—the mouse attacking the cat, for instance, which he transposes into the bird or the bug or the jun-

The Mayor grabbed a tom-tom. He started to smack it.
And, all over *Who*-ville, they whooped up a racket.
They rattled tin kettles! They beat on brass pans,
On garbage pail tops and old cranberry cans!
They blew on bazookas and blasted great toots
On clarinets, oom-pahs and boom-pahs and flutes!

Dr. Seuss, *Horton Hears a Who!*, 1954: Who-ville.

gle animals bullying an elephant. But instead of Herriman's bristling verbal barbarisms, spare, dreamlike landscapes, and abrupt schematic story lines, Dr. Seuss made his own work alluringly accessible, softly appealing, in word, image, and story. He learned from the comic mock-technology of Rube Goldberg in creating his own convoluted contraptions. He occasionally borrowed from the manifold mayhem of Palmer Cox's *The Brownies: Their Book* (1887), one of the delights of his youth,[35] but he did so in a less wearying and more tantalizing fashion, inviting us into pages less dense and more absurd than Cox's, pages that we can skip past in pursuit of the story and return to upon a rereading, alternating rapid visual assimilation and slow delighted exploration. He selectively but sparingly adopted some of the visual language of comics, its simplified hands and motion lines, but he avoided its speech-bubbles, its swift sequential framing, and its tough talk. He knew Lewis Carroll, Edward Lear, and

Hilaire Belloc, but his nonsense verse as the verbal obverse of a visual narrative was all his own.

Children's tales have existed for many millennia, cheap color reproductions of cartoons were possible from the 1890s, and comic masters like Winsor McCay and George Herriman were at work as early as the 1900s and 1910s. Without these universal *and* local conditions, Dr. Seuss would not have created the works he has; but even in these conditions, no one else responded as he did, creating children's books like his in manner or impact. The storytelling and comic traditions he encountered are enabling conditions, but insufficient to explain the special charge of his work. For that we need a finer-grained explanation: we need to take into account Dr. Seuss's own constitution, and the unique infectiousness his work has for other minds.

22

LEVELS OF EXPLANATION: INDIVIDUALITY AGAIN

MUCH OF THEORY, since Roland Barthes's 1968 announcement of the "death of the author," has sought—or professed—to downplay the individual, using the rhetorical strategy of referring not to authors but to texts, as if they were self-created or the product only of "systems of cultural production."[1] In fact even if they have nominally challenged the idea of the "single historically defined author,"[2] most critics have continued to discuss single historically definable authors in articles and books that they would be indignant not to have attributed to their own single historically defined selves.

We find individuality too important in ourselves, in others, and in authors to live without it; we are not slime molds or ants. The differences between individuals matter to us in life (people who fall in love with one identical twin will not fall in love with the other), in literature (we respect Hamlet for his singularity and dislike Rosencrantz and Guildenstern for their interchangeability), and in our appreciation of art (we may happen to value almost anything by Vermeer and to resist almost anything by his contemporary Poussin). But in modern literary Theory, individuality has been critiqued as a Western bourgeois notion or as an illusion of "the subject" (the subjective self) foisted on each of us by "the dominant ideology" (often conceived of in Theory, curiously, as unitary and self-directed, unlike individuals!) and as much less important in explaining literature than the local level, the historical or social or other context.

Biology has established more securely than ever the fact and the depth of individuality. The genetic recombination in sexual reproduction ensures so many dimensions for variation that the odds against two human parents' producing genetically identical children from different conceptions are seventy trillion to one. Without individual variation, evolution would never have begun, since there would be no differences for natural selection to select from. And variation extends beyond genetic difference. Development magnifies difference, so that even identical twins, the result of the separation of a single cell at the first cell division, differ in such important details as the folds of the cerebral lobes.

Modern developmental neuroscience has deepened our knowledge of neural plasticity, the mind's capacity to modify itself according to circumstance and experience. Shortly after birth, human minds have an oversupply of neural interconnections, more than they will need later in life. Connections not activated and strengthened by experience—necessarily unique for each individual—weaken and die off. Subsequent development strengthens initial biases. A higher or lower level of physical coordination, intellectual curiosity, or sociability will lead to different choices and actions and therefore unique microenvironments and ranges of experience even for children raised within the same family. Social pressures toward conformity will be partially counteracted by the situational pressures to establish a niche of difference for oneself.[3]

Individuality is no late Western invention but a biological and psychological fact. For any social animal, and especially for ultrasocial humans, the capacity to distinguish other conspecifics in fine-grained ways evolved because those who could do so better could more reliably predict the benefits and costs of future interactions.[4] Chimpanzees show marked individual differences not only in physique but also in intelligence, sociability, and temperament, and modify their responses to other chimpanzees on the basis of personal characteristics. Jane Goodall once reported a respected fellow scientist telling her that even if she was discovering individuality in chimpanzees, it might be better to sweep such knowledge under the carpet.[5] But why? She was certainly not imposing individuality on chimpanzees because of her own Western assumptions; in fact it was Japanese primatologists who first (apart from the chimpanzees themselves!) treated and heeded chimpanzees as individuals.[6]

The individuality of authors is no more a product of the West, the

Enlightenment, or the bourgeoisie than is the individuality of apes, and has no more reason to be hushed up. As readers of others and readers of authors, we have always had an intuitive grasp of individuality that we enjoy and rely on and need to articulate more clearly as part of literary theory, and that we can now trace to the capacity for discriminating individuals and intentions evident in many animal species.

What makes Theory's frequent denial of the importance of authorial individuality so odd is that we engage so naturally, personally, and pleasurably with particular artistic personalities. We have a rich sense of artists' individual "voices" and can easily savor them at a distance or recognize them close up, even if we cannot reproduce or describe or mimic them exactly. Audiences have always engaged with authors as individuals—even compulsive readers of Harlequin romances or other formulaic romance fiction have their favorite writers.

Since classical times, and in modern times since at least Erasmus in the sixteenth century, criticism has stressed the individual style of authors. In the eighteenth and nineteenth centuries this personal stamp was a major focus of criticism, even if the terms shifted from moral character in the age of Shaftesbury to personal imagination in the age of Shelley. In the twentieth century, even before recent Theory downgraded authors, critical theory had begun by the 1950s to understate the importance of authorial individuality. The New Criticism of the mid-twentieth century stressed authorial *intention* as crucial to interpretation, as indeed it is.[7] But in life we respond to others not only because of their actions and the intentions that we read into them, but also because of them as individuals, with their distinct looks, smells, sounds; with their personal range of movements, stances, gestures, glances, facial expressions; their vocal tone, timbre, pitch, and volume; their accents and register; their turns of word, phrase, and sentence; their social ease, responsiveness and awkwardness; their intelligence, curiosity, kindness, sense of humor; and much more. In the same way we respond individually to other artists as individuals, even before they consciously choose a certain artistic effect. In the words of Joan Didion: "The way I write is who I am, or have become."[8]

Children and adults sense Dr. Seuss's singularity from their first encounter with his books. He *made* himself even more singular while making the most of local conditions in the storytelling of his time, from Belloc to Herriman, or in the politics of his time, like Japan's impending election. He took what he needed from his local context to amplify his own individuality while appealing to our universal predilection for play.

Dr. Seuss's editor at Random House, Bennett Cerf, whose writers included eventual Nobel laureates Eugene O'Neill and William Faulkner, liked to insist that Dr. Seuss was the only genius among his authors.[9]

As part of Theory's dismissal of the individual, it has repudiated the notion of genius, both because it prefers to locate the source of outstanding works in the social energies of their time and because it rejects awe and reverence before genius. (With characteristic self-contradiction, Theory also spawned the star system in American academe, bestowing unprecedented adulation on its heroes such as Derrida and Foucault: genius was ousted, in theory, from literature, to be reinstated, in practice, in Theory.) But genius exists. Statistics confirm that a few people of exceptional creativity have an impact far beyond their time and place and a vastly disproportionate contribution to make to their field[10]—and even Theorists who repudiate the notion of genius prefer to write about the work of authors of genius. To acknowledge some individuals as phenomenally creative is not to consider them superhuman, however. Indeed, an evolutionary viewpoint can explain how genius arises in a perfectly naturalistic, perfectly Darwinian way. Differential success, after all, lies at the core of evolution.

In the past artistic genius has often been explained in transcendental terms like inspiration from the Muses. We can no longer accept supernatural explanations for exceptional creativity, but we do no better to deny genius and renounce attempts to explain it naturalistically. Dean Keith Simonton has offered an evolutionary account of genius as a third-order "Darwin machine."[11] Let me elaborate on this concept, introduced in Chapter 8.

The *first-order* Darwinian system is the evolution of life itself. Life, like other Darwinian systems, *generates* new genetic combinations, *tests* them in the environment, and allows those that do not fail the test (by dying before reproducing) to *regenerate*. This new generation produces new recombinations; some will fail, but others may fare better than any in the first generation. Gradually, success builds on success, by regenerating from the most successful available variants. Successive small improvements can accumulate and over time attain complex new functional design without planning or any preconceived goal. Of course whatever successes emerge in this way must also cope with the success of other organisms in the same environment, which also reproduce in the same Darwinian fashion, often to become better competitors, perhaps even predators or parasites. *Their* presence and developing capacities will therefore continually alter the conditions of the test phase of the cycle.

Second-order Darwin machines are systems within living organisms that themselves incorporate or deploy the same generate-test-regenerate principle. One essential second-order Darwin machine is the human immune system. Since it cannot predict exactly which pathogens individuals will happen to meet in their lifetimes, especially as bacteria and viruses evolve so fast, it *generates* a massive array of possible antibodies. Only when the environment *tests* positive, when antibodies meet pathogens they can fit and therefore disarm, are they selected for massive short-term *regeneration*. Another second-order Darwin machine is the human brain. Since evolution cannot predict in advance exactly what environments human brains will meet, brains begin by *generating* in infants more neural connections than they need, then *testing* them by experience, strengthening *(regenerating)* those interconnections that have been used, and killing off those that have not. New experience then tests these new connection networks, and alters *their* strengths, and new experience again alters *those* strengths, in a repetition of the cycle that lasts as long as the brains function.

Among *third-order* Darwin machines are our ideas and their concrete manifestations. The exceptionally creative rarely arrive at their best ideas easily. Responding to Cerf's acclaim, Geisel remarked: "If I'm a genius, why do I have to work so hard? I know my stuff looks as if it was all rattled off in 28 seconds, but every word is a struggle and every sentence is like the pangs of birth."[12] But genius as effortlessness is a myth. Even the precocious and seemingly insouciant Mozart and Pushkin were both painstaking revisers.[13] Exceptional creativity is possible only for those who apply themselves repeatedly and protractedly to particular kinds of problems, through cycles of generate-test-regenerate. "To produce a 60-page book," Geisel explained, "I may easily write more than 1,000 pages before I'm satisfied. The most important thing about me, I feel, is that I work like hell—write, re-write, reject, re-reject, and polish incessantly."[14] He kept in his studio what he called his "bone pile" of "the thoughts that got nowhere."[15]

Any original idea lies some way beyond the known. Its novelty usually makes it likely that the problem emerges ill-defined and therefore that the direction or range of possible solutions can hardly be anticipated in advance. Creative minds can only try to move blindly toward a still indefinite goal, and if some moves seem more promising than others, they then try new blind moves from these propitious positions. As experience at solving problems of a particular type accumulates, solutions can arrive a little more often, because previous efforts have produced so many partial successes, relatively complex moves or forms that have survived *some* tests that can then be recombined and regenerated to develop new, still ampler, degrees of success.

As a schoolboy, Geisel compared himself with others, recognized his special talent to make them laugh, and developed that as *his* competitive niche. He worked obsessively at his humor, in cartoons and parodies, in school and college magazines, in newspaper columns and advertisements. As he told his nephew: "You have to put in your hours, and finally you make it work."[16] From all those trials, he found ideas that particularly appealed to himself and others, like the especially benign visual features of Seussian fauna, his comic combinations of opposites or contrasts, his fantastic situations, his rhyming verse, rambunctious rhythms, and nonsense words. Each of these partial successes became the basis for further recombinations, regenerations, and tests: wacky new widgets on which he could ring endless new comic changes.

Take for instance *Horton Hears a Who!* It hatched out of *Horton Hatches the Egg,* which in turn developed out of the Seussian visual style, the Seussian situation of preposterous contrast, and the developing Seussian capacity for concentrated verbal and verse play. The first Horton story developed in particular from Dr. Seuss's personal predilection for heavyweight elephants and lightweight birds.

With his childlike attraction to the exceptional or the extreme, Dr. Seuss was fascinated by odd creatures, real or invented. Among real animals, he found elephants most fascinating of all. He liked playing with expectations by mixing opposites (large/small, heavy/light), transgressing probabilities (mixing heaviness and flight or fragility), and crossing boundaries (animate and inanimate, species and species or even biological order and order). Since at least his early twenties he had made comic capital of the contrast between elephants' bulk and the airy lightness of birds, as in a 1925 painting of a Horton-like "Elephant Presenting Flower to a Bird," which anticipates Horton and the Whos' clover of 1954.[17] He imagined many variations on elephants' size, weight, and shape: elephants in improbable flight (in 1930, with bird wings and bodies; in 1937, with bee wings and body) or sitting on elephant eggs that crack under their weight (1934) or, even less likely, on a bird egg that does *not* crack (1938), before Horton hatches *his* egg and sees an elephant-bird fly forth (1940).[18]

Evolving his comic elephants offered Dr. Seuss ready solutions to new comic problems, but even his talent and training offered no guarantee of success. In 1938's "Matilda, the Elephant with a Mother Complex" he combines two drawings with a short story exploiting the incongruous pairing of elephant and bird. He adds to the humor by mocking the whole mode of animal fables. But the audience is adult and the tone sourer than later Seuss. He portrays Matilda as absurd to feel maternal toward a chickadee egg. She deserves what she gets when the bird hatches from the egg, sights its elephantine incubator, cries out at her in terror, and flies away forever, leaving Matilda still roaming "the jungle, alone and friendless . . . a woebegone creature, with nothing at all to show for her pains but a very bad case of lumbago." There the story ends, except for a fingerposted "*Moral:* Do not go around hatching other folks' eggs." Seussian humor rarely limped so lamely, and the ungenerous tone was already uncharacteristic even of his 1930s work.[19]

Only two years later he turned around the elephant-as-incubator in *Horton Hatches an Egg*. Here for the first time he expands into a full-length story his perennial fascinations with elephants and with upsetting scale and crossing kinds: a huge heavy thing sitting down on a light frail one and thereby saving rather than destroying it; an elephant hatching a bird. Knowing that this premise will appeal to even the youngest, he then focuses their interest on heart and mind, on Horton's engagingly extravagant kindness and constancy, countering the mother bird's selfishness. Children understand the magical violation of biological kind at the end, when the egg yields a baby elephant-bird, as a just reward for Horton's patient parental care. *Horton Hatches an Egg* was his greatest success to date, but even after this, as we will see, Dr. Seuss had one more failure with Horton before the still richer success of *Horton Hears a Who!*

The Darwinian cycle-and-recycle of generate-test-regenerate can enable genius to emerge. All normal humans are born with two legs or two eyes, but in many other features determined by the interaction of multiple genes—and creativity seems more polygenic than any other biological feature discovered to date—there is a normal distribution of capacities (a bell curve), with successively fewer toward the low end of a continuous range, many in the central bulge, and successively fewer again toward the high end. Of those at the high end on some particular capacity, like the capacity to make others laugh, a smaller fraction will also be at the high end in terms of inclination, ambition, and determination, and may choose to work intensely at that particular capacity, since it allows them to excel and earn the admiration of others around them. Their ambition may cause them to make many trials of their own—and, often, motivate them to study the successes of others in their field—and to reject many of their own efforts as insufficiently successful, so that they generate further attempts from whatever has worked best in prior rounds.

Their exceptional capacity will become more singularly developed. Except in societies that discourage acclaim, their skill will produce a positive feedback loop that makes the value of distinction still greater, thereby amplifying ambition, multiplying effort, and raising the standards against which they test their own work. This process of repeated generation, testing, and regeneration of new efforts will provide the most talented with a highly personal set of partial solutions for recombination in future trials. And the mathematics of combination ensure that even a small number of elements can allow an almost infinite power of recombination, as the twenty-six letters of the English alphabet allow endless rearrangements from epics or epigrams to Eminem.

This continually iterated process explains why even highly creative persons create in distinctively personal patterns. Shakespeare learned from the opportunities and examples of the drama of his day—blank verse, rhetorical exuberance, multiple plots, the genres of tragedy, comedy, and history—but from the first extended them in his own way, becoming, as his work matured, more idiosyncratic in vocabulary, phrasing, imagery, meter, speech construction, characterization, scenic structure, plot development, plot parallelism, emotional change and range, and sheer artistic confidence. By working at their own kinds of problems intently, geniuses can build on their expertise, their peculiar neural networks, their own mental materials and methods, rather than reinventing elements and methods each time from scratch. Even writers with a high inclination or a high determination to maximize novelty will reach positions and discover practices distinctly their own that they continue to recycle and recombine in their own way.

This account of genius also helps explain Simonton's finding that the more exceptionally creative people are, the further they are likely to be from the norm of their milieu, the less likely they are to be representative of their era, the less adequately local contextual rather than individual explanations can account for their singular creativity.

Genius can arise, in short, in a perfectly naturalistic manner, through a nested hierarchy of Darwinian processes that compounds the high creative efficiency of such systems: the first-order system of the evolution of life; the second-order system of individual neural development; the third-order system of individuals' generating-testing-regenerating within a field of endeavor to which they are strongly motivated by an initial bias of unusual talent and ambition, and the positive-feedback effects of those efforts.

Most of our actions are simple and routine, small circumstantial variations on familiar patterns. We have many clearly defined and more or less automatically executed purposes: breathing, eating, walking, talking. When however we search for a novel solution to a novel problem, we may not know how to achieve it. When we do not know just what it is we want, we can only try blindly, and if some trials seem more promising than others, we then attempt new blind trials from these most promising new advance bases.

This logic suggests one reason why genius matters so much to us. It follows patterns similar to those we all use, but it seeks persistently, and often successfully, for the substantially new. Even genius does not know quite where it is going until it arrives there, usually after a long cycle of generate-test-regenerate. But it gradually builds on its partial discoveries to arrive at substantial and often lasting solutions to problems it could not formulate before reaching them. It becomes an efficient system for generating significant novelty.

A naturalistic account of genius explains rather than denies the existence of exceptional creativity, and does not encourage prostration before it. In stressing that it works by principles common through life at many stages, especially repeated cycles of blind variation and selective retention, it also registers the fact that genius, too, must often try out many ideas to find a single successful one, and will often have followed many wrong trails even in the course of reaching something wonderfully right.

23

LEVELS OF EXPLANATION: PARTICULAR

Success as Problem

Artists of genius will solve particular artistic problems in individual ways, using repertoires of materials and methods accumulated over years of creative effort. But past success itself ensures that each new project creates new problems: How to build on previous accomplishments yet not merely repeat them? How to make the most of audiences' high expectations and familiarity with one's creative aims and angles, and yet surprise them?

Horton Hatches the Egg had achieved great success in 1940, but Geisel had been diverted to other work by the urgencies of World War II. When in 1949 he wondered what children's book to try next, his Random House editor suggested he write a new Horton story. He did, but "Horton and the Kwuggerbug" added little to—almost subtracted from—*Horton Hatches the Egg*. Once again Horton "is suckered into helping a manipulative [and minuscule] animal during the month of May: . . . 'I know of a Beezlenut tree where some Beezlenuts grow!' . . . Horton . . . climbs a 9,000-foot-high mountain 'while the Kwuggerbug perched on his trunk all the time / And kept yelling "Climb! You dumb elephant, Climb!"'"[1] Dr. Seuss realized that this idea was not worth a whole new book, because the absurd premise of a tiny creature exploiting a much larger one who could squash it without noticing was merely recycled, and because the exploitation theme, amusing when it merely framed *Horton Hatches the Egg*, turned sour when persistently repeated. "Horton and the

Kwuggerbug" became no more than a two-page magazine story soon forgotten.[2]

Often an unusual constraint or problem can engender fresh ideas, whereas the mere suggestion of "the same again" fails to pose enough of a problem to fire the imagination. To express this in Darwin-machine terms, a mere sequel does not provide a severe and specific enough new test to motivate new recombinations of existing ideas—as a sudden climate change, by selecting from the existing flora and fauna according to new criteria, can prompt new life forms to develop.

But in 1953 a quite unexpected new problem fused for Geisel with the familiar challenge to think up the next Seuss book. During and after his trip to Japan, he felt a commitment to encourage participation in the democratic process there and to counter the indoctrination that he felt Japanese, like German, youth had been subjected to during their nations' expansionist years: in his words, "the worst educational crime in the entire history of the world."[3]

For many years he had been fascinated by the comic contrast between large and small. Often since his twenties that had taken the form of an elephant and a bird—an elephant falling in love with a bird or an egg, or an elephant exploited by a bird. That fascination had eventually led to *Horton Hatches the Egg,* and in "Horton and the Kwuggerbug" he intensified the contrast still further. In 1953 he saw a way to take the contrast of scale he had played with so often to a fantastic new extreme, by inventing creatures too small even to be seen, whose voices all needed to sound together if they were to be heard and saved: a whole community on a speck of dust, a wild exaggeration of Japan's population density and space shortage. As a contrast to these microscopic "persons" he could again resort to Horton, already established as an altruist in *Horton Hatches the Egg.* By making Horton the protector of a people whose existence only he could naturally hear, and by making others opposed to his protecting these invisible and inaudible people, until every last microperson spoke up as loudly and clearly as possible, Dr. Seuss had his story, his solution to the problem of encouraging democracy in Japan or elsewhere, and a series of visual and narrative opportunities to make the most of these creatures' absurdly small size but full "personhood."

He could satisfy the expectations built up by the endearing character of Horton in the earlier book, yet surprise his audience with something still more extreme. He could avoid the unpleasantness of Mayzic's or the Kwuggerbug's manipulation of Horton, or the parasites' exploitation in *Thidwick, the Big-Hearted Moose*. If his micropersons in some sense reflect Japan, then they should not be seen as exploiters reluctant to pull even their own microweight. Horton wants to help them almost unasked:

> Just a very faint yelp
> As if some tiny person were calling for help.
> "I'll help you," said Horton. "But *who* are you? *Where?*"

They then help themselves as much as they can, briskly rebuilding their city after the eagle drops it into the clover field—like the Japanese in their postwar building boom—and shouting together to make their tiny voices heard.

Dr. Seuss faced a number of recurrent problems in each of his books. He wished to appeal to children *and* their parents, to first-time readers *and* rereaders. He made his stories as much like play as possible, yet as artful as possible, saturating them with the cognitive toying with pattern that underlies all art. Especially after 1949, he writes his stories to invite their being read aloud[4] and to be comprehensible even to very young children who cannot understand all the words.[5] In this particular story, he also wants to appeal both to those who have and those who have not already read *Horton Hatches the Egg*. He therefore establishes Horton's agreeableness visually on the first page and his altruism verbally on he second.

And from the first he tries to maximize the benefits of attention and minimize the costs of comprehension. The narrative line of the story has a clarity that reduces comprehension costs while delivering maximum urgency and impact: Horton's goal of protecting a whole people; the obstacle of the other animals, who object to

Horton's eccentricity; Horton's and the Whos' and *our* urgent need to convince the other animals that the Whos really exist. The differences among the animal species simplify character distinctions, while their humanized expressions allow maximum emotional legibility.[6]

Each two-page spread moves the story along one narrative phase, with visual and verbal economy normally in crisp lockstep. Dr. Seuss times his story to fit a child's short attention span, ensuring that each spread ushers in a new stage of the story to keep interest regularly refreshed, with a single illustration, in simple outline, swiftly understood, with just one background color, a flat blue, and a red highlight, often reserved for the Whos' clover or some narrative key, like the Who-ville mayor's megaphone that allows Horton to hear him.

To vary the pace, and to plant deferred pleasures for the rereader, Dr. Seuss occasionally disrupts the usual spareness of detail and immediacy of effect, making the few pages with the Whos a protracted feast for the eye in their intricate comic profusion. Unlike his childhood favorite, Palmer Cox's *Brownies,* but like Breughel's *Hunters in the Snow,* even these crowded pages produce an immediate overall impression despite the centrifugal detail. These high-density Who-pages do not slow down first-time readers but allow rereaders to discover at leisure the local comic complications, like a crowded Who bedroom with three double beds and five-decker bunks.

Dr. Seuss plays again his frequent game of contrasting singularity and similarity, individuality and multiplicity, repetition and variation, concisely compacted in his 1950 story of the identical twins Tadd and Todd, one who loves being a twin, the other who hates it.[7] This features in several ways in *Horton Hears a Who!:* in the contrast between Horton, so much on his own (unlike the eight Hortonish elephants carrying a Zinzibar-Zanzibar tree in *Scrambled Eggs Super!*), and the other gregarious jungle animals, the kangaroo with her young mimic in her pouch, or the Wickersham Brothers with their seemingly endless repetitions; or the Whos, who in various different images seem to have an unceasing variety of tufts and stripes, as if each individual or family were its own species, yet also display comic repetitions of identical young or adults in identical poses, in a pram, a chorus, or an unstable stack.

Intensifying the Story:
The Confrontation Scene

Let us become still more particular and tighten our focus to the artistic problems, both typical and unique, facing Dr. Seuss on a single two-page spread within the story. We can dub this the confrontation scene: if the pages were numbered from the start of the story, these would be 38 and 39.

The first of the problems here faces storytellers anywhere: how to maintain and if possible heighten attention by moving the story along at speed. Dr. Seuss needs to make plausible and even alarming the situation of smaller animals threatening a much larger one, because by engaging readers' fear for Horton, he can amplify their attention: emotion, after all, directs attention.

The second is the problem of an already successful storyteller, and especially one as singularly successful as Dr. Seuss: how to redeploy the distinctive manner that has already won him such a following, how to give his readers more of the pleasures they have returned to his work to seek, yet also to surprise them, how to refresh their attention rather than let it fade into habituation.

A third problem is to minimize the comprehension cost for the audience, so that they can—literally—see the situation at once. The less effort they have to make, the more they will value whatever rewards they find.

Normally each double spread in *Horton Hears a Who!* moves strongly from left to right, usually with Horton, as protagonist, heading in that direction, from earlier to later, to the next page to be turned. In the confrontation scene the left-right movement continues but with a twist.

Horton has been searching for the clover with the Whos aboard, and several pages earlier (pages 30–31), exhausted, he has traced a comic Seussian serpentine down to where he faces the bottom right of the page, as if he will have to keep on meandering fruitlessly

"With the help of the Wickersham Brothers and dozens
Of Wickersham Uncles and Wickersham Cousins
And Wickersham In-Laws, whose help I've engaged,
You're going to be roped! And you're going to be caged!
And, as for your dust speck... *bah! That* we shall boil
In a hot steaming kettle of Beezle-Nut oil!"

> "*Boil* it? . . ." gasped Horton!
> "Oh, that you *can't* do!
> It's all full of persons!
> They'll *prove* it to you!"

Dr. Seuss: *Horton Hears a Who!*, 1954: the confrontation scene.

forever, on to the pages that follow. But on the next spread (pages 32–33), perked up, he stares cross-eyed at the speck he has been looking for on the one clover that he needs among these millions. Now his head faces us, so that we can see full-on his eyes staring in relieved surprise at the speck, but his visible left flank shows he has only made a three-eighth turn to his left. The next time we see him (pages 36–37) the kangaroos and the three Wickersham Brothers have come up from behind him, so he turns around to regard them from his position on page right, and faces almost completely left.

Now, in the confrontation scene, the rest of the animals advance further toward Horton, their target, while he faces decidedly left, as if backed into a corner with nowhere to go. Over the preceding scenes, Dr. Seuss has slowly turned Horton around, just over one half-turn, doing so perfectly naturally according to the needs of each page, but also so as to preserve the visual continuity for the reader and yet to offer change—like the surprise here of seeing Horton squeezed into the corner and almost cowering beneath all the other animals. Although he is larger than his jungle companions, Dr. Seuss places him here below them on the page, as if subordinate to the dominance their height allows them, thanks to the hillocks they stand on. Horton is on his own, and a fragment, not even a whole elephant. As if squashed flat, he looks up in wide-eyed and down-mouthed alarm at the cockily fierce kangaroos, mother and child comically reduplicating each other as they point to the threat of the cauldron where the Beezlenut oil boils.

This configuration may alert us retrospectively to another problem Dr. Seuss has had to face all along. Much of the humor of the story arises from discrepancies of scale between Horton and the other animals and between all the jungle animals and the Whos. But at the same time Horton must not seem so much larger than the other jungle creatures as to make it improbable that they can threaten and control him. Dr. Seuss quietly solves this problem by ensuring that on every other page in which Horton is present with another animal—until the last page, when all is resolved, when all are on the same side, and it makes sense to maximize Horton's hugeness—there is at least one other animal whose eye level is above his. (The sole exception is the page where the kangaroos dive away from him mockingly, almost aggressively, into the pool: here the position of the kangaroos, though lower than Horton, still exudes defiance as they hurtle into the water and splash him and his clover.) But no other page has the effect of the confrontation scene, with the kangaroo and her child glowering down at an abject Horton.

On the previous page we have seen the first three of the enfilade of jungle monkeys, the three Wickersham Brothers, who are all we have encountered earlier in the story. Now Dr. Seuss introduces a surprise as we turn the page to the confrontation scene and see the whole file, smilingly confident in their numbers, encircling Horton, to his right, left, and straight ahead, with no space visible behind him. The monkeys, though not identical in pose like the kangaroos, have comically similar, comically confident expressions, a vast proliferation against the lone Horton, matched in the repetition of their names in the verse:

> With the help of the Wickersham Brothers and dozens
> Of Wickersham Uncles and Wickersham Cousins
> And Wickersham In-Laws . . .

Dr. Seuss makes his singularity-versus-similarity or singleness-versus-multiplicity joke here work, both visually and verbally, not just for humor but also for danger, to add to the narrative threat suddenly posed by the overwhelming numbers facing the cornered Horton.

The kangaroos point toward the Beezlenut-oil cauldron. It stands out in bright red at the top left, threatening the only other patch of red, the clover flower at bottom right, with the speck of dust perched on it, almost squeezed to the edge of the page, as if in the tightest of tight spots. Horton tries to shield it by putting himself and the greatest visual distance the page allows between the clover and the threatening army. Highlighted by the emphasis of their final position in the line, and by their rhyming, the words "caged," "boil," and "oil" add to the threat, even as Dr. Seuss adds a playful twist in the invented Beezlenut (brazil nut and betel nut, with a bezel edge and a diesel flavor?).

The whole story has lurched forward, as often before, with the abrupt logic of nightmare. Horton's speech highlights the key surprise and threat, by starting with, and stressing, a repetition of the danger word: "*Boil* it?" The lines of his speech, visually separated from the kangaroo's though actually identical metrically, have been split in two, uniquely in this story, moved to the right so as to match the visual placement of kangaroo and elephant on the page and to reinforce the sense of the elephant's being cornered and suddenly cut down in size.

His last line repeats the story's key motif-word: "It's all full of persons! They'll *prove* it to you!" By signaling here the next phase of the plot, Dr. Seuss both reduces our comprehension costs (we know roughly where the story is heading) and whets our curiosity. As we are about to turn the page, we wonder: how *can* the Whos prove to others that they are there?[8]

All Dr. Seuss's decisions in the confrontation scene introduce and instantly amplify the danger and urgency of the situation, and hence intensify our attention, impel us on through the story, and prepare for the triumph of the Whos' proving their existence at the end— without ignoring the comic or the fantastic.

Noël Carroll characterizes the devices by which filmmakers ensure that viewers respond similarly to a scene—including soundtrack music, mise-en-scène, and editing—as "emotional prefocusing."[9] Like a master director, Dr. Seuss shapes his confrontation scene, in its place in the story, in its action, and in all its subsidiary choices, to prefocus our response, our attention, *and* our emotion. Each of his details, even if not consciously noticed, adds to the immediate impact of the story for first-time readers, and to the pleasure of rereading, the sense that everything counts, that nothing is wasted, and that there is still more there than first met eye and ear.

Everything in the confrontation scene bears the mark of Dr. Seuss's individual and invitingly playful style. The page builds on devices he has developed elsewhere, through many cycles of his Darwin machine: the two-page spread; the standardly Seussian verse; the nonce names and the rampant repetitions; the economical, unfussy, curving, loose-limbed Seuss line; the blithe, eyes-closed faces of the kettle-carriers, the guiltless grins of all the other Wickershams; the comic parent-child duplication, dating back to at least 1933; the elephant versus others, dating back to the 1920s; the elephant and flower (1925);[10] the Beezlenut name (1951); the singularity-versus-multiplicity jokes.

Yet despite the rich natural grain of his individuality, despite the personally tried and tested solutions, Dr. Seuss has also attended to the specific problems of *this* phase of the story and *this* two-page spread, and has found many small solutions, some of them unique, like the singularly pop-eyed Horton, with less of the curve of his head showing between eye and object than in any other profile.

Dr. Seuss has had to "write, rewrite, reject, re-reject, and polish incessantly," to generate, test, and regenerate, in order to make the elements of the confrontation scene work as swiftly and effectively as they do. He has done so more to tighten attention and minimize the costs of comprehension, even for a very young child, than to express any abstract meaning. And he has done so not by using arbitrary conventions but by calling on many features with deep biological roots, like the expressions on Horton's and the kangaroos' faces, height as a sign of dominance, and the number of other animals ranked against the lone elephant.[11]

Dr. Seuss had been thinking of elephants and tiny others for almost thirty years when he wrote *Horton Hears a Who!* He had considered reprising Horton since at least 1949, but it took a wholly unexpected problem, arising from his sojourn in Japan, to generate a new set of variations on these familiar themes and spark a new imaginative surprise for his readers: a whole populace on a speck of dust. Dr. Seuss solves a problem posed—to his unique sensibility, at any rate—by his local context, but transforms it into his own particular problem, to make of it a story appealing to readers of all kinds, young and old, and at all times.

A truly evolutionary approach to literature, I have tried to suggest through the simple example of *Horton Hears a Who!*, will apply a problem-solution model to multiple levels: the universal features of human nature that story in general and individual authors represent and appeal to; the local conditions available for earning an audience; the individual qualities that shape an author's unique range of likely problems and solutions, and that he or she develops and deepens over years of generating, testing, regenerating new artistic efforts; and the particular problems of a single story, especially of lowering the cost and raising the benefits of audience attention. An evolutionary analysis of any literary work can be sensitive to the moment *and* to the likelihood that past problems and solutions—within human nature, the local narrative tradition, the individual's experience, and the work's particular contours—will shape the new problems to be faced and the new solutions to be found.

24

MEANINGS

Events, Stories, Meanings

FOR ALL THE MODERN academic interest in narrative, a comprehensive understanding of the way we extract information from real events we witness, events we learn about, or stories we encounter seems some way off. Mark Turner suggests in *The Literary Mind* that human minds are unique because they can draw meaning from stories of all kinds, real or invented, through a process he terms "parable."[1] We do have an extraordinary capacity to express ourselves via oblique vignettes, but the ways in which we understand events and stories go beyond both causal explanation and parabolic implication. How we understand the Vietnam War or *War and Peace,* our parents' divorce or *Anna Karenina,* cannot be summed up as parable.

From an evolutionary standpoint, explaining how *we* understand events and stories necessitates also discovering how *other* animals learn from events. Although we still have much to discover about advanced animal cognition, creatures of many species appear to infer meanings from events. At the simplest this involves, say, the recognition that food can be obtained through a certain procedure, as a pigeon pecks at or a rat presses a lever in an experiment. Animals that observe others intently may acquire complex information about social opportunities and costs. Chimpanzees can find out not only the dangers of challenging a particular male or his coalition partners, but also, by observing over time how coalitions rise and fall, how to plan for their own strategic assault on the ruling alliance.

By monitoring the actions of others, animals can learn not only from their *own* actions—often a costly way to learn a negative lesson—and not only specific strategic information but also general information: *character* information about specific individuals or general types, and *plot* information, specific (A has *x*ed B) or general, about the costs and benefits, the risks and opportunities, of various kinds of behavior. What an animal assimilates—conditions, procedures, strategies, outcomes—will depend not just on what it encounters but also on its capacity to observe and infer. Events do not come with their implications labeled for bystanders, and animals that can extract more relevant information than their conspecifics will be in a position to make better decisions.

Events merely happen, and animals of all kinds may witness them or not. *Stories* on the other hand seem uniquely human, and whether factual or fictional get told because they are presumed worth their audience's attention either for the interest of the events themselves or for their implications.

Aristotle famously distinguished between the implications of history (actual events) and those of poetry (or, as we would say, literature, especially invented events), calling the former "particular" and the latter "universal."[2] But this often-repeated distinction seems wrong. We can derive "universal" or general implications from witnessed events or reported narratives whether factual or fictional. We cannot draw specific strategic information from fiction as we can from factual narratives: it is one thing to learn that Hannibal Lecter has behaved psychopathically, another to learn the same of your neighbor. But like other animals we can draw "universal" (general) implications from real events (in the human case, whether encountered directly or not), just as we can from fictional events. Plutarch's *Parallel Lives* used historical figures as moral exempla memorable enough to make him, in the Renaissance, the most popular of Greek classics.

Stories differ from events not only because they are indirect but also because they are told by storytellers. We interpret the actions of others all the time, and we often act (pull a face, roll our eyes . . .) only because we *want* others to interpret our actions in this or that way. Storytelling is a particular kind of action, and telling a particular story invites an audience to interpret not only the story's events but also the storyteller's action in telling it. Interpreting the teller's motives can be easy in casual conversation, when we may often see a story as a reminiscence or an example or a parallel triggered by the preceding conversation.[3] It may become more difficult when someone tells a preexisting story, since the story was not made for the occasion, though it may have been selected, and reshaped, to fit the situation of its retelling. It becomes still more difficult for a fictional rather than a factual story, and a printed rather than a spoken story, for now the narrative intentions can be inferred not from the situation or the teller but only from the story's events, manner, and impact.

Our capacity to infer meanings from stories emerges naturally from our capacities both to infer general meanings from specific events and to infer intentions from others' actions. But just as the meanings we choose to draw from events are not wholly determined by the events, so our reactions to others' actions may not be wholly determined by the actors' intentions. Someone might say, "It's hot in here" to prompt a window to be opened. The remark might indeed have its intended effect, but it could also elicit other reactions, spoken or unspoken, from "Yes, why do they always overdo the central heating?" to "Why don't you take your jacket off, then?" to "Is she having a hot flash? She looks that age" or "He always is one for stating the obvious." In the case of fiction, a storyteller like Dr. Seuss will try to elicit the kinds of responses he most wants, but he will also expect that he cannot entirely control the responses of the large and diverse audience he seeks. In trying to rule out some unwelcome reactions, he might irritate or bore others resistant to being told what to think and feel. He will have to balance intuitively the costs and benefits of each choice, and monitor for possible counterreactions to each attempt to direct audience response.

Dr. Seuss and Meaning

Like many storytellers, Dr. Seuss had an ambivalent attitude to the meanings or morals of his stories. He was successfully Seussian in advertising campaigns like the Flit bug-spray ads of the late 1920s or the Essomarine gas ads of the 1930s long before he cared for serious meanings at all. The story that sowed the seed of the Horton books mocked the idea of the animal fable, with its humiliating and frustrating conclusion for Matilda the elephant and the jeering "*Moral:* Don't go around hatching other folks' eggs." Dr. Seuss's first published children's book, *And to Think That I Saw It on Mulberry Street,* seemed remarkable in its time for allowing its hero to concoct fantastically tall tales and not be punished for doing so. It was rejected by publisher after publisher, on the grounds that they could find "'no moral or message' . . . nothing aimed at 'transforming children into good citizens.'" Reading such replies, Geisel roared to his wife across his studio: "What's wrong with kids having fun reading without being preached at?"[4]

Yet the rise of Hitler and the course of World War II made him a more serious and socially engaged writer.[5] He would later declare, apparently forgetting *Mulberry Street* or *The Cat in the Hat:* "It's impossible to tell a story without a moral—either the good guys win or the bad guys win."[6] And in a further partial contradiction, as if a moral was now not inevitable, but an infrequent and precious achievement, he remarked in his last year: "I think my stuff has become useful, not just amusing. That's important to me."[7] The antityrannical *Yertle the Turtle* (1958), the antiracist *The Sneetches* (1961), the antipollution *The Lorax* (1971), the antiwar *The Butter Battle Book* (1984) all have unmissable morals, despite their fantastic treatment.

In composing some stories, Dr. Seuss seems to have begun with a moral or message, and to have been guided by that throughout, although the local problems of making the story as wackily Seussian as ever dominate the composition of even these books. But the greatest Seuss stories cannot be reduced to any single moral.

Horton Hears a Who! began with an impulse to promote democracy in Japan but emerged quite independent of its origin. It has held audiences old and young, generation after generation, partly because it has too much life, like complex events in the real world, for us to extract only one neat meaning.

In "Matilda, the Elephant with a Mother Complex" Dr. Seuss mocked cozy moral meanings. In *Horton Hears a Who!,* however, he repeats four times "A person's a person, no matter how small": even the least have rights, and all have responsibilities even to the least of others. He notes: "In verse you can repeat. It becomes part of the pattern. To teach, you have to repeat and repeat and repeat."[8] The "moral of the story" recurs in variations on Horton's line or simply in numerous references to "small" and "persons." Horton, the mightiest animal in the jungle, not only voices the line but acts on it from the first, and we admire him for it. The other animals, as soon as they believe in the Whos' existence, instantly echo his sentiment and actions, playfully reprised at the end in the young kangaroo's "no matter how small-ish!" The story, its refrain, and its ending all suggest this as the natural and desirable attitude toward others.

Even a narrative "message" as explicit and reiterated as this nevertheless depends for its range of implication and application partly on audiences' attitudes and interests.[9] The problem-solution model after all should apply not only to artists but also to audiences. Selfish gene theory shows that we cannot expect organisms to work routinely for the benefit of others:[10] in this case they should seek to maximize the benefit *for themselves* of attending to a story. Audiences will therefore tend to seek what matters most to themselves, not necessarily to the artist, and even to appropriate the work in ways the artist did not intend.[11] Vegetarians could read *Horton Hears a Who!* as supporting their cause. An antiabortionist group *did* use "A person's a person, no matter how small" in its campaign until Geisel threatened legal action.[12]

But if audiences engage with art to serve their own purposes, rather than those of the artists, then artists, intuiting this propensity, will often try to make their interests coincide—or at least seem to—with those of their audiences, especially by promoting prosocial or group values, since we all benefit from associating with altruists or from living within thriving groups. In his two Horton books Dr. Seuss not only promotes sociality but does so with comic Seussian exuberance.

Yet most writers with literary rather than purely instructive ambitions prefer to open up rather than close off implications. Just as a live metaphor in poetry creates an aura of suggestion rather than a pinpoint illumination, meaning in story more open-ended than fable tends to radiate out. We can read *Horton Hears a Who!* as about the rights of all, even the smallest, or the duties of all, even toward those who can least speak up for themselves. In that sense, and given Dr. Seuss's primary audience and the diminutive size of the Whos, we may think specifically of the rights of children. The year before composing *Horton Hears a Who!*, Dr. Seuss had worked on his film, *The 5000 Fingers of Dr. T.* "The Kids' Song" there includes the lines

> Now just because we're kids,
> Because we're sort of small,
> Because we're closer to the ground,
> And you are bigger pound by pound,
> You have no right, you have no right
> To push and shove us little kids around . . .
> I'd hate to grow like some I know
> Who push and shove us little kids around.[13]

Horton Hears a Who!'s contrasts in size between Horton and the other jungle animals, and between the other animals and the Whos, can yield other implications: the weakness even of the strong if challenged by the rest (Horton for all his size cannot withstand the concerted opposition of the others); the strength of the comparatively weak if they are united (the other animals ranked against Horton; the Whos themselves in making themselves heard by the other animals).

Although it is a brilliant solution to Dr. Seuss's self-assigned problem of promoting democracy in Japan, *Horton Hears a Who!* is much more multivalent in its implications for the relationships among individual, group, and community. Every voice counts, even the smallest, we learn from Jo-jo's "Yopp!" making the crucial difference, so that the Whos can at last be heard and their interests taken into account. Yet this part of the story covers only the case of everybody speaking with *one* voice, when democracy allows people to speak freely with many conflicting voices—or to remain silent. As a case for democracy, a message that all should shout with one voice would be misguided if narrowly taken. But the rest of the story tries to debar such an inference. Horton acts on his own conscience, despite the concerted mockery and threats of the other animals. He resists their majoritarian pressure, and we commend him for doing so and resent the other animals for not tolerating dissent.[14]

Part of what makes the story so satisfying, indeed, is the delicate balance between our admiration for Horton's having the courage to stand up for himself despite the pressure of his entire jungle community and young Jo-jo's having the decency to respond to the pressure of *his* community. Both nonconformity and conformity have their claims. This kind of symmetry may not be consciously noticed even by most adult readers, yet it contributes naturally to our sense of the rightness of the story. In his case, Horton has good reason to resist the other animals; in his, Jo-jo has good reason to join his fellow Whos.

Other aspects of the story carry other implications. Our distaste for the other animals' initial intolerance of Horton's belief in the Whos encourages a tolerance of difference, including racial difference—the more focused moral of the story "The Sneetches," from the same year[15]—and differences in beliefs, and not just a tolerance for but an appreciation of the unusual, which Dr. Seuss seeks to stimulate throughout his work.

As in *Horton Hatches the Egg,* Horton's character and conduct serve as an example. He shows not only an instinctive readiness to help, even more than in the earlier book, but once again perseverance, in following the eagle and sifting through the clover, and now courage, in not yielding before the opposition of all the other animals. That encouragement of resilience and determination proves central in other Seuss stories like "What Was I Scared Of?" and *Oh, the Places You'll Go.*

We can reduce a story to any level of thematic abstraction, as Richard Levin forcefully pointed out when he debunked critical claims to have discovered supposedly ever deeper or wider "themes" in this or that Renaissance play, in what he called the "my theme can lick your theme" strategy.[16] Although sermonizers may start and stick with a theme, storytellers need first of all to have audiences attend. A tacit promise of eventual meaning, a take-home fortune cookie message, offers too little incentive if the fare itself lacks flavor.

This is not to deny that even longer stories have implications, especially through their power to engage our moral emotions. And indeed their implications will have a major bearing on the attention they receive. Not only do stories tend to be prosocial in their implications, but they are regularly selected for that reason by parents and communities.

Horton Hears a Who! encourages openness and independence of mind, resolution and courage in the face of obstacles, tolerance of others, compassion to those weaker than oneself, and efforts in support of the common good. Many a story with such laudable lessons nevertheless now languishes unread, and the success of *Horton Hears a Who!* depends less on its moral implications than on its appeal to attention in every line. Yet the fact that it can engage our moral emotions so simply and strongly, in such a rounded way, *adds* to its power to earn attention. Not only the artistry but also the moral tone of Dr. Seuss's elephant stories became progressively more engaging, from Matilda, derided for her desire to hatch the egg and unrewarded when the bird flies in horror from her intimidating bulk, to Horton, exploited by lazy Mayzie but rewarded by his hatching an elephant-bird, to Horton, eager to help the assiduous and resourceful Whos. Had the execution of *Horton Hears a Who!* remained as brilliant as it is but its story line been that of the mean and mocking "Matilda, the Elephant," it would hardly have become the favorite it remains.

We are likely to be wary of the precepts of others, especially if they seem more to the advantage of the instructor than to the advantage of the instructed. Yet we all love an altruist. As game-theory simulations of cooperation show, any participant in a social exchange benefits when the other partner is an altruist. And Horton's altruism is as colossal as his physique. He is ready to exhaust himself and risk his own freedom to save the lives of minuscule creatures who could never repay his kindness. There is an absurd Seussian extravagance here that appeals to us all, since we know that our concern for other creatures dwindles with size: we do not feel much solicitude for a fly, a mosquito, or an ant, let alone a bacterium, and we appreciate the extraordinariness of big Horton's concern for creatures microscopically small.

Dr. Seuss's comedy and his seriousness are the twin chambers of his story's huge heart. The fantastic extravagance of Horton's altruism makes him all the more attractive and makes us all the more readily sympathize with him, ally ourselves with his goals, and rejoice in the positive outcome for him and those he champions. We learn the moral lesson, in other words, in a memorable and pleasurable way, soaking up the admiration that Horton's own behavior earns. Our guard is down, our moral emotions are engaged,[17] our imaginations stirred. As novelist Philip Pullman writes: "'Thou shalt not' might reach the head, but it takes 'Once upon a time' to reach the heart."[18]

So far I have stressed the moral and social meanings of *Horton Hears a Who!* But there are more. As I and others argue, all art serves creativity. Creativity and imagination are favorite themes in Dr. Seuss from *And to Think That I Saw It on Mulberry Street* through *The Cat in the Hat* books and *Oh, The Thinks You Can Think!* to his unfinished *Hooray for Diffendoofer Day* (1998), completed by Jack Prelutsky and Lane Smith. A. O. Scott writes in the *New York Times*:

> That I think about the health of my children's imaginations at least as much as I worry about the strength of their characters, and that I picture their powers of perception as both resilient and fragile, probably owes more to Theodor Geisel than to any parenting manuals I've read since. I suspect that it's only a slight exaggeration—and exaggeration was one of his great gifts—to say that our current understanding of children, and of ourselves as former children, is the brainchild of Dr. Seuss.

He calls *Mulberry Street* "a hymn to the generative power of fantasy, a celebration of the sheer inventive pleasure of spinning an ordinary event into 'a story that no one can beat.'"[19]

A fine work of art not only expresses creativity but also inspires it in those who enjoy it. In children responding to Dr. Seuss, this appreciation may not be sophisticated, but it will be there, even in wonder at details: "Look! A whole city on a dust speck!" "See how he's drawn *them!*" Through the zaniness of his choices Dr. Seuss makes a good deal of the creative appeal of the craftsman evident even to the youngest children, but although he designs his stories for accessible and immediate effects, the care in his design allows rereaders to notice more details or implications each time they read—in itself a process of continued creative discovery.

While creativity and imagination are strongly implicit in the manner of *Horton Hears a Who!*, they are perhaps less central in its matter. Nevertheless, Horton's capacity to conjure up in vivid imagination—on two two-page spreads—one lone creature hanging on to the globelike sphere of the dust speck, and then a whole family and a token plant aboard the speck, is important in preparing for his immediate readiness to protect whoever is on the speck, and in opening our imaginations to this improbable scenario—soon to be outstripped, of course, by the direct glimpse of the happy Whos that the next, "objective" panorama of the speck offers us. And Horton's gift of imagination, and our imagination in merely enjoying a story so fantastical, also align us against the other animals, who stick stolidly and almost fatally to the normal and the probable. The whole story insists that without the capacity to think beyond what we can see and to explore the worlds of the possible our lives would be sadly restricted.

In discussing the *Odyssey,* I noted the role that the human understanding of false belief plays in its structure—in all the dramatic ironies surrounding the fate of Odysseus—and in the metaphysics that informs the story: the sharp awareness that the truth of a human situation may not be visible to individuals within it, that the truth may often need a larger, more comprehensive vision, stretching even all the way to the Olympian gods. Dr. Seuss, too, constructs his story around a simple, central, readily graspable instance of false belief, the other animals' assumption that there cannot be creatures living on a space as small as a dust speck. He simplifies the dramatic irony for his young readers, by including only one such irony, developing it gradually, and allowing the children to see first Horton's good but imperfect guesses and then the reality he cannot see. He also motivates his audience—by allowing us to see the happy Whos on Whoville directly, and by putting them and Horton in danger because the other animals refuse to believe they exist—to want the jungle animals to recognize the existence of the Whos. We not only want Horton to achieve his goal, to save the Whos; we also want the other animals to discover that their belief was false, that it is they and not Horton who acted on false belief.

The human understanding of false belief, I suggested in previous chapters, has been crucial for the art of story and for religion and science. Once we understand clearly that we may not know everything relevant to a situation, we will often wish to seek out a deeper explanation. In the past such an effort often entailed supernatural explanation, reading the seen in terms of unseen agency. In the modern world many of us seek to explain our world in strictly natural rather than in supernatural terms, without unseen agents. In *Horton Hears a Who!* the other jungle animals seem narrow-minded for rejecting the notion that there might be creatures smaller than they have conceived of on a mere dust speck. There are indeed unseen agents in this world—unseen by the jungle animals—but we can see them, and see that they are both fantastical, with their microcity, their microtennis, and microhockey, and real within the world of the story, perfectly natural, not in the least supernatural. They are also an emblem of the worlds within worlds that science has discovered—and an unwitting prediction of modern nanotechnology—and of the openness and imagination needed to discover the world through science.

In Parts 2 and 3 I suggested that the chief functions of art and story lie in improving human cognition, cooperation, and creativity. Even a story as playful and unassuming as *Horton Hears a Who!* explicitly and implicitly develops social cognition, encourages cooperation, and fosters the imagination, linking imagination with understanding our world more fully, with the creativity of human life as a whole, in religion and science as well as in art.

Mark Turner proposes the parable function of story. Story can be used in that way, in Aesop or La Fontaine, in the parables of Christ or the fables of Tolstoy. But even a simple children's story like *Horton Hears a Who!* can have myriad meanings radiating from its central core. These meanings are open, though not endlessly so, for they derive ultimately from the open-ended ways in which creatures can understand and emotionally assess not only specific strategic information but also the general implications of events. Dr. Seuss manages both to offer a firm happy close to his story and a clear explicit moral *and* to allow for a wide range of other audience responses, interests, and implications.

By focusing his efforts on securing our attention and activating our emotions, page by page, Dr. Seuss makes his simple story too rich to be compacted into a single moral, even the one he turns into a ringing refrain. Yet by focusing his efforts at the same time on Horton's helping the Whos, and winning others over to help them, he stirs up an indefinite set of meanings that make the story matter still more to us. Attention and meaning remain distinct from and irreducible to each other, but they feed off and into each other.

CONCLUSION: RETROSPECT AND PROSPECTS: EVOLUTION, LITERATURE, CRITICISM

> Criticism seems to be badly in need of a coordinating principle, a
> central hypothesis which, like the theory of evolution in biology, will
> see the phenomena it deals with as parts of a whole.
>
> Northrop Frye, *Anatomy of Criticism* (1957)

RETROSPECT

Evolution allows us a clearer view of art and literature. Like a lens
that can slide smoothly from macro to wide-angle to telephoto, it offers us more precision, breadth, and depth as we look at art in general or literature in particular, or the elements of human nature represented or appealed to in either.

Throughout *On the Origin of Stories,* I have developed a number
of key ideas:

Evolution

As I observed in Chapter 20, "Evolution generates *problems and solutions* as it generates life. Rocks may crack and erode, but they do not
have problems. Amoebas and apes do. Natural selection creates complex new possibilities, and therefore new problems, as it assembles
self-sustaining organisms piecemeal, cycle after cycle, by generating
partial solutions, testing them, and regenerating from the basis of

the best solutions available in the current cycle. In time, it can create richer solutions to richer problems."

Such *Darwinian processes* operate at multiple levels, not only at the level of genes or organisms, but also in secondary modes like the immune system. Particularly important for investigating art are the Darwinian processes within *individual brains,* where, as we grow, some of our neural connections atrophy through inactivity and others multiply through use;[1] within *culture,* where individuals together and over time engender and select new beliefs and practices;[2] and within *invention,* as individuals or groups construct new artifacts or ideas and, from the most promising platforms, build still higher.[3]

Adaptations are complex biological systems, physiological or behavioral, which through the cumulative Darwinian process of *blind variation and selective retention* have developed a *design* that reliably serves some *function,* in other words provides a sufficient solution to some problem a species faces to improve chances of survival and reproduction.

Art as Adaptation

I have proposed that we see *art* as a form of *cognitive play* that appeals to our intense human appetite for the rich inferences that *pattern* allows. Art in this broad sense is a human *adaptation,* its chief functions being (1) to *refine and retune our minds* in modes central to human cognition—sight, sound, and sociality—which it can do piecemeal through its capacity to motivate us to participate again and again in these high-intensity workouts; (2) to raise the *status* of gifted artists; (3) to improve the coordination and *cooperation* of communities, in our very social species; and (4) to foster *creativity* on an individual and social level.

The pleasures of *play* entice animals with flexible behaviors to enjoy expending energy rather than resting, so that they learn and overlearn key motor and social skills. We humans owe our competitive success, among ourselves and against other species, to intelligence. The appeal of the cognitive play in art makes art as compulsive for us as play, enticing us to forgo mental rest for mental stimulation that helps us to learn and overlearn key cognitive skills, especially our capacity to produce and process information *patterns.* Art entices us to engage our *attention* and activate our minds in ways that we find most pleasing, and allows the most gifted individuals to earn *status* by their power to command the attention of others.

Cooperation offers many *benefits*, although like all biological features it also incurs *costs*. It has not been easy for a competitive process to produce cooperation, but evolution has done so, especially through a series of social and moral *emotions*—like all emotions, nature's motivators for particular adaptive behaviors. Nevertheless the spread of cooperation through large communities of mostly unrelated individuals with inevitably competing interests needs the support of numerous evolved adaptations *and* cultural inventions that build on them. Among these are the community pride and coordination possible through shared design, music, dance, and story: tattoos, anthems, legends, and much more.

Creativity matters to us as humans, but may seem difficult to account for in terms of hard biological costs and benefits. Yet like sex, wasteful in that we lose half of our variable genes in each offspring (whereas asexual cloning transfers all genes to the next generation), creativity ultimately benefits us in producing a wider array of behavioral options, some of which will survive better than others under unpredictable selection pressures. Although many artistically creative choices do not lead directly to survival benefits, they strengthen dispositions that do so, through preparing human minds to think beyond the given and to feel confident in controlling aspects of the world on their own terms, and can increase motives for trade and invention.

Storytelling appeals to our social intelligence. It arises out of our intense interest in monitoring one another and out of our evolved capacity to understand one another through *theory of mind*. Our capacity to *comprehend events*, many facets of which we share with other animals, underlies our capacity for story but should not be confused with *narrative*, with *telling* events, an effortful process we undertake only to direct the attention of others to events real or imagined.

Stories, whether true or false, appeal to our interest in others, but fiction can especially appeal by inventing events with an intensity and surprise that fact rarely permits. Fictions foster *cooperation* by engaging and attuning our social and moral emotions and values, and *creativity* by enticing us to think beyond the immediate in the way our minds are most naturally disposed—in terms of social actions.

Explanation

I have proposed four levels of explanation we can focus on for any story or literary work: the universal or specieswide, the features of human nature *represented* or *appealed to* in literature; the local, in time, place, or culture, which may modify human behavior and interests in more or less substantial ways; the individual powers, interests, and experience of storytellers or their audiences; and the particular features of this or that story or audience situation.[4]

At the level of particular stories, the *problem-solution model* allows us to analyze closely the choices we infer behind a work of art.[5] Storytellers' first problem is to capture *attention*. They may choose from among many different kinds of audience, but whichever audiences they seek, they will attempt to appeal through both common human predispositions and the fine-tunings of local culture. Each attempted solution to the problem of holding audience interest will involve different costs and benefits for both storytellers and their audiences. Storytellers will aim to reduce the costs and raise the benefits of their own composition efforts and their audiences' responses.

Nature has shaped us to observe events and to infer information from them, both specific (about this person, situation, or action) and general (about this type of person or the likely outcomes of that sort of situation or action). When we respond to fictions, we need not believe in the actuality of *particular* individuals, situations, or actions in order to infer general *implications*.

We also interpret the actions of others in terms of their intentions. Telling a story is itself a complex action. So long as a storyteller holds our interest, we will infer significance both from the story world—from characters and events—and from the storyteller's intentions in recounting these events in just this way. Both the fictional world's implications for our world and the storyteller's inferred intentions in telling us the story contribute to our sense of its import. But attention precedes meaning, although an emerging intuition of meaning may also feed back into our interest in the story.

PROSPECTS

What are the consequences for literature of seeing storytelling as a human adaptation?

A first consequence should be that we realize that fiction is even more central to human life than we feel from the moment we first engage in pretend play to the moment we finish our last story. Although people continue to write and read literature with pleasure and passion, academic literary study over the last few decades has often felt on the defensive, at least about literature, as if it were a peripheral indulgence justifiable only by making it a stalking-horse for political reform. But if storytelling sharpens our social cognition, prompts us to reconsider human experience, and spurs our creativity in the way that comes most naturally to us, then literary studies need not apologize.

Evocriticism versus Theory

The prevailing mode of literary theory in academe often calls itself simply "Theory," as if theories like those of gravity, evolution, and relativity were nugatory compared with a single subset of literary theory. Capital-T theory has become, as even some of its proponents admit, an initiation ritual, a graduate student boot camp for the intellectual officer corps. It has isolated literary criticism from the rest of modern thought and alienated literary studies even from literature itself.[6]

A biocultural or evolutionary approach to fiction can reverse these trends. Let me call it "evocriticism," until a better term establishes itself or, still better, until all literary criticism accepts that our evolutionary past provides a natural framework for understanding our present and we can dispense with the distinguishing label. Evocriticism lets us link literature with the whole of life, with other human activities and capacities, and their relation to those of other animals, as they compete, cooperate, and play, as they observe, understand,

and empathize with others. It can reconnect literature with the whole range of human experience, and especially with what matters most in literature and life—and not just the features of life closest to animal existence.

Evocriticism can offer a literary *theory* both theoretical and empirical, proposing hypotheses against the full range of what we know of human and other behavior, and testing them. Though compatible with much earlier literary theory and criticism, it will reject some possibilities, such as assumptions of radical disjunction between human minds of different eras or cultures based on a general cultural constructivism or particular "epistemic shifts."

It can connect literature, for so long our best repository of information about human experience, with ongoing research of various kinds that can refine and challenge our understanding of human nature and thought.[7] In this, it stands opposed to Theory, which cuts literature off from life by emphasizing human thought and ideas as the product of only language, convention, and ideology—although Theory then tries to compensate for severing literature from three-dimensional life by insisting that it is always political or ideological. Theory also cuts literature off from research by eschewing empiricism, the testing of ideas against factual evidence.

An evolutionary approach rejects taken-for-granted, locally unquestioned assumptions about human nature. To understand stories at all requires at least an implicit theory of human nature. But most literary criticism has little more: either an intuitive theory of human nature as self-evident, or recently an explicit and untenable theory of it as entirely socially constructed. By adopting a biocultural perspective, literature can both inform and be informed by science by engaging with the inquiry into human nature that a grasp of evolution makes possible.

The Western tradition has customarily linked humans to gods or God, and literature to the inspiration of the divine, to the Muses or God. Daniel Dennett refers to such supernatural explanations as "skyhooks," hoisting the human above the earth by hanging us on to some celestial support.[8] Postmodern thought has tended to define the human in terms of culture or convention, no longer suspending us from the sky but keeping us somewhere above ground, as if culture allowed minds to hover like a head-high fog. A biocultural explanation builds from the ground up, from single-celled organisms to humans. It interprets human *society* partly in light of the many forms of sociality that have evolved, under sometimes similar pressures, throughout the animal world; human *culture* in light of culture in other animals; and human *conventions* in light of the conventionalizing of behavior in animal ritual.

Postmodern thought has been a last bastion of human exceptionalism. Partly through sheer ignorance of culture and conventions in the animal world, Theory has decreed the world of human life to be entirely shaped by culture and convention and therefore distinct from the rest of reality. A biocultural account of the human, by contrast, sees us as having evolved under pressures often similar to those other animals face, but with a heightened capacity for sociality, social learning, and flexible thinking, which all helped, and were in turn helped by, the evolution of language, and which may have allowed us to pass through another major transition in evolution.

Evocriticism makes possible genuine and valid interdisciplinarity, through a connected, coherent, cumulative and relentlessly self-critical body of knowledge, and not the dilettantish smorgasbord (a dash of chaos theory here or Lacanian pseudopsychology there) exposed in the Sokal hoax.[9] It can call on a model of human nature tested *cross-culturally* from hunter-gatherers to modern industrialized societies; tested *comparatively, across species,* within and beyond the primate and the mammalian lines; tested not shallowly but in *real historical depth,* over the millions of years that shaped the human mind and that account for the similarities among people of very different cultures, without neglecting immediate or midrange causal factors; and tested in the *neurophysiological* terms that are now becoming available through brain imaging technology.

An evolutionary approach to literature can encourage literary scholars to learn from the strengths of science without abandoning their own expertise. Science advances by subjecting its hypotheses to hard testing against possible counterevidence. Literary studies can work in the same way.

As we read fiction we generate ideas about the story, about *why* events so far have happened, about what we think will happen, and about why the author has told *this* story in *this* way. We revise our judgments as we read, and as we *re*read we may *re*-revise them, noticing new details incompatible with earlier impressions or new patterns suggesting deeper explanations. This process closely resembles the scientific advancing and testing of hypotheses.[10]

Yet in literary criticism too many still share the naïve conviction that only positive evidence counts (that X had a motive, and access to the murder weapon; that work *W* can be explained as "about" theme *T*), when negative evidence counts for more (that X was elsewhere at the time of the murder, or Y was seen plunging the knife into the victim; that in almost every line of work *W* we can find details with no relation to theme *T*). We learn more when evidence *against* a reading surfaces, since it forces us to account for a richer stock of information. Literary critics can only gain from learning to test their own interpretations, their "readings," against possible counterevidence, from within a text or beyond. They need to abandon the axiom that evidence amassed *for* a conclusion suffices as proof of good faith, never mind how much may count against.

In recent literary and film criticism scholars have often applied to fictional works some subset of Theory, ideas hallowed by their provenance from, say, Derrida, Foucault, Lacan, or Althusser: "one thought-system is taken over as setting the standards by which [authors] should be read. Critics derive their assumptions about language and literature, their methodology (in some cases their renunciation of method), their attitudes to life even, from a law-giving individual or system."[11] Such critics assume that if they can "apply" the theory, if they can read a work in its light, they thereby somehow "prove" it, even if the criteria of application and evidence are loose.

Partly for that reason, I resist the term "literary Darwinism" favored by some evolutionary critics. Darwin's ideas have been far more fertile and valid than those of Marx or Freud, but precisely for that reason they belong to a widespread research program that has in many ways outstripped what Darwin proposed. Evolutionary critics, unlike those who have over the years labeled themselves as Marxist

random

or Freudian, should appeal not to a founding father but to a live and empirically accountable research program. Marxism and Freudianism by contrast remain popular in literary studies despite having been rejected by their home disciplines, and despite their reluctance or refusal to countenance negative evidence and their suspect, self-serving flexibility in their "search" for positive "evidence."

Evolutionary anthropology, biology, economics, and psychology are research programs, not bodies of doctrine. Critic Frank Lentricchia dismissed Theory—which he had championed until he realized how thoroughly it quashed literary pleasure—by saying: "Tell me your theory and I'll tell you in advance what you'll say about any work of literature, especially those you haven't read."[12] Such predictability will not be possible in evolutionary literary criticism. Evocriticism will not replace one grand theory with another, although it will attempt to provide a basis for literary theory. It can investigate works at multiple levels, from the universal to the particular, or specific problems or topics in human life as represented in literature—love, cooperation, sexual difference, life history, family conflict, and much more—or specific ways of engaging audience attention or rousing audience response, in specific traditions or contexts, or within or across periods and genres.[13]

Evidence and "Proof"

Unlike current Theory, evocriticism prefers proposals concrete enough to be subjected to potential falsification by evidence, in both the theories used for literary analysis and the analyses themselves. Aspects of my own proposal for art as adaptation, for instance, would be falsified if repeated engagement in particular arts did not improve the processing of information patterns in the relevant cognitive modes, or if there were on average a negative correlation between individual artistic success and status or between the artistic richness of societies and independent measures of social cohesiveness.

Claims within an evolutionary approach to literature can hardly be "proved" by application to literary works. Yet the fertility of such an approach should be tested both on particular works and on literary theory. If evocriticism cannot generate literary theory and criticism richer than those possible through other methods, it does not warrant the considerable effort of learning enough about human evolution to apply it to literature.

In *On the Origin of Stories* I have discussed only two stories in considerable detail. I chose the *Odyssey* and *Horton Hears a Who!* not because they bear out evolutionary themes in which I have a prior investment, but simply as supremely successful stories as near as I can get to narrative's phylogenic and ontogenic origins. But it is not only narratives near the historical or individual origins of story that can be explained with help from evolution. I had intended to include other highly successful stories, a Shakespeare tragedy *(Hamlet)* and comedy *(Twelfth Night)*, a classic novel *(Pride and Prejudice)*, a modernist novel (Joyce's *Ulysses*), and a postmodernist text (Spiegelman's *Maus*), the last two by authors associated with the avant-garde.[14] Space forced me to defer those additional examples.

I have wanted to show, among other things, that whatever else evocriticism may be, it need not be reductive and can be expansive. It can offer the widest possible explanatory perspective, on, say, the evolution of cooperation (the *universal* level) without neglecting the historical or regional circumstances (the *local* level: the institution of *xenia* in the *Odyssey* or the reintroduction of democracy in Japan in *Horton Hears a Who!*), or the individuality of the author (the *individual* level: problematic in Homer's case, given our uncertainty about his identity, pervasive and intense in Dr. Seuss's), or the *particular* problems and solutions the author faces in *this* work, to secure an audience both by building on the appeal of other works in its tradition and by ensuring the many surprises *this* work offers within that tradition.

An evolutionary approach can be more sensitive than others to artistic detail and design. It can show the problem situations of storytellers aiming to engage the attention of wide audiences, their attempts to minimize costs while maximizing benefits, and the Darwinian process of composition itself. It takes seriously the idea that the complex emerges out of the simple, mostly building slowly, by minute increments, in the design of species (the universal), in cultural tradition (the local), in personal development (the individual), and in artistic composition and comprehension (the particular). It allows us to see art as part of an unbroken series of nested problems and solutions, from species to local to individual to particular impacts and implications.

My discussions of Homer and Dr. Seuss have often drawn on scholarship of a nonevolutionary kind, historical, cultural, biographical, and bibliographical. Evocriticism should not exclude such detailed work, and in fact can offer a theoretical justification for employing it more secure than Theory's recent rejections of the individual and particular.

An evolutionary approach should not seek to subvert commonsensical readings. After all, works of art would fail if their meanings were integral to their overall effect yet had remained unrecognized until the latest critical findings. Occasionally, however, evocriticism will challenge common interpretations in the light of the best empirical information; and it will consistently call into question readings from biologically untenable positions, such as those that insist that reality is only culturally constructed, or that the human mind can be explained by psychology in the speculative Freudian tradition. (Naomi Goldenberg, for instance, reads *The Cat in the Hat,* in Philip Nel's words, "as a masculine attempt to overcome womb envy: the Cat gives birth to little alphabetic cats (invents language) in order to erase the role of the mother (removing a menstrual stain from mother's dress and bed).")[15] Yet I would also stress (with David Bordwell) that meaning is only part of the effect of works of art.[16]

Although it is informed by science, evocriticism does not limit itself to scientific reduction, since art by its nature invites a creative and original response, to some degree subjective and open-ended, an aura of implication rather than an exact transfer of information. Evocriticism can account for the range of responses and appropriations in terms of individual and local variation, a resultant diversity of purposes and interests, and creativity as an adaptive function of art.

A biocultural approach should also be able to account for the existence of the critical tradition. Attention continually shifts in both intensity and direction, and noticing in a new way a single detail or pattern in a work of art may begin to modify one's response to the whole. Even artists who expertly appeal to audiences and monitor their own reactions as they compose cannot predict the fluctuations of attention and response in all possible audiences. Most of these fluctuations will make little difference to overall judgments, but a new recognition of any feature of a work of art *may* transform what we heed in the rest, whether we register it ourselves or see it highlighted by others. Since we are primed for social learning, we are predisposed to learn from the responses of others—around the campfire, in the next seat in the theater, in the review pages, in the library stacks—whenever we engage with stories.

Evocriticism should not be easy, should not impose a template, should not arrive at a priori conclusions. It is a research program, not a set of prefabricated conclusions or all-purpose methods. It is open-ended at both ends, in terms of the evidence in literary works and the evidence about human nature or culture arriving from anthropology, biology, economics, history, psychology, and sociology.

Evocriticism does not dispense with literary expertise, the sensitive response of trained readers. It does not replace expert readers with a standard checklist of evolutionary questions and answers, but opens up to good readers a new range of problems and possibilities.

Critical expertise, too, should be explicable in evolutionary terms, and it is. Since we enjoy sharing attention and since we are predisposed to learn from others, we like to discuss works of art—designed, after all, to maximize sharable attention and response. Since stories, especially, often communicate values important in the community, parents or elders explain fables, stories, or mythologies to the young. But the flexibility of human minds and behaviors allows us to specialize. In a society in which not all need be generalists, those more inclined or able can devote sufficient time and energy to mastering a particular skill so that it becomes, as we say, second nature. Or in neurological terms: the expert processes relevant information rapidly and automatically in a more posterior region of the brain, leaving more room in the most advanced and explicit part of the mind, the neocortex, for higher-level processing, deeper analysis, more novel and searching response.[17]

To become an expert reader, there is no substitute for reading literature widely and intensely, but someone who does so *and* adopts an evolutionary and cognitive perspective can ask new questions about literature and the human nature seen in or stirred by fiction.

ATTENTION

The evolutionary theory of art and fiction I have proposed stresses *attention*. Other evolutionary theories do not. I have several reasons for foregrounding attention.

1. *Attention is important to all art, literature, and story,* as artists and audiences—and Aristotle—have known for millennia. Alternative evolutionary explanations, like narrative as a source of information (Michelle Scalise Sugiyama)[18] or as a source of cognitive order (Joseph Carroll),[19] do not cover all cases. Information can be important in some stories, as even in the *Odyssey,* where the rituals of hospitality or setting sail convey information directly (albeit information likely to be already known to the original audience). So, too, can cognitive order, like Homer's confident overview of the cosmos, the worlds of the immortals, the mortals, and the dead. But the commonest stories, and the fastest to circulate, are jokes. Neither jokes nor children's stories from a poo-pooing dragon to Dr. Seuss's kindly elephant constitute valid sources of information or cognitive order, but they are highly effective at earning attention for different audiences and durations.

2. *Art's effects on human minds depend on its power to compel attention.* Like play, art can reshape minds piecemeal because its high doses of pattern engage us compulsively and over time, over many experiences, alter our capacity to produce and process pattern.

3. *Attention capture can explain the* design *features of stories,* as other explanations cannot.

Information as a key to narrative can account for factual narrative, but not for fiction. Why invent a form of story that makes it an error to infer specific information (that a prince called Hamlet really existed, that Elsinore was the capital of Denmark, that it perched on precipitous clifftops) and a puzzle to know what general information to infer (that ghosts walk and talk, that daughters should resist their fathers' meddling in their love lives, that sons should not berate their mothers)? In a Tolstoy fable the story's implications may be succinctly spelled out, but what *information* do we extract from *Anna Karenina*? Where do we start? Where do we stop?

Cognitive order, vague though it is as an explanation for fiction, fares no better.[20] Even an implied story as brief as "Ralph, come back, it was only a rash" calls on our cognitive capacities to decode, which we can do in a flash. It catches our eye and provokes a sudden explosion of amused emotion, but what cognitive order does it provide? Stories from childhood pretend play, like the chaotic dinosaur and duck examples in Chapter 12, to adult fiction like Donald Barthelme's bewildering "The Indian Uprising"[21] employ our capacity for rapid makeshift cognitive inference but do not supply cognitive order—although *some* stories may.

But the aim of securing attention can explain the design features of stories from jokes or children's pretend play to timeless masterpieces or avant-garde challenges. Because of our ultrasociality, we have a huge appetite for learning about social events. Fiction can compel our attention for hours on end. As I showed in discussing the *Odyssey* and *Horton Hears a Who!,* the large- and small-scale choices that successful stories continually make about genre, character, plot, medium, structure, character contrast, irony, and much else can be explained in terms of attention, in terms of offering enticements—especially emotional enticements—to keep audiences distracted from responding to their immediate surrounds as they half-immerse themselves in the far-fetched world of story.

4. *Attention can explain the social effects of art.* Attention to one another comes so naturally to us all that we rarely think about its biological significance. In social animals in general, and especially in primates, the ability to command attention is closely correlated with status, which indicates priority of access to resources and therefore normally correlates highly with survival and reproductive success.[22]

And as all of us in our singularly social species know, there are few ways of holding others longer than by telling a vivid story. Humans, moreover, have a uniquely intense motivation and capacity to share attention—and this proclivity has recently been proposed as the key factor in the singular development of human intelligence and culture.[23]

The attention-earning power of art can lead not only to a primary market in status for artists but also to a secondary market in acceptance or status for audiences. We may wish to belong to or excel within some audience, whether large—those interested in the pop charts or best-seller book lists or new movie releases—or small and select—those with cult or elite tastes, for 1930s blues music, or Julian Barnes's "Dan Kavanagh" detective fiction, or avant-garde film. We may earn the attention of others through our access, especially when restricted, to artists or their works: a first edition, a signed copy, a manuscript, an original painting, a commissioned performance, an expensive seat at the opera, a night in a rock star's hotel bedroom.

5. *Attention explains artistic kinds.* Artists can seek many different audiences with their art, even within the art of fiction. "Whose attention does it invite?" may be the first question we ask about a story: does it define its audience by age (child, adolescent, adult), by literary level (pulp, popular, literary, experimental), by genre (romance, mystery, thriller, Western, sci-fi, fantasy)? Does it promise immediate, low-effort attention-worthiness, or longer-term and higher-effort engagement, to last over time and repeat rereadings? Is it the sort of book we would discard after reading or keep and return to?

6. *Attention offers a guide to appreciation and evaluation.* An approach to story that focuses on attention will respect the achievement of low-culture fiction that appeals successfully to its chosen audience, although it will also rate differently fiction that can hold interest over many readings and many generations.

7. *Attention can also explain the role of artistic reputation.* In all social species, individuals learn from others, especially from attending to what others attend to, even if it is no more than a threat like

a predator or an opportunity like food. As a species, humans have evolved to be highly disposed to and able to learn from others.[24] If others value something, it may be worthwhile for us, too, unless we know their values to be at odds with ours. If others clamor for *The Da Vinci Code* or the latest Harry Potter book, or have long held Cervantes or Stendhal in high regard, perhaps we should try what they have enjoyed.

In capital-T Theory, critics rarely turn their critiques on themselves. A Theorist claims, in effect: "Meaning is endlessly irresolvable, but I expect you to understand precisely this point when *I* make it"; or "*The author* is an outmoded concept, but please quote *me* on that." In this book I, too, am competing for attention (for a certain kind of audience: for you, dear intelligent educated reader), and I can do so partly by latching on to major works, as Homer himself chose the major story cycle of his time. If I thought I could command audiences for generations by doing so, I might concoct a new story and compete for still wider notice, but by riding on Homer's classic status and Dr. Seuss's perennial popularity, I hope to avoid the fate of novelists who find themselves altogether ignored.

ATTENTION, INTENTION, AND PROBLEMS

Evolutionary considerations strengthen and deepen the recognition, common among artists and audiences though not among academics, that works of art must first catch and hold the interest of audiences. This need presents the first problem situation for artists, although the problem will be further specified according to their intended audiences, their medium, the local expectations within that medium, and the ways in which they wish to direct audiences' responses.

Most of our behavior is routine and quickly decided on. Art is not, to the extent that it is original. A Darwinian approach helps us realize that we rarely reach thoroughly novel outcomes in a single move. Novelty will most likely result from repeated cycles of variation and retention, including, in the case of artists, the accumulation of artistic experience over a tradition or a lifetime that can provide strategies or intuitions to be newly recombined. Within a particular work's composition, another trial-and-error process occurs: an approximate intention modified, expanded, abandoned, redirected, or elaborated as particular elements become concrete and suggest new solutions or new problems.

But intention does not explain all. In evolutionary analysis and in everyday behavior, individual variation matters and can be analyzed apart from particular choices. We (and many other species) distinguish individuals in terms of not only their physical characteristics but also their personality, for the good reason that evidence of others' past behavior makes possible better predictions of their future behavior. While intentions help explain the particular actions people choose, the range of ideas to choose *from* will not be the same for one individual as for others, because of differences in capacity, experience, and inclination.

Evolutionary biology uses sophisticated cost-benefit analysis to explain observed behavior. Despite specieswide traits, individuals differ considerably from one another, genetically, environmentally, and developmentally. Part of their strategy will therefore be to recognize their own powers relative to the powers of others of their kind, and to act to suit their talents and interests: toward one or other of the arts, in certain cases, and toward certain directions within it.

We can see authors as problem-solvers with individual capacities and preferences making strategic choices within particular situations, by shaping different kinds of appeals to the cognitive preferences and expectations of audiences—preferences and expectations shaped at both specieswide and local levels—and balancing the costs against the benefits of authorial effort in composition and audience effort in comprehension and response. Such an approach allows us to make tentative reconstructions of the problem situations authors face as they seek audiences now and in more or less unimaginable futures. Clarifying the primacy (but not necessarily the ultimacy) of attention helps also to reconstruct a major part of an author's problem situation and eliminates or makes less likely many alternative constructions of intent or implication.

Where From? Where To?

Evolutionary literary criticism offers no set questions, let alone set answers. It requires simply a readiness to accept that humans evolved and a commitment to empirical research to find out just what we are and why. Like evolutionary approaches to human life, evolutionary approaches to literature will suggest new questions and be more or less pertinent to old ones.

As children most of us will have occasionally written our address not as the bare minimum, "204 Ferguson Street, Palmerston North," say, but as an expansive maximum, adding something like "North Island, New Zealand, Pacific Ocean, Southern Hemisphere, the World, the Solar System, the Milky Way, the Universe." When we consider a famous textual crux in *King Lear,* a long evolutionary view of human nature will probably be as superfluous as the outer orbits of our childhood address. But when we discuss the play, we *may* find a larger context relevant even when we consider textual detail, like the different kinds of attention Shakespeare aims for in the Quarto and Folio versions. When we examine aspects of the play's impact and implications, we will *very* likely find that a view of human nature within a larger natural world can help explain *King Lear's* power across centuries and countries. We may compare Lear's rage with the fury of an alpha male chimpanzee deposed from dominance, or note the sudden spike in levels of the stress hormone cortisol in animals that suffer loss in rank, or the strength of family attachment *and* intergenerational and intragenerational conflict across species, or the vulnerability that all social animals feel when they lose their place within the shelter of their group. These contexts may do as much to explain *King Lear* as studies of the notion of service or inheritance in 1605, and more to explain the play's visceral universality.

Evocriticism may help to explain the force of a metaphor, the handling of a theme, the depth of psychological representation. It may help to show that claims like the unthinkability of x before time t_1 or the unfeelability of y before time t_2, or other claims that we suppose differ radically from culture to culture or era to era, founder both on the evidence of texts and on cross-cultural and cross-species comparisons. But precisely what questions an evolutionary criticism will open up in what works or traditions depends as much on the imagination of good readers as on the researches of cognitive and evolutionary anthropologists, economists, psychologists, and sociologists. And what answers they will suggest remain to be discovered.

AFTERWORD: EVOLUTION, ART, STORY, PURPOSE

> [Darwinism] seems simple, because you do not at first realize all that
> it involves. But when its whole significance dawns on you, your heart
> sinks into a heap of sand within you. There is a hideous fatalism
> about it, a ghastly and damnable reduction of beauty and intelli-
> gence, of strength and purpose, of honor and aspiration . . .
>
> George Bernard Shaw, Preface to *Back to Methuselah* (1921)

DOES EVOLUTION BY NATURAL SELECTION rob life of a sense
of purpose, as many have feared? I suggest that, on the contrary,
Darwin has made it possible to understand how purpose, like life,
builds from small beginnings, from the ground up. Art, including
the art of storytelling, and science, including the theory of evolution,
have played key roles in the recent expansion of life's purpose.

A recurring element of *On the Origin of Stories* has been the theme
of problems and solutions, and the power of a Darwinian system—a
cycle of generating, testing, and regenerating—to attain highly complex solutions to new problems that cannot be fully formulated, let
alone planned for, in advance. This idea allows us to generalize and
expand on the core concern of *On the Origin of Species* (1859).

There Darwin showed how new species could evolve through a
process of blind variation and selective retention. He transformed at
a stroke our understanding of natural design. Living things manifest
complex design but can be produced by a mindless process that does

no more than passively register, in terms of survival and reproduction, the advantages of particular variations. In *The Blind Watchmaker* (1986) Richard Dawkins explains how nature is, indeed, like a blind watchmaker, and thereby overturns the famous argument of theologian and naturalist William Paley. Paley opened his *Natural Theology—or Evidences of the Existence and Attributes of the Deity Collected from the Appearance of Nature* (1802):

> In crossing a heath, suppose I pitched my foot against a *stone*, and were asked how the stone came to be there; I might possibly answer, that, for anything I knew to the contrary, it had lain there for ever: nor would it perhaps be very easy to show the absurdity of this answer. But suppose I had found a *watch* upon the ground, and it should be inquired how the watch happened to be in that place; I should hardly think of the answer which I had given, that for anything I knew, the watch might always have been there . . . [The precision and intricacy of its mechanism would have forced us to conclude] that the watch must have had a maker: that there must have existed, at some time, and at some place or other, an artificer or artificers, who formed it for the purpose which we find it actually to answer; who comprehended its construction, and designed its use.

Nobody could reasonably disagree with this line of reasoning, Paley adds, yet this is tantamount to what an atheist does, for "every indication of contrivance, every manifestation of design, which existed in the watch, exists in the works of nature; with the difference, on the side of nature, of being greater or more, and that in a degree which exceeds all computation."[1] After reminding us that the complexity of natural organisms surpasses that of the most intricate watch by far more than science could guess in Paley's time, Dawkins goes on to show how simple processes of variation and selective retention can, over many cycles, create products with even this degree of design.

As we have seen, other processes working *within* natural selection have been found to follow the same principle: the human immune system, the synapses in the young human brain, culture, and invention. Such "Darwin machines"—Dawkins uses the term "universal Darwinism"[2]—allow genuine novelty to be achieved without advance knowledge of what will work best in an unpredictable and open world.[3] The common principle of blind variation and selective retention allows for a deeply indeterministic process that explores patches of possibility space in multiple directions and pursues any direction that is provisionally more promising than others. It tracks through the vastness of the possible in ways that surprisingly often lead to rich solutions, as it compounds immediate advantages and retains achieved complexity into the next round of variations. Such Darwinian processes occur perhaps anywhere we find deeply original novelty.

Darwin's explanation of evolution by natural selection shocked, and still shocks, because it appears to deny purpose. We think of purpose as prior to decision and action: I want to raise my arm, and unless I am paralyzed or restrained, I do. But in fact purpose emerges slowly, phylogenically and ontogenically. My capacity to move my arm in as many ways as I can depends on things like the evolution of forelegs into arms early in the primate line, the evolution millions of years later of a rotating socket in the shoulder of great apes to enable arboreal brachiation, and the further freeing up of arm movements after early hominids became fully bipedal. Babies flail their arms uncontrolledly and purposelessly for months before they can direct them in a particular way for a particular purpose.

Paley's example of the watch assumes a purpose we already understand: the intricate integration of material objects into an instrument for telling the time. But humans did not become capable of constructing multipart mechanisms until within the last 50,000 years or so. Until their manual control reached a high level and their stone-flaking attained the precision of the Levallois technique, they would have been unable to conceive of such mechanisms or, if confronted with them, to recognize anything of their construction or purposes. The idea of telling the time precisely would have been unknown and meaningless to our human ancestors even at the end of the Stone Age.

Purposes can emerge only piecemeal; problems cannot even define themselves until many of the elements are already in place. The position of the sun in its daily sweep can indicate the phases of the day, but nothing more precise. Sundials and the sticks in the ground that preceded them afford more finely determined divisions of time during daylight. These in turn made it possible to imagine coordinating common actions in advance. The first water- and sand-clocks enabled even tighter coordination. Mechanical clocks and bells to chime the hour or even the quarter-hour took social synchronization still further. Not until European navigation and mapping flourished in the fifteenth century did anyone consider a portable clock to ascertain longitude, yet maritime clocks still remained highly inaccurate for another three centuries. The first watches, in the early sixteenth century, could register only the hour, and to tell the time down to the minute on a portable mechanism took more than another century. By Paley's day, the recent invention of the duplex escapement made it possible to keep time accurately on a pocket watch, but only over the course of the nineteenth century did highly accurate timing allow new degrees of precision measurement and new research options in physics and psychology. Not until well into the twentieth century did cesium clocks attain the exactitude and reliability of timekeeping needed for quantum physics or space flight.

As instruments and standards for measuring finer time intervals developed, new purposes have been discovered, each largely inconceivable a stage or two previously. Purposes arise not in advance, but as possibilities materialize. Of course, when the purpose becomes established, it can *then* be implemented ahead of time: I can choose to move my arm in such a way as to put on a sweater, or measure mental responses down to the millisecond in a psychology laboratory to assess the complexity of the neural processing involved. But each of these purposes, although it precedes a definite action, has a long history that precedes *it,* in the species, the culture, and the individual, a history of prior trials and prior errors, before the purpose could be conceived and fully defined, let alone specified in advance.

Life could become established only once matter organized itself in a way complex enough to sustain and reliably reproduce itself. Maintaining such a highly improbable and functional arrangement of matter became life's first purpose. As species continued to evolve, so did the purposes of their organs and behaviors. New behaviors, like new organs, begin uncertainly, with slight modifications of existing structure, but become defined over time, and their function or purpose clearly specifiable in advance: a certain kind of spider, for instance, will spin a certain kind of web to catch a certain range of insect prey under certain conditions.

As creatures began to act in more complex and flexible ways, nature evolved emotions to motivate better decisions. Satisfying these emotions—escaping fear, appeasing hunger, fulfilling desire, sustaining love, and so on—became important purposes in themselves for much of the animal kingdom. As behaviors standardize, as purposes define themselves, social animals can learn to understand not only their conspecifics' actions but even their desires and intentions before they act. We not only learn to infer others' intentions, but in social species that benefit from cooperation we also evolve to empathize with or emotionally react against others' purposes. And without this capacity, stories would be impossible.

stories emerge from cooperation

Yet we should not forget that despite our thinking of purposes as prior to actions, they have emerged over long stretches of evolutionary and experiential time. Intentions are efficient routes to objectives clearly defined only after many preliminary stages of variation and selection within animals' collective and individual pasts. Like design, purpose *emerges* rather than precedes.

like we have evolved abilities to walk, we have evolved purposes

Purposes evolve, and Darwinian processes extend them. Intelligence, cooperation, and creativity are all purposes that have emerged in the course of life on earth. Stephen Jay Gould famously argued that if we could rewind and replay the tape of evolution, humans and human intelligence would not reappear.[4] Probably not; no one disputes that contingency strongly inflects evolution. But as Simon Conway Morris has stressed, certain capacities have evolved again and again, because of the singular advantages they offer: senses, locomotion, minds, emotions, sociality, intelligence, cooperation, to name those that concern *us* most.[5] Let us start late in this lineup.

Intelligence allows us to respond flexibly to circumstances, to solve problems not only according to successful old routines (prior purposes, if you like) but also in novel and context-sensitive ways. Because it can sometimes find new solutions, intelligence is highly advantageous, yet hard to evolve. Although mind has been necessary for all motile creatures, more advanced intelligence has emerged in relatively few lineages, though quite diverse ones: invertebrates like octopi and cuttlefish, vertebrates like crows and parrots among birds, and cetaceans and primates among mammals.

Intelligence has large benefits but also incurs costs. Out of the pressure to develop social intelligence, humans have increased in self-awareness, so that we can model ourselves as others see us in competitive and cooperative scenarios. That greater self-awareness offers real advantages in anticipating others' actions and reactions, but it also carries costs, including the ability to envisage our own death and absence from the ongoing world. For humans this awareness has raised the question of our purpose in the face of our ultimate lifelessness, one we have answered most frequently by imagining that we continue in some form after death. To judge from grave rituals dating back at least 70,000 years, and from evidence of the fear of death and the hope of immortality in the records of early civilizations, the preoccupation with death has loomed large ever since the appearance of a distinctly human culture.

Cooperation allows individuals to gain access to resources they could not obtain by themselves. It, too, has large benefits, and has evolved in many species. It, too, has been difficult for natural selection to evolve, but nevertheless it *has* developed through a whole series of partial solutions like mutualism, inclusive fitness, reciprocal altruism, and the emotions that motivate such behavior (trust, gratitude, shame, indignation, and the like) and the cultural complements that make it still more robust (social monitoring and punishment, and even the punishment of nonpunishers).

Creativity is the capacity to produce significant and valuable novelty. This seems the most difficult capacity of all for evolution to evolve, and for good reason. Significant and valuable by what criteria?

Human creativity matters for human beings, but creativity hardly matters for evolution. Single-celled organisms reproduce themselves readily, and life would go on—did go on, for billions of years on Earth—with barely more complexity. Life persists through reproduction, through transmitting accumulated complexity to subsequent generations. If inherited design were radically changed each time an organism reproduced, the hard-won gains of natural selection would rapidly be lost. Life can evolve new possibilities only slowly, through variations small enough not to threaten existing functions. But although evolution has thereby spawned many new species and even major new forms of life, it does not *need* or *pursue* creativity.

Yet organisms vary, even if only through imperfections in reproduction, and conditions change. Over enough time conditions will always alter, including competition with other organisms in the environment. Since any organism can become a source of energy for others, each has to find ways of exploiting others more efficiently and to avoid being exploited by them, including predators, parasites, and pathogens. In species with a wide range of variations, some individuals will be able to exploit changing opportunities or avoid changing threats more effectively than others. Variation in itself— enough to gain advantage, but not so large as to imperil existing design—therefore offers a measure of security against unpredictable circumstances. For this reason, sex has evolved many times: by recombining genes unpredictably but reliably, it generates a range of initially viable variations from which conditions will select.[6] Some species even toggle between reproducing sexually or asexually according to the degree of environmental instability.

Just as natural selection has evolved sex as a means for amplifying genetic variation, I suggest, it has evolved art in humans—first as a means of sharpening minds eager for pattern, but gradually also for creativity, for amplifying the variety of our behavior. Gene-culture coevolution has reconfigured art for creativity.

The immune system and the infant brain, as we have seen, naturally overproduce options to cope with as many unpredictables as they can, then pare back whatever is not activated by experience, and regenerate from whatever *has* been stimulated by experience. These second-order Darwinian processes allow an extra level of flexibility beyond first-order genetic variation, a still more sensitive adjustment to even shorter-term unpredictability. Art in turn has become a subsidiary Darwin machine that generates not natural but unnatural variations or options. By "unnatural" I mean only that art involves highly deliberate human choices, both individual and cultural, even if our choices derive ultimately from nature.

Creativity as a principle, as a Darwinian process, solves no particular prespecifiable problem, but offers another way of generating new possibilities that *may* prove to solve problems, even significant ones, provided there is a consistent pressure toward a solution—whether over generations, as in natural selection, over weeks or months or years, as a storyteller, say, posits and revises a story, or in only minutes or seconds in the spreading neural activation in a poet or a scientist seeking a new image or idea, in a mind prepared over many years, by many trials.

New evolutionary solutions often themselves spawn new problems. When our brains allowed us to become superpredators, to dominate our environments and earn the food we needed in much less time than our waking hours, we did not solve the "problem" of spare time, as did other top predators, lions, tigers, or bears, by sleeping the extra hours away to conserve energy. Even at rest our large brains consume a high proportion of our energy, and since they offer us most of our advantages against other species and other individuals, we benefit not from resting them as much as possible but from developing them in times of security and leisure. The cognitive play of art could engage us in a self-rewarding way that entices us to improve our mental processing power in the kinds of information that matter most to us.

But if art solves the problem of maximizing the advantages of the human brain in situations free of other pressures, it also sets up a Darwinian system for producing variation—and does so very efficiently. Because art appeals to our own cognitive preferences, we have built-in incentives to *generate* art: its effects should be pleasing in themselves. Since art's criteria for success are human preferences, since the testing mechanisms come preinstalled in our minds as we sing, or tell a story, or dance, or daub, we can readily adjust our actions to produce more satisfying effects: we can easily *select* from what we do, as we do it, and try out new variations, or stop when interest flags.

In most societies art has been collective and active, and even in modern cities dance and song often still remain so. Where art tends to be more individual than communal, those with talent enough to spark the interest of others have a strong extra incentive to develop their skill for the attention, admiration, and status it can earn them. Although professional artists may not have appeared until agriculture and permanent settlement allowed resources to accumulate and labor to specialize, the quality of some of the earliest art suggests that *some* individuals long before agriculture had the luxury of developing singular skills. Creative concentration and feedback during composition could work like a speeded-up version of natural selection as these artists rapidly generated, discarded, and regenerated new variations.

Even in societies in which art has become individualized and professionalized, it remains highly social. Art not only activates our private preferences but also adjusts and amplifies them through our sociality. From the first, mothers and others engage with infants in multimodal social play involving fine-grained attunement and interaction. We instinctively make learning enjoyable for children by making it social, by making it play, and by making it art, by appealing to the cognitive preferences that art animates. Throughout life, participation in artistic activities in group settings, whether actively, in performance, or more or less passively, as audiences, continues to amplify the emotional charge of art.

Art's social nature can not only help motivate our participation but also provide ready-made models that reduce the costs of invention and increase the benefits of response. Nursery rhymes, for instance, appeal to children's interest in pattern, extend their capacity to play with rhythm and rhyme, and may prompt them to subvert or refresh an existing ditty with a new word, intonation, or elaboration, or even to invent new chants or taunts of their own. Tribal arts like weaving, classic forms like the sonnet, and modern arts like filmmaking all depend on the existence of shared norms to provide prompts and challenges. As David Bordwell observes of film: "Norms help unambitious filmmakers attain competence, but they challenge gifted ones to excel. By understanding these norms we can better appreciate skill, daring, and emotional power on those rare occasions when we meet them."[7]

Art can engender variations through other factors present else-where in nature—like randomness, "an intrinsic part of brain func-tion" and "nature's way of exploring unforeseen possibilities,"[8] or copying errors—and through distinctively human factors. Art does not need to start from scratch but can recombine elements already developed in the same or a different art or tradition. Just as sex, by recombining genes, and hybridization, by recombining lineages that have had time to separate, can engender novel forms, art, too, can readily recombine, from the animal-human blends in cave art to the Minotaurs of the ancient Greeks or the modernist Picasso.

In *any* species attention diminishes with persistence or reiteration, but humans are especially curious. And as we have seen, attention, especially shared or commonly focused attention, has become un-precedentedly important in our species. To attract attention art ex-plores variation, even in traditional societies, and all the more so in societies in which professional art and a highly competitive market for attention act as incentives to discover new variations within ex-isting forms or entirely novel combinations.

Art is *uniquely* well designed as a Darwinian system for producing new variations. It is not well designed to generate *useful* ideas. The markings on faces, costumes, or containers in tribal societies, early empires, or modern cities serve no immediate practical function, only the function of appealing to human preferences—but they may make individuals to whom they belong more striking, or those who create them more admired.[9] They may have consequences in terms of sexual or general social selection (an individual effect); they may be invested with social significance, in terms of group identity or re-ligious observance, and thereby promote cooperation (a group-level effect); and they may inspire new efforts in processes like ceramics or glassmaking or weaving, and thereby boost industries and trades, from silk manufacture to video games (a multileveled indirect ef-fect).

Art offers a system of "unnatural" variation. It can produce sig-nificant and valuable novelty, often by the slow accumulation of small variations, like natural selection itself, though much faster, over generations or even lifetimes; sometimes, especially in societies favoring novelty and complex enough to permit dedicated specialization, by intense individual efforts (Joseph Haydn and the string quartet, Louis Armstrong and the jazz solo, Wu Man and the pipa). The selection occurs at two levels: artists judge their own new variations and combinations, selecting and recombining them until the result satisfies; audiences then select them according to their own preferences, either awarding them attention or not, perhaps even preserving a favored few by remembering, retelling, repeating, or reworking them.

Art may produce results not directly *useful* except in terms of mattering to other humans, but in a species as highly social as ours this itself makes a difference. Our extremely large-scale sociality seems inconceivable without the stimulus and relaxation that art provides. And art benefits individuals both in the creative play of childhood and in raising adult creativity, in artistic traditions and models we can readily adopt or adapt, in encouraging us not just to accept our world but to remake it to suit our preferences.

Creativity has become increasingly a purpose of art. It seems not to have been so, at first, if we judge by the comparatively slow development and proliferation of styles in traditional societies. Nevertheless we can recognize awesome creativity already evident in work like the Mas d'Azil spearthrower from the Upper Paleolithic (14,000 B.C.E.?) or Nok sculptures from almost two and a half millennia ago in what is now Nigeria. The *deliberate* pursuit of novelty and originality has further intensified in highly differentiated societies around the world, and has become still more insistently the norm over the last two centuries. A work like the *Odyssey,* although it celebrates the inventiveness of an Odysseus or a Hephaestus, asserts the authority of tradition. Dr. Seuss represents his more modern age in embodying, celebrating, and promoting originality on every page.

Not that the conscious pursuit of novelty ensures successful creativity, even in solely artistic terms. Thousands of galleries open in hundreds of cities every week, but few will offer new works that posterity "selects" as worthy of continued attention. Yet here art operates like any other Darwinian process. Many variations are generated—in this case, thousands or millions of attempts to produce new variations on existing visual art models—but far fewer will survive selection. But the very strength of selection, even if only through human

Nok sculpture
(Nigeria), ca.
400 B.C.E.

preferences, produces some remarkable results, just as consistent selection pressures in nature lead to extraordinary new species and adaptations. In art the pressure to earn attention as few other things can do produces some of the triumphs of the human spirit.

And the process of generation itself spawns new ideas. Although these may individually be small-scale, they may feed not only into major new artistic developments but also into new styles, tastes, and products. When settled existence and agricultural surpluses allowed the first accumulations of wealth, artisans or artists who could attract the attention of the wealthy could raise their standards of invention and execution, and thereby provide an impetus for new processes that could produce earth, metal, or glass vessels in new forms, or with new durability, luster, or malleability. Highly crafted small-scale wares offered a first impetus to long-distance trade, and today new forms of computer graphics, music sharing, and narrative organization help drive the development of computer software and hardware.

Evolution does not *aim* at creativity. It aims at nothing. But purpose ⟩ emerges, and creativity as a purpose has emerged more and more ⟨ distinctly in human life, through the arts' appeal to our innate and acquired preferences, and through the essential part that generating and responding to art, including stories, plays in our development and interests as we mature. Of course there will be wide variations in individuals' inclinations to art, and for all a trade-off between cost and benefit, effort and reward. For many, whether in a compound of Balinese carvers or a Canadian editor's condominium, the trade-off may involve relaxing with a television soap opera. For others, it may require the intense creative struggle, the arduous process of trial and error, of bold variation and selective retention, of inventive art.

Not a well-defined problem for natural selection, creativity has not been easy to evolve. But it offers another solution to the open-ended problem of unpredictability. Stories in particular foster our ability to multiply options and imagine possible actions in the face of any eventuality.

If art is "unnatural" variation, science is "unnatural" selection. Art appeals to our species preferences and our intuitions, often as they have been modified by local culture. Science rejects our species preferences and our intuitions, *even as* modified by local culture. It tests ideas not against human preferences but against a resistant world, and its methods of testing, by logic, observation, and experiment, encourage us to reject even ideas that seem self-evident and apparently repeatedly confirmed by tradition.

Because science exposes itself to falsifying evidence, in the long run it retains and builds on only the most rigorously selected ideas. This winnowing strategy does not prove all science's ideas correct, but it improves on the ratio of tested to untested ideas attainable by any other procedure.

Art could evolve as an adaptation because it appealed to our deep-grained species preferences. Science could not. It appeals to one strong human inclination, our curiosity, but otherwise goes against the grain of our intuitions. All assumed, with Aristotle, and until Galileo, that a heavy object fell more rapidly than a lighter one. Information gathering, invaluable for all kinds of animals and even for plants, has mattered especially for humans, but the knowledge gained has mostly been in the form of heuristics, partly right, but not necessarily so, like our hunches about falling objects or the sun's motion around the Earth. And although accurate information is invaluable, indecision is fatal, and no organism can always afford time to search for correct information. It was not possible to devote effort to a time-consuming, difficult-to-imagine, and increasingly resource-expensive process of testing ideas until in Renaissance Italy the right conditions happened to converge: a considerable buffer of security and overproduction, opportunities for intense specialization, and the availability of information and conflicting explanations that the printing press made possible.

Science still calls for qualities that are unnatural.[10] Information sponges that they are, children soak up what they need in order to understand the basics of their world or their language. They need not be *taught* how to speak or pretend play. But they do need slow formal instruction to read, write, or calculate, and they need even more training, and the help of externalized information (books, diagrams, models), to master the knowledge on which science builds. If they undergo the intensive training required to become scientists, they will still need imagination to find new ways of testing or reexplaining received knowledge. Even for those with training, looking for potential refutations of cherished ideas is both emotionally difficult and imaginatively draining, especially since we have an inbuilt confirmation bias. And whereas art appeals to human preferences, science has to account for a world not built to suit human tastes or talents.

Unnatural selection though it is, science allows for the accumulation of advantageous variations and the rapid "evolution" of complex intellectual and technological design. Art works very differently. Many of its unnatural variations may not be intrinsically deeply valuable, even if they aim to appeal to human preferences. But some art *is* deeply valuable, and speaks profoundly to many, over long stretches of time or life, and across many cultures. Because art is primarily a process of variation—although artists and audiences also select—

412

there is not the same ratcheted accumulation of better design that we meet in science. Hence art of thousands of years ago, like Homer's or the Nok sculptors', can be superior in many ways, as examples of creativity, to most works generated now, simply because Homer or the Nok craftsmen could appeal to preferences that they understood deeply and that have not changed radically since their day.

Art can produce major examples of creativity, as can science, but in a different and complementary way. It produces "unnatural" variation in us all, as we clap, dance, and sing, or as we engage in pretend play, with our dinosaurs and ducks, our monkeys falling down sky. It makes us realize that we can vary what we are given, that we need not accept things as they are, and that we can enter a new possibility space, including a counterfactual one, and populate it as we choose. True, our predilection for "unnatural" variation will make us much more likely to invent a melody or a metaphor than a cure for cancer, which depends on an external reality we cannot easily discover rather than on human predispositions we cannot help feeling. But precisely because art appeals to us, it offers something we can share and have evolved to create and consume.

Religion partakes of elements of both art and science. It could not have begun without the understanding of false belief—our awareness that we may not have all we need to understand a situation—that has amplified our curiosity and spurred us to science. Nor could it have begun without the capacity for story that grew out of our theory of mind and our first inclinations to art, like chant and bodily decoration. Storytelling launched a thousand tales. The tales most often told not only involved agents with memorably exceptional powers but also helped to solve problems of cooperation by suggesting that we are continually watched over by spirits who monitor our deeds and punish or reward them.[11]

At the same time that religious stories helped to solve problems of securing attention and fostering cooperation, they could also allay the unease that arose from our awareness of false belief. The social intelligence from which our grasp of false belief emerged also allowed us to imagine being dead and to foresee the world without us. It brought with it a new anxiety about the possible purposelessness of our lives, although this could be allayed to some extent by stories of spirits guaranteeing purpose *prior* to human life or promising continued existence afterward.

Religion and power commandeered art, not entirely, but substantially, for millennia. Not that art as play did not persist, between parent and child, or among children, or among adults letting off steam. But where they could, religion and power appropriated to their own ends art's ability to appeal to human imaginations.

Only when science began to offer alternative naturalistic explanations of the world did religion and art start to split right apart. When science offered a detailed explanation of natural design without the need for a designer—the theory of evolution by natural selection—*that*, more than any other single idea, stripped us of a world made comfortable by a sense of purpose apparently underwritten by beings greater than us.

Nevertheless if we develop Darwin's insight, we can see the emergence of purpose, as of life itself, by small degrees, not from above, but by small increments, from below. The first purpose was the organization of matter in ways complex enough to sustain and replicate itself—the establishment, in other words, of life, or, in still other terms, of problems and solutions. With life emerged the first purpose, the first problem, to preserve at least the improbable complexity already arrived at, and to find new ways of resisting damage and loss. As life proliferated, variety offered new hedges against loss in the face of unpredictable circumstances, and even new ways of evolving variety, like sex. Still richer purposes emerged with emotions, intelligence, and cooperation, and most recently with creativity itself, pursued naturally, and unnaturally, through human invention, in art, and pursued unnaturally, through challenging what we have inherited, in science.

Art at its best offers us the durability that became life's first purpose, the variety that became its second, the appeal to the intelligence and the social emotions that took so much longer to evolve, and the creativity that keeps adding new possibilities, including religion and science. We do not know a purpose guaranteed from outside life, but we can add enormously to the creativity of life. We do not know what other purposes life may eventually generate, but creativity offers us our best chance of reaching them.

NOTES

BIBLIOGRAPHY

INDEX

NOTES

Epigraph: Nabokov 1962: 289 (Kinbote).

Introduction

1. For general refutations of the charge of reductionism in evolutionary approaches to the human, see, for instance, A. Campbell 2002: 16; and, in evolutionary approaches to literature, J. Carroll 1995: 223 and Gottschall 2003: 262. Rather than arguing in general terms, I try here to demonstrate through particular cases the nonreductiveness of evolutionary explanations of literature.

2. See, for instance, in psychology: Barkow, Cosmides, and Tooby 1992; Pinker 1997, 2002; Buss 2005; in anthropology: D. Brown 1991; Sperber 1996; Cronk 1999; in sociology: Sidanius and Pratto 1999; in economics: Frank 1988; Fehr and Henrich 2003; in history: Smail 2007; in linguistics: Pinker 1995, 1999, 2007a; in law: Law 2004; O. Jones and Goldsmith 2005; in politics: Axelrod 1990, 1997; Thayer 2004; in religion: Boyer 2001; Atran 2002; D. S. Wilson 2002; J. Barrett 2004; Dennett 2006.

3. Bonner 1980; Dehaene 1997; Hauser 2000; de Waal 2001a.

4. Among recent writings on evolution and literature, see especially J. Carroll 1995, 2004; Storey 1996; B. Boyd 2001; Gottschall and Wilson 2005; Nordlund 2007; Gottschall 2008a, 2008b; B. Boyd, Carroll, and Gottschall in press.

5. See, for instance, Critical Inquiry 2004; Patai and Corral 2005; B. Boyd 2006b.

6. Gould and Lewontin 1979; see also Segerstråle 2000; A. Brown 1999.

7. Fears misplaced: Radcliffe Richards 2000; Pinker 2002; basis of free-

417

dom: Cosmides and Tooby 2000; basis of culture: Tooby and Cosmides 1992; D. S. Wilson 2005.

8. Marten et al. 1996; White 2001.

9. Fossey 1983: 184–185.

10. Jolly 1999: 290–291.

11. Hansen 1999: 353.

12. Chauvet, Deschamps, and Hillaire 1996.

13. D. Morris 1962; Lenain 1997.

14. See Schank and Childers 1984; Schank 1995; Duchan, Bruder, and Hewitt 1995.

Book I. Evolution, Art, and Fiction

Epigraphs: Thompson 1942; Darwin [1859] 2003: 397.

1. Tooby and DeVore 1987.

2. D. Campbell 1983.

Part 1. Evolution and Nature

Epigraph: Appiah 2006a: xi.

Chapter 1. Evolution and *Human* Nature?

1. Headlam Wells 2005 ably and amply refutes recent critical claims that Shakespeare himself had no concept of or interest in human nature, that Shakespeare himself was antiessentialist.

2. Midgley 1995: 166.

3. McGrew 2003: 420.

4. See the Introduction, note 2.

5. Gould 1981.

6. Lewontin 1972; Cavalli-Sforza, Menozzi, and Piazza 1994; Dawkins 2004: 331–343; A. W. F. Edwards 2003.

7. Cosmides and Tooby 1998, echoing the postulate originally proposed by the nineteenth-century German ethnographer Adolf Bastian.

8. Ridley 2003: 20.

9. Barthes 1972a: 9.

10. For a recent example, see Menand 2005 and the reply in B. Boyd 2006b.

11. D. Brown 1991; Tooby and Cosmides 1992; Cronk 1999; Cronk, Chagnon, and Irons 2002; Barkow 2006.

12. Fanaai Le Tagaloa, quoted in Cronk 1999: 36.

13. Appiah 1992: 42.

14. Bordwell 2004: x.

15. de Waal 2001a; Gaulin and McBurney 2001; Cronk 1999; Nordlund 2007.

16. Gaulin and McBurney 2001: 9.

17. Ridley 2003.

18. Sapolsky 2006.

19. Iacoboni 2008: 175.

20. Dawkins 1985: 59.

21. Ridley 2003: 247.

22. Dawkins 1989b: 332, 331.

23. Radcliffe Richards 2000: 129.

24. M. Wilson, Daly, and Scheib 1997: 437.

25. D. S. Wilson 2005, 2007a.

26. D. Brown 1991: 147–149; J. Carroll 1995: 270–271.

27. "...without a word of the book itself needing to be changed" (Dawkins 2004: 158).

28. Pinker 1995: 427; see also Dissanayake 1988: 28.

29. D. S. Wilson 2007b: 16.

30. Radcliffe Richards 2000: 121.

31. Singer 1999; Radcliffe Richards 2000.

32. O. Jones and Goldsmith 2005.

33. Boehm 1993, 1999; Tomasello et al. 2005; D. S. Wilson 2007a.

34. D. S. Wilson and E. O. Wilson 2007.

35. Maynard Smith and Szathmáry 1995; Tomasello et al. 2005; D. S. Wilson 2007a; D. S. Wilson and E. O. Wilson 2007.

36. This kind of critique often calls itself antifoundationalism (a rejection of claims to a firm foundation for knowledge): see, e.g., Menand 2005. For a critique of the limitations of the kind of antifoundationalism common in modern literary studies, with a defense of a more valid antifoundationalism, common in modern science and almost a

necessary consequence of accepting human evolution, see B. Boyd 2006b.

37. Keesing 1981, quoted in Eibl-Eibesfeldt 1988: 62.

38. See Bordwell 1985: 104–110 for a lucid critique of the notion that visual perspective reflects the ideology of the unified subject.

39. Cf. Gombrich 1982: 282: "In fact the widespread view has recently been challenged that the conventional elements in photographs bar naïve subjects such as unsophisticated tribesmen from reading them."

40. There are minor individual differences, such as color blindness in 8 percent of the population and, recent research suggests, a fraction of women who may have partially tetrachromatic vision: Hollingham 2004.

CHAPTER 2. EVOLUTION, ADAPTATION, AND ADAPTED MINDS

1. Nordlund 2007: 32.

2. Dawkins 1996: xiii.

3. Fromm 2003: 317.

4. Dawkins 1997: 66.

5. Dawkins 1997: 2.

6. Dawkins 1997: 148–154, reporting on Nilsson and Pelger 1994.

7. Gould offered the strongest critique of what he called the "adaptationist program," but no one disputes the role of other factors than adaptation, and Gould never showed or even tried to show how complex design could be the result of chance. See J. Carroll 2004: 227–244.

8. Atran 2002: 22.

9. Cain 1964: 51, quoted in Dawkins 1989a: 31.

10. See the Introduction, note 2; and Sosis 2000, 2004; Irons 2001. For an evolutionary view that religion is not adaptive but a byproduct, see Kirkpatrick 2005.

11. For other accounts of art as adapted, see Dissanayake 1988, 1995, 2000; G. Miller 2000b; Coe 2003. For literature as adaptive, see J. Carroll 1995, 2004; Scalise Sugiyama 2001a, 2001b, 2004, 2005.

12. Dawkins 1989a: 31.

13. Dissanayake 2000: 153.

14. G. Williams 1966: 16.
15. Pinker 1997: 525 and 2007b.
16. Daisies: A. Campbell 2002: 29; elephants: Dennett 1996a: 276.
17. Atran 2002: 33.
18. Radcliffe Richards 2000: 1; Richerson and Boyd 2005.
19. G. Williams 1992: 77.
20. Hernadi 2001: 62; Špinka, Newberry, and Bekoff 2001: 163.
21. Deacon 1997: 351.
22. Baron-Cohen 1995; Kobayashi and Koshima 2001.
23. Atran 2002: 21.
24. Aiello and Wheeler 1995; G. Miller 2000b: 134: 15 percent of our oxygen intake, 25 percent of our metabolic energy, 40 percent of our blood glucose.
25. Plotkin 1997: 173.
26. Plotkin 1997: 174–175.
27. Meltzoff and Moore 1995; Meltzoff 1996.
28. de Waal 2001b, 2002.
29. Tinbergen 1963.
30. de Waal 2002: 187, 190.
31. Pinker 2005; de Waal 2002: 187; D. S. Wilson, Near, and Miller 1996: 285.

Chapter 3. The Evolution of Intelligence

1. For an accessible general overview, see Pinker 1997; for technical details, see Buss 2005.
2. Lorenz 1978; Pinker 2002: 40–41.
3. For summaries of the controversies, see A. Brown 1999; Segerstråle 2000; Buller 2005; and the replies to Buller noted by Cosmides and Tooby at http://www.psych.ucsb.edu/research/cep/buller.htm.
4. Gould and Lewontin 1979; Gould 1991.
5. See especially Barkow, Cosmides, and Tooby 1992.
6. Symons 1992: 142.
7. Geary 2005.
8. Fodor 1983.
9. See also Bonner 1980 on flexibility.

10. Geary 2005: 10. As Geary notes, he refines here the model proposed by Rochel Gelman.
11. Geary 2005: 132.
12. Humphrey 1976; Alexander 1989; Dunbar 1998a; de Waal and Tyack 2003.
13. Alexander 1989; Geary 2005: 60–66; Flinn, Geary, and Ward 2005.
14. Geary 2005: 105.
15. Geary 2005: 230–233.
16. Geary 2005: 232.
17. Baddeley and Hitch 1974; Baddeley and Logie 1999; Baddeley 2000.
18. Geary 2005: 170.
19. Cf. Bonner 1980: 147.
20. Geary 2005: 184, 216–217, 272.
21. Similar conclusions have been reached by Tooby and Cosmides 2001.

Chapter 4. The Evolution of Cooperation

1. W. D. Hamilton 1963, 1964; G. Williams 1966.
2. Radcliffe Richards 2000: 174.
3. D. S. Wilson and E. O. Wilson 2007: 335 and passim.
4. D. S. Wilson and E. O. Wilson 2007: 338.
5. Maynard Smith and Szathmáry 1995; D. S. Wilson and E. O. Wilson 2007: 339–344.
6. Tomasello et al. 2005; D. S. Wilson and E. O. Wilson 2007.
7. Richerson and Boyd 2005; D. S. Wilson 2005.
8. Pinker 1997: 428.
9. See Dugatkin 2002 for an overview.
10. Maynard Smith 1982a: 2 notes that this was often overlooked in early sociobiological accounts of cooperation.
11. Butler 1954; E. O. Wilson 1975/2000: 7.
12. Barash and Barash 2004; Leary and Downs 1995: 128: "human beings appear to possess a fundamental motive to avoid exclusion from important social groups."
13. Hamilton 1963; Dawkins 1989a: 80; Gaulin and McBurney 2001: 320. Nevertheless multilevel selection theory reinterprets Hamilton, as Hamilton himself came to accept: "When Hamilton reformulated his

theory in terms of the Price equation, he realized that altruistic traits are selectively disadvantageous within kin-groups and evolve only because kin-groups with more altruists differentially contribute to the gene pool" (D. S. Wilson and E. O. Wilson 2007: 335).

14. Wright 1996: 190.

15. Reeve 1998: 48: genetic "relatedness is not a measure of the absolute genetic similarity between two individuals, but of the degree to which this similarity exceeds the background similarity between randomly chosen individuals from the population."

16. Trivers 1972, reprinted with discussion in Trivers 2002a: 123–153.

17. See Wright 2000 for the implications of non-zero-sum games in human life.

18. Singer 1999: 47–48.

19. Trivers 1971, reprinted with discussion in Trivers 2002a: 3–55.

20. Trivers 1971; Axelrod and Hamilton 1981; Axelrod 1990.

21. Trivers 1971, 1972, 2002a; Frank 1988; cf. Pinker 1997: 404–405.

22. Pinker 2002: 53.

23. Wilkinson 1984.

24. de Waal 2001a: 366n17.

25. Packer 1977.

26. de Waal 1998: 203.

27. Axelrod 1990.

28. D. S. Wilson 2002: 193.

29. Wason 1966; Cosmides 1989; Cosmides and Tooby 1992.

30. Cosmides and Tooby 1995: 90.

31. Cosmides and Tooby 1989: 91.

32. Cosmides and Tooby 1995: 91.

33. See Cosmides and Tooby 2005 for the most recent discussion.

34. Cosmides and Tooby 2005; Buller 2005; see http://www.psych.ucsb.edu/research/cep/buller.htm.

35. Ermer et al. 2007.

36. Fehr and Henrich 2003.

37. Henrich et al. 2001.

38. Brosnan and de Waal 2003.

39. D. S. Wilson 2002: 8; D. S. Wilson and E. O. Wilson 2007: 336 and passim.

40. See also R. Boyd and Richerson 1992; Fehr and Gächter 2002; D. S. Wilson and E. O. Wilson 2007: 339.

41. Flesch 2007 unwarrantedly makes this almost the sole biological reason for storytelling.

42. Boehm 1993, 2000.

43. D. S. Wilson 2002: 64–65.

44. See Chwe 2001.

45. J. Carroll et al. under submission.

46. No to our genes: Dawkins 1989b; Pinker 1997. Our genes say no: de Waal 1996, 2006.

PART 2. EVOLUTION AND ART

Epigraph: Cain 1964, quoted in Dawkins 1989a: 31.

CHAPTER 5. ART AS ADAPTATION?

1. For fuzzy categories, see Rosch 1973 and 1975.

2. D. Brown 1991; Dissanayake 1988, 1995, 2000.

3. S. Davies 2000: 200, 201.

4. Dutton 1993: 21 and 1995: 35.

5. Dürer 1971: 24–25, quoted in Dutton 1977: 392; Goethe, conversation with Eckermann, 31 January 1827, quoted in Wright 2000: 155; Japanese: Pinker 2002: 408.

6. For animal culture, see Bonner 1980; de Waal 2001a. For the circularity of explaining culture by culture, see de Waal 2001a and Nordlund 2007.

7. Dickie 1997: 19.

8. Novitz 1996: 162: "We expect that works of art will make us feel more at home in our world."

9. Dissanayake 1988: 67, summarizing John Dewey, Herbert Read, and Suzanne Langer.

10. Shklovsky 1990.

11. For an appreciative survey, from someone interested in both the wide functions of art and an evolutionary perspective on the arts, see Dissanayake 1988: 64–106 and 1995: 10, 84. For brief dismissive catalogues of proposed functions, see G. Miller 1999: 79 and 2000b: 262.

12. Popper 1976: 61–62.
13. S. Davies 1997: 19; N. Carroll 1999: 227 ff.; for the charge of circularity, N. Carroll 1999: 261.
14. See D. Brown 1991; Dissanayake 1988, 1995.
15. "The greater the number of generations in which a cultural behavior has been replicated, the greater the probability of evidence of design," that is, of adaptation (Thornhill and Palmer 2000: 27).
16. Dissanayake 1988, 1995.
17. Griffiths 1997: 59.
18. Darwin [1871/1874] 2003.
19. G. Miller 2000b.
20. G. Miller 2000b; Cronin 1991.
21. Trivers 1972; G. Miller 2000b: 85–88.
22. G. Miller 2000b: 260–261.
23. Pinker 1997: 536; Darwin [1871/1874] 2003: 572.
24. Including Frisch 1974; Hamilton 1975; and Fox 1972.
25. G. Miller 2000b. For a more detailed critique of Miller, see B. Boyd 2005a.
26. J. Carroll 2004: xx.
27. Dissanayake 1995: 10–11.
28. G. Miller 2000b: 35.
29. Lane 1989: 272. Mothers singing: Sandra Trehub in Glausiusz 2001; infant preferences: Trehub 2000: 440.
30. G. Miller 2000a: 334.
31. Bahn 1998: 283.
32. Steven Brown 2000a: 245–246.
33. See de Waal 2002, 2001a: 26: "one could define language so narrowly that the babbling of a toddler does not fall under it, but does this mean that babbling has nothing to do with language? Narrow definitions neglect boundary phenomena and precursors, and they often mistake the tip of the iceberg for the whole."
34. Steven Brown 2000a, 2000b.
35. Fukui in Glausiusz 2001. See also B. Boyd 2008b for a discussion of the different selection pressures on music and visual art.
36. Dutton 2000: 513.
37. Darwin [1871/1874] 2003.

38. Lewis-Williams 2002: 98.
39. Jolly 1999: 97–98 notes that human hair appears to have coevolved with preferences for hair styling.
40. Coe 2003.
41. Kohn and Mithen 1999; see also G. Miller 2000b: 289–291; Bahn 1998: 86. Kohn and Mithen were not the first to feel that flints were knapped to the level of a "virtuoso's elegance" (Eiseley 1979: 270–271, cited in Dissanayake 2000: 132).
42. G. Miller 2000b: 29.
43. Thiessen and Umezawa 1998: 302.
44. See Sidanius and Pratto 1999.

CHAPTER 6. ART AS COGNITIVE PLAY

1. G. Williams 1966.
2. Pinker 1997: 525; see also Pinker 2002, 2007b. For another account of Pinker's byproduct hypothesis, see J. Carroll 1998.
3. Pinker 1997: 525, 539, 526.
4. Pinker 1997:528
5. Pinker 1997: 523.
6. Dennett 1996a: 339–340 wryly questions why we are ever awake, since action is costly.
7. Dawkins 2004: 160.
8. Rabkin 1999: 47.
9. V. Johnson and Tjapaltjarri 1994; V. Johnson 2003; Petyarre 2001.
10. Wright 2000: 5.
11. *OED*, s.v. "pattern," 8c.
12. Gazzaniga 2008: 215–216.
13. Weiss 1955: 286.
14. Haidt 2006: 7.
15. Quoted in Gould 1992: 268.
16. Weiss 1955: 286, 290.
17. Atkins 2003: 139.
18. Gould 1985: 199.
19. Solso 1994: 52.
20. Habituation is "the gradual loss of interest in repeated stimuli. Habit-

uation is a universal property of all nervous tissue. We find it in sea slugs. We find it in rats" (Martindale 1990: 11).

21. Lehrer 2008: 130–131.
22. Sloboda 2003: 38.
23. Dissanayake 1988: 74–78 for a detailed and sound consideration of art as adult play; and Dissanayake 1995: 43–33, for Schiller, Spencer, Freud, and Huizinga on art and play. Friedrich Schiller linked art and play in [1794] 1967. For animal play, Bekoff and Allen 2002.
24. Smith 1984: 67; Fagen 1995: 25; Mather and Anderson 1999.
25. van Hooff and Preuschoft 2003: 266.
26. Pinker 2007b: 170.
27. Bekoff 2007: 100.
28. Renner and Rosenzweig 1987; Doidge 2007: 43.
29. Jenkins et al. 1990; Merzenich et al. 1999.
30. Gordon et al. 2003; Goleman 2006: 178.
31. Špinka, Newberry, and Bekoff 2001.
32. E. O. Wilson 1975/2000: 164.
33. Pinker 1995: 340.
34. Siviy 1998.
35. Stopping rules: Gigerenzer, Todd, and ABC Research Group 1999. As a biological adaptation, curiosity has built-in stopping rules, when it exposes an animal to too much risk in novel environments, or when the rewards of new effort cease to justify continued searching. Science manages to overcome those stopping rules by cultural means.
36. Sutton-Smith 1997: 194 summarizes Gerstmyer 1991, who took videotapes of his/her daughter from twenty-one to thirty-two months, and concluded that play is typically a highly condensed representation of whatever it enacts, which means that much can be covered with great brevity.
37. Quoted in Doidge 2007: 306.
38. Merzenich et al. 1999; Doidge 2007: 68, 87, 111, 155.
39. D. Morris 1962; Lenain 1997.
40. Nyreröd 2004.
41. Martindale 1990: 11.
42. Carey 2006.

43. Kobayashi and Koshima 2001; Bekoff, Allen, and Burghardt 2002: 205; D. S. Wilson 2007a: 166.

44. Provine 2000: 94.

45. Hauser 2000: 211; Provine 2000: 93.

46. Stern 1977: 35–37; Morgan and Morgan 1995: 104; Meltzoff and Moore 1995; S. Johnson, Booth, and O'Hearn 2001: 639.

47. Jolly 1999: 331.

48. Corballis and Lea 1999: 340.

49. Trehub 2000: 437.

50. Winkielman et al. 2006.

51. Goleman 2006: 36.

52. Dissanayake 2000: 29.

53. Keller, Schölmerich, and Eibl-Eibesfeldt 1988; Trevarthen 1993; Trevarthen, quoted in Goleman 2006: 36.

54. Tomasello et al. 2005; D. S. Wilson 2007a.

55. Stern 1977: 39; J. Barrett, Richert, and Driesinga 2001: 51; Moore and Dunham 1995; Tomasello and Call 1997: 405, citing Stern 1985 and Trevarthen 1979; Dissanayake 2000: 7, 29; Povinelli and Preuss 1995: 422–423.

56. Michael Tomasello, quoted in D. S. Wilson 2007a: 169.

57. Deliège and Sloboda 1997: 121.

58. Hrdy 1999: 157, 288.

59. Stern 1977: 73.

60. Morgan and Morgan 1995: 102 and passim.

Chapter 7. Art and Attention

1. Jolly 1999: 291–292.

2. Aristotle 1951; among those outside an evolutionary approach who have recently foregrounded attention are Bordwell 1997; N. Carroll 1998; within such an approach: Aiken 1998; Bedaux and Cooke 1999; Cooke 1995; Coe 2003; Eibl-Eibesfeldt 1988; Nettle 2005.

3. For a signal exception, see Dissanayake 2000.

4. D. Campbell 1983.

5. D. S. Wilson 2002, 2007a; Richerson and Boyd 2005; D. S. Wilson and E. O. Wilson 2007.

6. Plotkin 1997: 204.
7. Gaulin and McBurney 2001: 338.
8. Amotz Zahavi, Bernd Heinrich, and John Marzluff call raven roosts "mobile information centers" (Ward and Zahavi 1973; Marzluff, Heinrich, and Marzluff 1996; H. Bloom 2000); Whiten and Byrne 1997: 162.
9. de Waal 2006: 25.
10. Bekoff 2002b: 36.
11. H. Bloom 2000: 89; Matsuzawa 2003: 387; Buss 1999a: 15.
12. Nettle 1999: 222.
13. R. Wells 2003: 56.
14. Kingdon 2003: 212–215.
15. Sugiyama 1972; Goodall 1986; Marker 2000: 42; Geissman 2000: 116; Sebeok 1979.
16. Tomasello et al. 2005; D. S. Wilson 2007a.
17. Iacoboni 2008.
18. Goleman 2006: 25, 29–30.
19. Pellis 2002.
20. Imitation: Meltzoff and Moore 1995, esp. 51; but see de Waal 2001a, esp. 216–238; de Waal and Tyack 2003. 3/1 for a strong but justified refutation of the denial of the capacity to imitate in other animals than humans; rehearsal: Donald 1998 and 1999, esp. 141, 144.
21. Hewlett and Cavalli-Sforza 1986; Scalise Sugiyama 2001b: 238.
22. Goleman 2006: 30–34; Oakley 2007: 148.
23. Goleman 2006; D. S. Wilson 2007a.
24. B. Boyd 2008b.
25. Alcorta and Sosis 2005.
26. McNeill 1995: 2; Haidt 2006: 237.
27. Dunbar 1996: 146.
28. Appiah 2008.
29. Chwe 2001.
30. Jolly 1999: 390.
31. Gould 1980: 267.
32. Researchers distinguish between two forms of status: domination (deference enforced by the more powerful party), in any animal species; and prestige (deference freely afforded by others), in humans

alone. See Henrich and Gil-White 2001; and R. Johnson, Burk, and Kirkpatrick 2007.

33. R. Hogan 1983, quoted in Buss 1999b: 31.

34. Darwin [1859] 2003: 142.

35. Richerson and Boyd 2005 discuss at length the reasons for the anomalous "demographic transition" over the last century, for high-status couples' now tending to have smaller families than lower-status couples.

36. Wright 2000: 26.

37. Boehm 1993, 1997, 2000.

38. For prestige, see Henrich and Gil-White 2001.

39. de Waal 1996: 76.

40. G. Miller 2000b: 320.

41. Chance and Larsen 1976, esp. 2, 13, 18, and 153.

42. Chance and Larsen 1976: 194.

43. Sloboda 2003: 38, 40.

44. Morgan and Morgan 1995: 149.

45. Knight 1998: 78; Dessalles 1998.

46. Hatfield 2002.

47. H. G. Wells 2007: 397.

48. D. S. Wilson 2002: 172.

49. Richerson and Boyd 2005.

CHAPTER 8. FROM TRADITION TO INNOVATION

1. David Bordwell, private communication.

2. D. S. Wilson 2002.

3. Haidt 2006.

4. Durkheim 1915: 381: "the principal forms of art seem to have been born out of religious ideas" (quoted in Dunbar, Knight, and Power 1999: 165).

5. Coe 2003: 27.

6. Atran 2002.

7. Pinker 2007a.

8. Boyer 1994, 1996, 2001; J. Barrett 1996, 2004; Kinderman, Dunbar, and Bentall 1998; Atran 2002.

9. For false belief, see Dennett 1978; Astington 1990; Perner 1991; Carruthers and Chamberlain 2000: 186.

10. Cosmides and Tooby 1994; Spelke 1995; Caramazza and Shelton 1998; Wellman and Gelman 1998. For a review of the large literature on theory of mind, see Baron-Cohen 2000.

11. Eibl-Eibesfeldt 1988: 54.

12. For a discussion of the human capacity to "time-travel," including envisaging the future, see Suddendorf 1999; Suddendorf and Corballis 2007.

13. Atran 2002: 12, 144–145.

14. Quoted in D. S. Wilson 2002: 64.

15. D. S. Wilson 2002: 17, citing Ostrom, Gardner, and Walker 1994; Richerson and Boyd 2005.

16. Atran 2002; Boyer 2002; D. S. Wilson 2002.

17. D. S. Wilson 2002: 42.

18. Zahavi and Zahavi 1997; G. Miller 2000b.

19. Sosis and Bressler 2003; D. S. Wilson 2002; Alcorta and Sosis 2005: 330–332.

20. Coe 2003.

21. Atran 2002; Boyer 2002; D. S. Wilson 2002.

22. Dissanayake 2000: 216.

23. Jackson 1982.

24. Coe 2003 offers an adaptive explanation of visual art in terms of the cooperative effects that ancestor-worship can create among the ancestors' descendants, even arguing that art's function is entirely one of social cohesion, and that art in traditional societies is therefore rigidly traditional. But she does not account for either the origins or the features of visual art, and her claims about the rigidity of traditional art do not square with the inventive diversity of forms across cultures or the diversity within cultures, both diachronically and synchronically. Many traditional cultures encourage innovation: in Sepik carving, for instance (Denis Dutton, private communication), or Ojibwe verbal play (Erdrich 2003). Maori art provides a good case study of a traditional culture with known parameters. In the mere five hundred years between Maori settlement in New Zealand and European occupation, Maori art diversified rapidly, regionally and locally.

25. Dawkins 1997: 279; G. Williams 1966.
26. Pinker 1997: 428.
27. Boden 1991; Csikszentmihaly 1996; Simonton 1999.
28. See Dawkins 1989b: 269 on the evolution of evolvability.
29. D. Campbell 1987; Dennett 1996a; D. S. Wilson 2005.
30. Edelman 1987; also Patricia Churchland, Terrence Deacon, Terrence Sejnovksi (see Sterelny 2003: 164).
31. Plotkin 1994.
32. Or, in evolutionary biological language, no local optima: see Bettinger, Boyd, and Richerson 1996.
33. P. Davies 1988: 55–56; J. Hobson 1994: 214.
34. Donald 1998.
35. Dawkins 1999: x.
36. See the discussion of prefocusing in N. Carroll 1998: 261.
37. Martindale 1990.
38. Stern 1977: 83–84.
39. Plotkin 1994: 166: "Put crudely, in the case of instruction [the Lamarckian process] the environment rules; in the case of selection [the Darwinian], internal or organismic states lead."
40. As Plotkin notes, as long as intelligence "operates, even if in only small part, by Darwinian processes involving the unpredictable generation of variants, then . . . intelligent behaviour . . . cannot be reductively explained by genetics or genetics and development" (1994: 176).
41. Haydon 1990; Wright 2000: 54, 192.
42. See Root-Bernstein 1989 and 1996.
43. Mumford wrote "dreams," not "stories." Dreams take their imagery from the waking world, including the world of story, which may be what Mumford actually meant (quoted in Allmann 1989).
44. Root-Bernstein 2003: 267.
45. Root-Bernstein 2000: 134.
46. Sharecroppers: Agee 1941.
47. Gombrich 1979: 130.
48. Bjorklund, Gaultney, and Green 1993: 97. The feeling of control has considerable survival value, neuroscientist Michael Gazzanigga observes, and must have been selected for (Gazzanigga 1994: 186).

PART 3. EVOLUTION AND FICTION

Epigraph: L. Carroll [1893] 2006: chap. 23 ("The Pig-Tale"), 301–302.

CHAPTER 9. ART, NARRATIVE, FICTION

1. The term "metarepresentation" was introduced in Pylyshyn 1978.
2. Tooby and Cosmides 2001: 12.
3. Boyer 2001: 152.
4. Schaller 1991.
5. Meir Sternberg quotes Wendy Lehnert, a former student of Roger Schank, the leading researcher in artificial intelligence and story: "Thus far, 'no natural language [computer] system can claim to tackle general language in an open-ended domain or task orientation,' such as would qualify it to 'read fiction'" (Sternberg 2003a: 359n; Lehnert 1994: 170–171).
6. The approach therefore accords with that of Joseph Carroll's in, and since, his *Evolution and Literary Theory* (1995), and with David Bordwell 1985, 1989, 1996. Where Carroll argues that understanding the world in general cannot be a matter of convention, I apply the same arguments to the specific field of comprehending events, including the reported events of narrative.

CHAPTER 10. UNDERSTANDING AND RECALLING EVENTS

1. P. Hobson 2004: 30, 31; Meltzoff and Moore 1995.
2. Garcia and Koelling 1966; Garcia, Ervin, and Koelling 1966; Garcia and Ervin 1968.
3. Frank 1988: 149.
4. Boyer 2001; Sperber and Wilson 1988: vii.
5. H. Damasio et al. 2004.
6. See pp. 39–40 above for a discussion of the "frame problem" in artificial intelligence.
7. Dennett 1996b: 57.
8. Gigerenzer, Todd, and ABC Research Group 1999: 5, 128, 21; Boyer 2001: 162; Sperber and Wilson 1988: 45.

9. Wittgenstein, *Remarks on the Philosophy of Psychology* 2:570, quoted in P. Hobson 2004: 243.

10. Emmott 1997: 26.

11. Boyer 2001: 100.

12. For the implications for the development of art, see Martindale 1990.

13. Arterberry 1997: 51. For a contrary view, see Sirois and Jackson 2007.

14. Gopnik and Meltzoff 1997.

15. Meltzoff and Moore 1995.

16. Daly and Wilson 1998a: 433.

17. Quine 1960: 6.

18. Baillargeon 1986.

19. Premack and Premack 2003: 117; Pinker 1995: 157.

20. Gopnik and Meltzoff 1997: 181.

21. Lea 1999: 26–27.

22. Caramazza and Mahon 2003.

23. Caramazza 2000: 1038.

24. Gelman 1990.

25. Gergely et al. 1995: 167; Michotte 1963.

26. Santos, Hauser, and Spelke 2002: 205–210.

27. Brothers 1997: 40.

28. Gergely et al. 1995; Caramazza and Shelton 1998.

29. Baron-Cohen 1995.

30. New, Cosmides, and Tooby 2007.

31. Atran 1990; Astington 1990.

32. J. Barrett 2004.

33. Michotte 1963.

34. Heider and Simmel 1944.

35. Premack and Premack 1995; P. Bloom and Veres 1999.

36. Gigerenzer, Todd, and ABC Research Group 1999; Johansson et al. 1994.

37. Nelson 1996: 197.

38. Schank and Abelson 1977; Schank 1995.

39. Premack and Premack 2003: 146, 171–172.

40. Tomasello 1995; Gergely et al. 1995: 174; Premack and Premack 2003: 145.

41. Wellman quoted in Gopnik and Meltzoff 1997: 125; Wellman 1990; Perner 1991: 118.

42. Darwin [1871/1874] 2003, chap. 4; de Waal 2001a: 352.
43. de Waal 2001a: 352; Bekoff 2002a: 35.
44. de Waal 2001a: 326.
45. de Waal 2001a: 352, 357.
46. de Waal 2001a: 355.
47. D. Griffin 2001: 216.
48. Boyer 2001: 122; J. Harris 2006.
49. Gaulin and McBurney 2001: 193; Buss 1999b: 51.
50. Smuts 2000: 94; de Waal 1996: 73–74; Klein et al. in press.
51. Goleman 2006.
52. Baron, Keller, and Burnstein 2002.
53. Kenrick, Sadalla, and Keele 1998: 504.
54. Mock 2004: 22.
55. See Frank 1988 and Katz 2000.
56. Darwin [1871/1874] 2003, chap. 5.
57. Brosnan and de Waal 2003.
58. Douglas 2001; Fehr and Henrich 2003.
59. D. S. Wilson 2002: 101, 191.
60. Wright 2000: 262; Pinker 2002: 256–257.
61. Cosmides and Tooby 1992, 2005.
62. Pinker 2002: 188.
63. Brosnan and de Waal 2003; de Waal 2006.
64. Frank 1988: 65.
65. Jung 2003; Nunez 2001.
66. B. Boyd 1998; Zunshine 2006, the first book on theory of mind and fiction, demonstrates a shaky hold on the scientific literature and its application to fiction; see B. Boyd 2006a for a review and discussion, and Zunshine 2007 and B. Boyd 2007a for an exchange.
67. Jolly 1966; Humphrey 1976; Byrne and Whiten 1988; Alexander 1989; Whiten and Byrne 1997; de Waal and Tyack 2003; Geary 2005; Zuber-bühler and Byrne 2006.
68. D. S. Wilson 2007a; Tomasello et al. 2005.
69. Iacoboni 2008: 139.
70. See Baron-Cohen 1995 for the first, P. Hobson 2004 and to some extent Baron-Cohen 2005 for the second.
71. Rakoczy, Tomasello, and Striano 2005; Tomasello et al. 2005.
72. See Astington and Baird 2005.

73. Tomasello et al. 2005: 675.
74. Baron-Cohen 1995, 1998: 181; Klin, Schultz, and Cohen 2000: 363; Boyer 2001: 104.
75. Sacks 1995: 272.
76. Baron-Cohen 1995: 3. Baron-Cohen 2005 has revised his explanation of autism as theory of mind deficit by including empathy as a component of theory of mind.
77. Vinden and Astington 2000: 513; Cosmides and Tooby 1994: 102; Heyes 2003: 721; Saxe 2004: 2.
78. Astington 1990: 154–157; Perner 1991; Baron-Cohen 2000: 5–14; Saxe 2004: 1.
79. Perner 1991: 270; Gopnik and Astington 2000: 194.
80. Perner 1991: 161, 162.
81. Perner 1991: 153.
82. Perner 1991: 158–159. This natural human concern with the sources of knowledge even shapes the grammar of certain languages, such as Turkish, in which grammatical markings are needed to indicate the source of one's knowledge about past events: if you witnessed them, heard about them, or had to infer them (Perner 1991: 149).
83. Perner 1991: 136.
84. Baron-Cohen et al. 1999: 410–411.
85. Baron-Cohen et al. 1999: 407–408.
86. Kinderman, Dunbar, and Bentall 1998.
87. Dunbar 1998b: 103. Something as complex as theory of mind naturally engages many brain areas: current neuroimaging studies suggest medial, orbitofrontal, and lateral areas of the prefrontal cortex, especially left medial frontal (Brodmann's areas 8 and 9), Brodmann's area 32, the medial temporal area, and the temporo-parietal junction: see L. Barrett, Henzi, and Dunbar 2003; Haan and Johnson 2003; Saxe 2004.
88. See Emmott 1997.
89. Baker-Ward, Ornstein, and Principe 1997: 91; Travis 1997: 119; Trabasso and Stein 1997: 243–244; Bourg, Bauer, and van den Broek 1997: 393–396.
90. Berman and Slobin 1994; Travis 1997: 114; Trabasso and Stein 1997: 238, 241.
91. Nelson 1997: 2; Baker-Ward, Ornstein, and Principe 1997: 93; Bourg, Bauer, and van den Broek 1997: 395.

92. Bauer and Mandler 1992; Travis 1997: 133, 136; Fivush and Haden 1997: 172.
93. Bauer 1996.
94. Hauser and Spaulding 2006, cited in Pinker 2007a: 217.
95. Nelson 1997: 21; Travis 1997: 114; van den Broek 1997: 335; Bourg, Bauer, and van den Broek 1997: 397.
96. Trabasso and Stein 1997: 242.
97. Berman and Slobin 1994: 67.
98. Stein and Liwag 1997: 201.
99. Gergely et al. 1995; Bourg, Bauer, and van den Broek 1997: 388, 399.
100. Travis 1997: 113.
101. Travis 1997: 114.
102. Doidge 2007: 265–266, 278–279.
103. Trabasso and Stein 1997: 248.
104. See Chapter 13, notes 27 and 28 and text.
105. Tulving 1985.
106. Long, Oppy, and Seely 1997: 373.
107. Fletcher, Briggs, and Linzie 1997: 344–345.
108. C. Brown 2001: 109.
109. Kendrick 2004: 48.
110. Clayton and Dickinson 1998; L. Barrett, Henzi, and Dunbar 2003: 496.
111. Bauer and Wewerka 1997: 141–142.
112. Bauer 1996; Bauer and Wewerka 1997.
113. Bauer and Wewerka 1997: 146; Fivush and Haden 1997: 177; Bauer 1996.
114. Nelson and Ross 1980: 96.
115. Baker-Ward, Ornstein, and Principe 1997: 84, 89; Trabasso and Stein 1997: 246.
116. Anderson and Schooler 1991.
117. Pinker 1997: 142–143.
118. Baker-Ward, Ornstein, and Principe 1997: 88. Goldinger 1998, however, suggests a multiple-trace memory model that stores aspects of surface detail as well as deeper structure.
119. Bartlett 1932.
120. Fivush and Haden 1997: 175.
121. Premack and Premack 2003: 238; Fivush and Haden 1997: 181; Trabasso and Stein 1997: 247.
122. Klein et al. 2002.

123. Klein et al. 2002, in press.
124. Emmott 1997.
125. Schacter and Addis 2007b: 331.
126. Barsalou 2008 offers a review. Evolved: Barsalou 2008: 620: "there are no a priori reasons why simulation cannot have a strong genetic basis. Genetic contributions almost certainly shape the modal systems and memory systems that capture and implement simulations"; 622, citing Barsalou 2005: "Barsalou . . . further proposed that nonhumans have roughly the same simulation system as humans but lack a linguistic system to control it."
127. Barsalou 2008: 618–619.
128. Barsalou 2008: 624, citing Tucker and Ellis 1998.
129. Barsalou 2008: 624, citing Pulvermüller et al. 2006.
130. Barsalou 2008: 625, citing Franklin and Tversky 1990.
131. Barsalou 2008: 627 reviews the relevant studies.
132. Schacter and Addis 2007a: 778, quoting Atance and O'Neill 2001; also Buckner 2007; Suddendorf and Corballis 2007; Schacter, Addis, and Buckner 2008.
133. Schacter and Addis 2007a, 2007b: 332.
134. Buckner 2007: 318.
135. A. Damasio 1996.
136. Barsalou 2008: 633.
137. Schacter and Addis 2007a: 779.
138. Bower 2008 reviews this tradition of research.
139. Barsalou 2008: 623.

Chapter 11. Narrative

1. Aristotle 1951: 1447a: "Epic poetry and Tragedy, Comedy also and Dithyrambic poetry . . . are all in their general conception modes of imitation"; 1448a: "the objects of imitation are men in action"; 1449b: "Tragedy, then, is an imitation of an action that is serious, complete, and of a certain magnitude."
2. See N. Carroll 1996b and 1998 on musical prefocusing, Bordwell 1997 and 2005 on aspects of visual focusing in narrative film, McCloud 1993 and Spiegelman 1999 on comics.

3. See Strawson 2004 for a rare critique of these assumptions.

4. The patient known as David, discussed in A. Damasio 2000: 119.

5. Dawkins and Krebs 1978. See also J. Krebs and Dawkins 1984; Sperber 2006.

6. J. Krebs and Dawkins 1984.

7. Noble 2000: 57. Paul Grice 1975 famously stressed the cooperative principle of conversation.

8. Frisch 1950, 1967; D. Griffin 2001.

9. Seyfarth, Cheney, and Marler 1980; Cheney and Seyfarth 1990.

10. E. O. Wilson 1975/2000: 474; D. Griffin 2001: 229.

11. Meltzoff 1988: 59.

12. Meltzoff 1988; P. Hobson 2004.

13. de Waal 2001b: 4.

14. Donald 1998.

15. Tomasello et al. 2005.

16. Dennett 1993: 144; 1996b: 57.

17. Boyer 2001; Barkow 1992: 629.

18. de Waal 1998: 169.

19. P. Hobson 2004: 75.

20. Marco Iacoboni, quoted in Oakley 2007: 104–105.

21. P. Hobson 2004: 54.

22. P. Hobson 2004: 247.

23. Pinker 1997: 207.

24. King, Rumbaugh, and Savage-Rumbaugh 1999: 112.

25. Eibl-Eibesfeldt 1982: 194.

26. Cf. Nettle 2005: 45.

27. A. Damasio 1996, 2000.

28. Haidt 2006: 54.

29. Ibid.

30. Dunbar, Marriott, and Duncan 1997.

31. Dessalles 1998.

32. Schank 1995.

33. Sperber 2006.

34. Leach 1954: 266.

35. Langellier and Petersen 1992: 172.

36. Jolly 1999: 110.

37. Maltz and Borker 1982: 208, quoted in A. Campbell 2002: 107.
38. For a general view of a wider sense of "parable" in narrative, see M. Turner 1996.
39. Boyer 2001.
40. See N. Carroll 1998: 206.
41. J. Carroll 1995, 2004; Sternberg 2003a, 2003b.
42. As claimed in Dunbar 1996.
43. Pinker 1995, 1997, 2007a.
44. Pinker 2007a: 149, 178, 187, 430, 208, 431.
45. I adapt the example and simplify the more elaborate and rigorous terminology of Fauconnier 1997: 73.
46. For negation, see Verhagen 2007, 2008b.
47. Scalise Sugiyama 2001b and 2004 makes much—probably too much—of the importance of narrative warnings about predators in hunter-gatherer societies.

CHAPTER 12. FICTION

1. Lorenz 1971.
2. Bekoff 2000; Bekoff and Allen 2002; Špinka, Newberry, and Bekoff 2001: 142.
3. Allen and Bekoff 1997; Bekoff 2002a, 2002b: 113.
4. Fagen 1995: 35; see also Stuart Brown 1998.
5. Stuart Brown 1998: 248; B. Sutton-Smith 1997: 40.
6. Sutton-Smith 1997: 169–173.
7. Corballis 2002: 79.
8. Bekoff 2002a: 124–125.
9. Allen and Bekoff 1997: 99–100.
10. van Hooff and Preuschoft 2003: 267.
11. Bateson 1955; Jolly 1999: 287.
12. Allen and Bekoff 1997: 109.
13. P. Hobson 2004: 76.
14. Rakoczy, Tomasello, and Striano 2005: 78; Baron-Cohen 2000: 8; Wellmann and Lagattuta 2000: 26.
15. Perner 1991: 51.
16. P. Harris 1991: 285; Dunbar 1996: 68.
17. Astington 1990: 151–152.

18. Sutton-Smith 1981: 15.
19. Nelson 1996: 197. See also Sutton-Smith 1997: 163.
20. Suddendorf and Whiten 2001: 633.
21. Leslie 1987, 1994; Boyer 2001: 130.
22. Taylor 1999; Boyer 2001: 149.
23. Nelson 1996: 193.
24. Sutton-Smith 1986: 67.
25. Sutton-Smith 1981: 53–54.
26. Baillargeon 2002; P. Bloom 2004: 12–13.
27. Ridley 2003: 179.
28. Waelti, Dickinson, and Schultz 2001.
29. Sawyer 1997: 118.
30. Sutton-Smith 1997: 158–160; Göncü and Klein 2001: 4; Sutton-Smith 2001: 162.
31. Peterson and McCabe 1983.
32. Pellegrini 1995: 154; Sutton-Smith 2001: 162.
33. Stuart Brown and Sutton-Smith 1995: 40.
34. Sutton-Smith 1981: 110–111.
35. Sutton-Smith 1981.

Chapter 13. Fiction as Adaptation

1. Fauconnier 1997: 121.
2. Erdrich 1990: 76.
3. Baron-Cohen 1995, 2000.
4. Bekoff 2007: 100.
5. Hrdy 1999: 157.
6. Green and Bavelier 2007; Rosser et al. 2007.
7. Levitin 2006: 193.
8. Levitin 2006: 190.
9. M. Johnson 1998: 428, 440.
10. Fischler 1998: 394.
11. Goleman 2006.
12. Dimberg, Thunberg, and Elmehed 2000.
13. Aziz-Zadeh et al. 2006.
14. Dennett 1996b: 152.
15. Cf. N. Carroll's clarificationism, 1998: 320–324.

16. As dreams may also do?
17. Nettle 2005: 44 discusses story as superstimulus.
18. Gerrig 1993.
19. Cosmides and Tooby 2000: 89.
20. Dawkins 1989a: 4.
21. J. Carroll 2004: 116.
22. J. Carroll 2004: 198.
23. M. Turner 1996 discusses the parable function of stories, our ability to extract sense relevant to us; see also Schank 1995.
24. L. Ellis 1995; de Waal and Tyack 2003 (hyenas); Pusey, Williams, and Goodall 1997; de Waal 1998: 163, 165 (chimpanzees); Wright 2000: 26, 37 (humans).
25. Allport and Postman 1946: 46.
26. Sidanius and Pratto 1999; Baumeister 2007.
27. Cf. P. Hogan 2003.
28. Daly and Wilson 1998b.
29. Haidt 2006: 193–200; Algoe and Haidt under submission.
30. R. Boyd and Richerson 1992; Fehr and Gächter 2002.
31. Oakley 2007: 259–260.
32. See Chapter 7, note 30 and text.
33. Laylā and Majnūn, quoted in P. Hogan 2003: 147.
34. Pretend play: Dunbar 2000: 239; modeling: Perner 1991: 87.
35. Cosmides and Tooby 2000: 65.
36. Lewis 1973; Collins, Hall, and Paul 2004.
37. The tilting of the pelvis as we moved from quadrupedal to bipedal locomotion is the major reason human births are so difficult. Richerson and Boyd 2005 show that maladaptation offers even better evidence of evolution as opposed to "Intelligent Design" than does perfect adaptation.
38. Boyer 2001: 13.
39. Atran 2002: viii–ix.
40. P. Harris 1990.
41. Mead 1932; Kohlberg 1969; Wellmann and Gelman 1998: 562.
42. Boyer 2001; Atran 2002.
43. Horton 1993 stresses the importance of the sense of control that religion imparts, especially in preindustrial societies.
44. J. Barrett 1996, 2000, 2004; Boyer 2001; Atran 2002.

45. Dennett 1996a.
46. Gigerenzer, Todd, and ABC Research Group 1999; Sperber and Wilson 1988.
47. Plotkin 1994: 191.
48. Richerson and Boyd 2005.
49. Atran 2002: 94; J. Barrett and Keil 1996; J. Barrett 2004.
50. Atran 2002: 88.
51. Boyer 2001: 153.
52. Quoted in D. S. Wilson 2002: 64.
53. D. S. Wilson 2002: 41.
54. D. S. Wilson 2002: 42.
55. Horton 1993.
56. Horton 1993: 231.
57. For evolutionary discussions of humor, see B. Boyd 2004 and Gervais and Wilson 2005.

Book II. From Zeus to Seuss

Epigraph: McEwan 2004: 4.

1. For an excellent but very different evolutionary approach to Homer (especially the *Iliad*), see Gottschall 2008b.
2. Popper 1999, 1976; Gombrich 1960, 1982, 1995; Bordwell 1985, 1989, 1997, 2005; Bordwell and Carroll 1996.

Part 4. Phylogeny

Epigraph: Sidney [1595] 1966: 40.

Chapter 14. Earning Attention (1)

1. Tracy 1990: xi.
2. van Wees 1999: 11.
3. de Jong 1999a: 7.
4. M. Edwards 1987: 1.
5. Kirk 1993: 293.
6. M. Edwards 1987: 20.
7. M. Edwards 1987: 44.

8. Gunn 1970.

9. M. Edwards 1987: 88.

10. J. Griffin 1977, 1980.

11. Scodel 2002: 34–36.

12. Frank Turner 1997: 137.

13. W. H. D. Rouse, cited in Sternberg 1978: 56.

14. Labov 1972.

15. Aristotle 1951: 14.1453b.22–26, p. 51.

16. de Jong 2001: 215.

17. Budgen 1960.

18. Sternberg 1978: 92.

19. Barnouw 2004: 21.

20. Lakoff and Johnson 1999, 2003.

21. Felson and Slatkin 2004: 106.

22. From the Murray/Dimock translation, Homer 1995: 193.

23. Stanford 1999: 201.

24. Bowlby 1988; Hrdy 1999.

25. See P. Hogan 2003.

26. de Jong 2001: 313.

27. On the preference for happy endings see Aristotle 1951: 13.1453a30–35, p. 47; and Lowe 2000: 129.

28. Olson 1995: 24 notes that since at least 1895 this has become something of a critical commonplace.

29. For an evolutionary analysis of this contrast in terms of "agonistic structure," see J. Carroll et al. under submission.

30. As Ruth Scodel aptly summarizes, Sale 1987 and 1989 convincingly argues "that the *Iliad* avoids the sharply hostile traditional epithets of the Trojans and that the epic language is weak in formulae for movement from Troy or for being in Troy. In other words, the full, sympathetic treatment of the Trojans in the *Iliad* is a relatively recent development in the tradition" (Scodel 2002: 52–53).

Chapter 15. Earning Attention (2)

1. Levin 1971, 1979, 2003; Bordwell 1989.

2. Brann 2002: 109 notes that epic, "if ever there was a genre, is meant for the second and subsequent readings."

3. Sternberg 1978, 2003a, 2003b. Note Scodel 2004: 52: "Perhaps the most common misconception about traditional stories is that they do not allow for suspense. This is not true—a good story-teller can create suspense even when the audience knows the outcome . . . In Homeric epic, the situation is especially complicated, since many details of a traditional story could vary among different versions. So the audience could never know just how the end would be reached. The poet carefully keeps the audience aware of how the plot will turn out in general, but uncertain about specifics."

4. Scodel 2002: 48–49.

5. Aristotle 1951: 23.1459a.30–34, p. 89.

6. He reports the first Greek killed in the war, nine years earlier (2.701); he has the leading warriors on the battlefield identified to Priam by Helen as they stand together on the battlements of Troy, as if the war were just starting; he opens the fighting between named warriors with Menelaos and Paris, Helen's Greek and Trojan husbands, in single combat, again as if hostilities had just begun; and he grimly foreshadows, though he does not show, the death of Achilleus and the fall of Troy.

7. Scodel 2002 offers a masterly account of these tactics.

8. Felson and Slatkin 2004: 93.

9. Lukács 1971: 67.

10. Scodel 2002: 49.

11. There are two distinct ways of dating the events in the *Odyssey,* depending on whether the council of the gods in book 5 is seen as being a second version of the council in book 1 or as a new event. In the latter case, the duration of the action would change from about twenty-seven to forty days.

12. Cf. Knox 1996: 3.

13. Tracy 1990: 3.

14. Goethe in Cook's translation, Homer 1993: 303; G. Finsler (1914) in de Jong 1999a: 4; Olson 1995: 140; Scodel 2004: 52.

15. P. Jones 1991: 9; Walsh and Merrill 2002: 18.

16. de Jong 2001: 34; J. Griffin 1980: 57.

17. The evidence is inconsistent. Kalypso serves Hermes ambrosia and nectar on her own, and seems otherwise to have no one else with her. Nevertheless, when she serves Odysseus food, "her serving maids set

nectar and ambrosia before her" (5.198–199). Yet there is no other reference to these serving maids or to anyone else on Ogygia. Odysseus, in an otherwise accurate report to Arete, describes Ogygia as an island where "subtle Kalypso lives, with ordered hair, a dread goddess, and there is no one, neither a god nor a mortal person, who keeps her company. It was unhappy I alone whom destiny brought there to her hearth" (7.245–249). Circe does have maidservants, who are nymphs and therefore immortal.

18. See de Jong 2001: 171 for the critical reactions.
19. Frog: Silk 2004: 35; knight: Woodhouse 1930: 54–65.
20. For both Circe and Kalypso, see 9.29–32.
21. Bordwell 1989.

CHAPTER 16. THE EVOLUTION OF INTELLIGENCE (1)

1. Barnouw 2004: 54 and passim; Maehler 1999: 9–10, 17.
2. Snell 1975: 31; Barnouw 2004: 163.
3. Jaynes 1976: 75, 273, 276.
4. B. Williams 1993; Barnouw 2004.
5. For a superb demonstration of how modern artists and especially writers have anticipated the findings of psychology, see Lehrer 2008, which came to my attention too late to incorporate more fully.
6. B. Williams 1993.
7. Premack and Premack 2003: 145; Gergely et al. 1995: 174; Gopnik and Meltzoff 1997: 125; Barkow, Cosmides, and Tooby 1992: 89; Saxe 2004.
8. Geary 2005: 14, 195, 215, 297; Pinker 2007a: 129.
9. Geary 2005: 195, 13, 14, 211, 232.
10. Bernard Williams discusses a version of the formula from the *Iliad:* "'Yet still, why does the heart within me debate on these things?' Hermann Fränkel had already noticed that remarks made to the *thumos* are then represented as made by it. But the word translated as 'debate on,' *dielexato,* refers, unsurprisingly, to a two-way discourse, and the formulae capture the idea that in talking to a *thumos* a man is talking to himself. What happens in both cases is that the character pulls back from a course of action he has been considering in favour of a course of action with which he is more identified" (B. Williams 1993: 350). Homer's use of *thumos* seems to reflect the idea of working memory

as a mental workspace where one *can* talk to oneself more or less explicitly.

11. P. Jones 2003: 18.

12. Barnouw 2004: 12.

13. de Jong 2001: 305.

14. Barnouw 2004.

15. For a fine discussion of these ironies, see de Jong 2001: 347–348, 388, 396.

16. This has been denied by some Homerists, but their denial is amply refuted by J. Griffin 1980: 51–65.

Chapter 17. The Evolution of Intelligence (2)

1. See Perner 1991; Cosmides and Tooby 2000.

2. Auerbach 1957: 5, 9. Barnouw 2004: 319 notes that Auerbach's reading is both "much celebrated and much criticized." See, e.g., Cook in Homer 1993: 332; M. Edwards 1987: 4.

3. J. Griffin 1980: 64.

4. Auerbach 1957: 2–3.

5. See, e.g., Barnouw 2004: 321.

6. Auerbach 1957: 5.

7. de Jong 1985, 2001.

8. Barnouw 2004: 322.

9. Tracy 1990: 117.

10. Sternberg 1978: 76.

11. Whiten and Byrne 1997: 7.

12. Bekoff 2002a: 56.

13. Suddendorf and Whiten 2001: 640; Plotkin 1997: 217. But see Hare, Call, and Tomasello 2001; and de Waal 2004.

14. See Tomasello et al. 2005; D. S. Wilson 2007a.

15. Rutherford 1999: 286.

16. Barnouw 2004: 343.

17. Heyes 2003: 721.

18. But see de Waal 2004.

19. For a book suggesting that an evolutionary understanding of "come-uppance" is the key to all fiction—surely an oversimplification—see Flesch 2007.

20. Stanford 1999: 201.
21. Barnouw 2004: 327.
22. Classicist Barry Powell (1991) has even advanced strong arguments that the alphabet, the most efficient writing system yet known and the *other* cultural precondition for the great Greek tradition of critical thinking, was invented specifically in order to record Homer's epics.
23. Horton 1993.
24. Dennett 1993: 144 and 1996b: 57.
25. J. Griffin 1980.

Chapter 18. The Evolution of Cooperation (1)

1. This seems to be woven into Penelope's name: *pene,* thread or woof, and *elopeaiai,* from *olopto,* to pluck out, hence "weaver/unraveler" (Barnouw 2004: 338).
2. Cummins 2005: 688.
3. Cummins 2005: 689.
4. A. Damasio 2003: 47.
5. A. Damasio 2003: 46; see especially the work of Frans de Waal.
6. Jolly 1999: 264.
7. Tooby and Cosmides 1996.
8. D. S. Wilson 2002: 119.
9. The system of *xenia* has been ably described in, e.g., Finley 1979.
10. Cf. de Jong 2001: 223.
11. Most 1999: 489 ff.
12. Knox 1996: 29–30.
13. Walsh and Merrill 2002: 27.
14. D. Krebs 2000: 143.

Chapter 19. The Evolution of Cooperation (2)

1. D. S. Wilson 2002: 8.
2. de Waal 1998; Fox 1995.
3. de Jong 2001: 199, 417.
4. D. S. Wilson 2002: 194.

5. Frank 1988.
6. Nisbett and Cohen 1996.
7. Cummins 2005: 681; Clutton-Brock and Parker 1995.
8. Atran 2002: 28.
9. Atran 2002: 114.
10. Sosis 2000, 2004; D. S. Wilson 2002; Alcorta and Sosis 2005.
11. D. S. Wilson 2002.
12. de Waal 1996, 2006.
13. A. Campbell 2005: 641.
14. Richerson and Boyd 2005.
15. Cf. Radcliffe Richards 2000: 201.

Chapter 20. Problems and Solutions

1. Morgan and Morgan 1995: 248.
2. Sheff 1987: 55, quoted in Minear 1999: 263, who notes that this was only Japan's first *post*war election.
3. MacDonald 1988: 75.
4. Minear 1999: 263.
5. J. Harris 2006: 239.
6. Morgan and Morgan 1995: 233.
7. Morgan and Morgan 1995: 88.
8. Cohen 2004: 110. The magazine was *Printer's Ink*.
9. Nel 2004: 97.
10. Cohen 2004: 204.
11. Boyer 2001; Atran 2002; J. Barrett 2004.
12. Cohen 2004: 341.
13. Darwin [1872] 1998: 56.
14. Cf. Aiken 1998; Solso 1994.
15. B. Boyd 2004; Gervais and Wilson 2005.
16. Cohen 2004: 110, 117.
17. Frederick Turner and Pöppel 1988.
18. Morgan and Morgan 1995: 7.
19. Morgan and Morgan 1995: 81.
20. Quoted in Nel 2004: 6.
21. Nel 2004: 88.

22. Morgan and Morgan 1995: 82.

23. Morgan and Morgan 1995: 176.

Chapter 21. Levels of Explanation: Universal, Local, and Individual

1. Sheff 1987: 55 in Minear 1999: 263.

2. Appiah 2006b: 41.

3. Minear 1999: 264.

4. Nel 2004: 57.

5. Morgan and Morgan 1995: 119.

6. Quoted in Nel 2004: 57.

7. Barthes 1972b: 135.

8. Braunmuller 1992: 78.

9. Menand 2002: 148.

10. See Levin 2003, chap. 2.

11. Morgan and Morgan 1995: 72; Cohen 2004: 105.

12. Nel 2004: 60.

13. Morgan and Morgan 1995: 125.

14. Quoted in Nel 2004: 29.

15. Menand 2005: 14.

16. Menand 2005: 13.

17. Menand 2005: 14.

18. E. O. Wilson 1999.

19. Menand 2005: 12.

20. Menand 2005: 15.

21. Ibid.

22. D. Brown 1991.

23. Gombrich 1991: 18.

24. Pollock 1983: 23, critiquing this position.

25. Nordlund 2007: 55.

26. Howard Bloch, quoted in Nordlund 2007: 25.

27. Nordlund 2007: 53–62; P. Hogan 2003; Diamond 2003; and Fisher 2004.

28. Kenan Malik, quoted in Nordlund 2007: 19.

29. Bordwell 1997: 44.

30. Bordwell 1997: 43.

31. Richerson and Boyd 2005: 166.
32. Bordwell 1997: 150–151, 156.
33. Morgan and Morgan 1995: 72.
34. Cohen 2004: 26–27.
35. Nel 2004: 77–79.

CHAPTER 22. LEVELS OF EXPLANATION: INDIVIDUALITY AGAIN

1. Critiqued in Levin 2003, chap. 2; and in Burke 1998.
2. Braunmuller 1992: 78.
3. J. Harris 2006.
4. Klein et al. 2002.
5. Goodall 1990.
6. de Waal 2001a.
7. Despite the so-called Intentional Fallacy of Wimsatt and Beardsley; see Livingston 2005.
8. Didion 2005: 7.
9. Morgan and Morgan 1995: 170.
10. Simonton 1999.
11. Ibid.
12. Nel 2004: 35.
13. For Mozart, see Steptoe 1996: 33. Pushkin's heavily revised manuscripts feature in many Russian editions of his work.
14. Nel 2004: 32.
15. Stofflet 1986: 65.
16. Morgan and Morgan 1995: 279.
17. Seuss 1995: 18–19 (the dating may be incorrect).
18. See Cohen 2004: 8–10.
19. *Judge,* April 1938: 17; reproduced in Cohen 2004: 10.

CHAPTER 23. LEVELS OF EXPLANATION: PARTICULAR

1. Cohen 2004: 306.
2. Seuss 1951.
3. Cohen 2004: 299.
4. In June 1950 *Redbook* magazine carried Dr. Seuss's "Gustav, the Gold-

fish," which was "advertised as 'An Amazing Adventure story for Youngsters and Other People of Imagination'" and introduced a new series of stories "'presented with a technique that's new. To get the best results . . . read it aloud to your youngsters.' Up to this point, Ted had published eight books, half of which rhymed and half of which did not. Having had the experience with the boy who was too young to read but who had memorized the sounds of the words in *Thidwick*, Ted was now planning to use the sounds of his words to reach youngsters at an early age" (Cohen 2004: 303).

5. Morgan and Morgan 1995: 82.
6. For the last of these, cf. Nel 2004: 67.
7. Cohen 2004: 305.
8. N. Carroll 1988: 170–181 analyzes this question-and-answer narrative under the term "erotectic."
9. N. Carroll 1998: 261–275.
10. See Cohen 2004: 8–10.
11. See Dehaene 1997 for the biological roots of mathematics, including the evolved capacity for appraising quantity common to many animals.

CHAPTER 24. MEANINGS

1. M. Turner 1996.
2. Aristotle 1951: 1451b, pp. 34–35.
3. Schank 1995.
4. Morgan and Morgan 1995: 81.
5. Morgan and Morgan 1995: 98.
6. Morgan and Morgan 1995: 211.
7. Morgan and Morgan 1995: 286.
8. Morgan and Morgan 1995: 155.
9. Schank 1995.
10. As noted in Chapter 4, this does not mean that cooperation cannot evolve, only that it requires special conditions.
11. See Bordwell 2008 for a model of narrative reception that moves from perception (substantially the same for all) to comprehension (more variable) to appropriation, the value a given audience assigns to the story (relatively open). Bordwell does not invoke selfish gene theory but would accept its relevance.

12. Morgan and Morgan 1995: 277.

13. Cohen 2004: 289.

14. In this story written in late 1953, we might suspect that another "local" context has played its part: Senator Joseph McCarthy's campaign against supposed Communists in the federal government was approaching its peak.

15. See Cohen 2004: 220.

16. Levin 1979: 28–41.

17. See Haidt 2006 and Algoe and Haidt under submission for discussions of the emotion of elevation as an inspiration to emulate the prosocial dispositions and actions of others.

18. Quoted in L. Miller 2005: 54; http://www.newyorker.com/archive/2005/12/26/051226fa_fact.

19. Scott 2000.

Conclusion

Epigraph: Frye 1957.

1. Edelman 1987.

2. D. S. Wilson 2005.

3. D. Campbell 1965, 1974; Hull 1988; Simonton 1999.

4. Cf. J. Carroll 1995, which distinguishes the first three levels.

5. Popper 1999, 1976; Gombrich 1995, 1960, 1982; Bordwell 1985, 1989, 1997, 2005; Bordwell and Carroll 1996.

6. Patai and Corral 2005; B. Boyd 2006b, 2006c.

7. Among those who have recently argued for the role that literature can play as a source and a testing-ground for psychology are Livingston 1988; Sternberg 2003a, 2003b; J. Carroll 2004; Gottschall 2008b; Slingerland 2008; J. Carroll et al. under submission.

8. Dennett 1996a: 74.

9. Sokal 1996, 2008; Sokal and Bricmont 1998; Editors of Lingua Franca 2000. Frustrated by what he saw as the meaninglessness of postmodern Theory's language and the emptiness of its claims, physicist Alan Sokal wrote a parody of Theory-speak (1996), awash in scientific quotations but actually meaningless. After it was published by the editors of *Social Text*, whose ideological preconceptions it flattered, in its special issue "Science Wars," Sokal exposed and explained his hoax.

10. Spitzer 1948; J. Ellis 2005: 100–101.

11. Vickers 1993: 417.

12. Quoted in Patai and Corral 2005: 6.

13. Cf. Bordwell and Carroll 1996 for a discussion of the midrange investigations possible after the demise of Theory.

14. While *Maus* is mainstream, I discuss another pointedly avant-garde Speigelman comic, *The Narrative Corpse,* in B. Boyd 2008a.

15. Nel 2004: 254. Nel's source is Steinfels 1995: 9: "The pink mess on the white dress reveals that menstrual taboos have also been violated [and] conjures up fears about the improper loss of virginity." The stain also expresses "the children's resentment over the loss of their mother and the playful protective environment she represents . . . the little cats . . . mark the advent of the alphabet," the power of language to "transform the pink pollution into structures of culture."

16. Bordwell 1989.

17. See Wolf 2007 for the neuroscience of reading, including the effects of reading on the plastic brain.

18. Scalise Sugiyama 2001a, 2001b, 2005.

19. J. Carroll 2004; J. Carroll et al. under submission. For a critique of Carroll's position, see B. Boyd 2009b.

20. For a more detailed critique of Carroll's proposal of the adaptive role of literature in terms of cognitive order, see B. Boyd 2008d.

21. Barthelme 1968.

22. Exceptionally, high status in humans in the last hundred years has begun to correlate with smaller families. Richerson and Boyd 2005 discuss the reasons for the "demographic transition," which from a purely biological point of view would in normal circumstances be maladaptive.

23. Tomasello et al. 2005.

24. Richerson and Boyd 2005.

Afterword

1. Dawkins 1991: 4, 5.

2. Dawkins 1983, 1989b, 1991; Plotkin 1994.

3. D. Campbell 1965; Hull 1988.

4. Gould 1991: 48–52.

5. S. Morris 2003.

6. Michod, Wojciechowski, and Hoelzer 1988; H. Bloom 2000: 91–95.

7. Bordwell 2006: 18.

8. J. Hobson 1994: 214; P. Davies 1988: 55–56.

9. See Bettinger, Boyd, and Richerson 1996 for an evolutionary discussion of style and function.

10. Wolpert 1993.

11. Boyer 2001; J. Barrett 2004; D. S. Wilson 2002.

BIBLIOGRAPHY

Adolphs, R., L. Cahill, and R. Schul. 1997. Impaired declarative memory for emotional material following bilateral amygdala damage in humans. *Learning & Memory* 4:291–300.

Adolphs, R., H. Damasio, D. Tranel, G. Cooper, and Antonio Damasio. 2000. A role for somatosensory cortices in visual recognition of emotion as revealed by three-dimensional lesion mapping. *Journal of Neuroscience* 20:2683–90.

Agee, James. 1941. *Let Us Now Praise Famous Men.* Boston: Houghton Mifflin.

Aiello, Leslie C., and P. Wheeler. 1995. The expensive tissue hypothesis: The brain and the digestive system in human and primate evolution. *Current Anthropology* 36:199–121.

Aiken, Nancy. 1998. *The Biological Origins of Art.* Westport, Conn.: Praeger.

Alcorta, Candace, and Richard Sosis. 2005. Ritual, emotion, and sacred symbols: The evolution of religion as an adaptive complex. *Human Nature* 16:323–359.

Alexander, Richard D. 1989. Evolution of the human psyche. In *The Human Revolution: Behavioural and Biological Perspectives on the Origins of Modern Humans,* ed. P. Mellars and C. Stringer. Princeton: Princeton University Press, 455–513.

Algoe, Sara B., and Jonathan Haidt. Under submission. Witnessing excellence in action: The "other-praising" emotions of elevation, gratitude, and admiration.

Allen, Colin, and Marc Bekoff. 1997. *Species of Mind: The Philosophy and Biology of Cognitive Ethology.* Cambridge, Mass.: Bradford/MIT Press.

Allman, William F. 1989. *Apprentices of Wonder: Inside the Neural Network Revolution.* New York: Bantam.

Allport, Gordon W., and Leo Postman. 1946. *The Psychology of Rumor.* New York: Holt, Rinehart, and Winston.

Anderson, John R., and Lael J. Schooler. 1991. Reflections of the environment in memory. *Psychological Science* 2:396–408.

Anderson, Joseph, and Barbara Anderson. 1996. The case for an ecological metatheory. In Bordwell and Carroll 1996, 347–367.

Appiah, Kwame Anthony. 1992. *In My Father's House: Africa in the Philosophy of Culture.* London: Methuen.

———. 2006a. *Cosmopolitanism: Ethics in a World of Strangers.* New York: Norton.

———. 2006b. Whose culture is it? *New York Review of Books,* 9 February: 38–41.

———. 2008. *Experiments in Ethics.* Cambridge, Mass.: Harvard University Press.

Aristotle. 1951. *Poetics: Aristotle's Theory of Poetry and Fine Art.* Trans. S. H. Butcher. 1911. Reprint, New York: Dover.

Arterberry, Martha E. 1997. Development of sensitivity to spatial and temporal information. In van den Broek, Bauer, and Bourg 1997, 51–78.

Astington, Janet Wilde. 1990. Narrative and the child's theory of mind. In Britton and Pellegrini 1990, 151–171.

Astington, Janet Wilde, and Jodie A. Baird, eds. 2005. *Why Language Matters for Theory of Mind.* Oxford: Oxford University Press.

Atance, C. M., and D. K. O'Neill. 2001. Episodic future thinking. *Trends in Cognitive Sciences* 5:533–539.

Atkins, Peter. 2003. *Galileo's Finger: The Ten Great Ideas of Science.* Oxford: Oxford University Press.

Atran, Scott. 1990. *Cognitive Foundations of Natural History: Towards an Anthropology of Science.* Cambridge: Cambridge University Press.

———. 2002. *In Gods We Trust: The Evolutionary Landscape of Religion.* New York: Oxford University Press.

Auerbach, Erich. 1957. *Mimesis: The Representation of Reality in Western Literature.* 1946. Trans. Willard R. Trask. Reprint, Garden City, N.Y.: Doubleday.

Axelrod, Robert. 1990. *The Evolution of Cooperation.* 1984. Reprint, London: Penguin.

———. 1997. *The Complexity of Cooperation: Agent-Based Models of Competition and Collaboration.* Princeton: Princeton University Press.

Axelrod, Robert, and William D. Hamilton. 1981. The evolution of cooperation. *Science* 211:1390–96.

Aziz-Zadeh, Lisa, Stephen M. Wilson, Giacomo Rizzolatti, and Marco Iacoboni. 2006. Congruent embodied representations for visually presented actions and linguistic phrases describing actions. *Current Biology* 16 (19):1818–23.

Baddeley, A. D. 2000. The episodic buffer: A new component of working memory? *Trends in Cognitive Sciences* 4:417–423.

Baddeley, A. D., and J. G. Hitch. 1974. Working memory. *Psychology of Learning and Motivation* 8:47–90.

Baddeley, A. D., and R. H. Logie. 1999. Working memory: The multiple-component model. In *Models of Working Memory: Mechanisms of Active Maintenance and Executive Control,* ed. A. Miyake and P. Shah. Cambridge: Cambridge University Press, 28–61.

Bahn, Paul G. 1998. *The Cambridge Illustrated History of Prehistoric Art.* Cambridge: Cambridge University Press.

Bahn, Paul G., and Jean Vertut. 1997. *Journey through the Ice Age.* Berkeley: University of California Press.

Baillargeon, René. 1986. Representing the existence and the location of hidden objects: Object Permanence in 6- and 8-month-old infants. *Cognition* 20:191–208.

———. 2002. The acquisition of physical knowledge in infancy: A summary in eight lessons. In *Blackwell Handbook of Childhood Cognitive Development,* ed. U. Goswami. Cambridge, Mass.: Blackwell, 47–83.

Baker-Ward, Lynne, Peter A. Ornstein, and Gabrielle F. Albert Principe. 1997. Revealing the representation: Evidence from children's reports of events. In van den Broek, Bauer, and Bourg 1997, 79–107.

Barash, David, and Nanelle Barash. 2005. *Madame Bovary's Ovaries: A Darwinian Look at Literature.* New York: Delacorte.

Barash, Nanelle, and David Barash. 2004. Biology, culture, and persistent literary dystopias. *Chronicle of Higher Education* 51 (15):B10.

Barkow, Jerome H. 1992. Beneath new culture is old psychology: Gossip and social stratification. In Barkow, Cosmides, and Tooby 1992, 627–637.

———, ed. 2006. *Missing the Revolution: Darwinism for Social Scientists.* Oxford: Oxford University Press.

Barkow, Jerome H., Leda Cosmides, and John Tooby, eds. 1992. *The*

Adapted Mind: Evolutionary Psychology and the Generation of Culture. New York: Oxford University Press.

Barnouw, Jeffrey. 2004. *Odysseus, Hero of Practical Intelligence: Deliberation and Signs in Homer's Odyssey.* Lanham, Md.: University Press of America.

Baron, Andrew Scott, Matt Keller, and Eugene Burnstein. 2002. Character information enhances memory for faces. Paper presented at the annual meeting of the Human Behavior and Evolution Society, Rutgers University.

Baron-Cohen, Simon. 1995. *Mindblindness: An Essay on Autism and Theory of Mind.* Cambridge, Mass.: Bradford/MIT Press.

———. 1998. Does the study of autism justify minimalist innate modularity? *Learning and Individual Differences* 10 (3):179–191.

———. 2000. Theory of mind and autism: A fifteen-year-review. In Baron-Cohen, Tager-Flusberg, and Cohen 2000, 3–20.

———. 2003. *The Essential Difference: Male and Female Brains and the Truth about Autism.* New York: Basic Books.

———. 2005. The empathizing system: A revision of the 1994 model of the mindreading system. In Ellis and Bjorklund 2005, 468–492.

Baron-Cohen, Simon, Michelle O'Riordan, Valerie Stone, Rosie Jones, and Kate Plaisted. 1999. Recognition of faux pas by normally developing children and children with Asperger syndrome or high-functioning autism. *Journal of Autism and Developmental Disorders* 29:407–418.

Baron-Cohen, Simon, Helen Tager-Flusberg, and Donald Cohen, eds. 2000. *Understanding Other Minds: Perspectives from Developmental Cognitive Neuroscience.* 3d ed. Oxford: Oxford University Press.

Barrett, Justin L. 1996. *Anthropomorphism, Intentional Agents, and Conceptualizing God.* Ithaca, N.Y.: Cornell University Press.

———. 2000. Exploring the natural foundations of religion. *Trends in Cognitive Sciences* 4:29–34.

———. 2004. *Why Would Anyone Believe in God?* Lanham, Md.: Altamira.

Barrett, Justin L., and Frank C. Keil. 1996. Conceptualizing a nonnatural entity: Anthropomorphism in God concepts. *Cognitive Psychology* 31:219–247.

Barrett, Justin L., Rebekah Richert, and A. Driesinga. 2001. God's beliefs versus mother's: The development of nonhuman agent concepts. *Child Development* 72:50–65.

Barrett, Louise, Robin Dunbar, and John Lycett. 2002. *Human Evolutionary Psychology.* Princeton: Princeton University Press.

Barrett, Louise, Peter Henzi, and Robin Dunbar. 2003. Primate cognition: From "what now?" to "what if?" *Trends in Cognitive Sciences* 7:494–497.

Barsalou, Lawrence W. 2005. Continuity of the conceptual system across species. *Trends in Cognitive Sciences* 9:309–311.

———. 2008. Grounded cognition. *Annual Review of Psychology* 59:617–645.

Barthelme, Donald. 1968. *Unspeakable Practices, Unnatural Acts.* New York: Farrar, Straus and Giroux.

Barthes, Roland. 1972a. *Mythologies.* 1957. 2d ed. 1970. Trans. A. Lavers. New York: Hill and Wang.

———. 1972b. To write: An intransitive verb? In *The Languages of Criticism and the Science of Man: The Structuralist Controversy,* ed. R. Macksey and E. Donato. Baltimore: Johns Hopkins Press, 134–156.

Bartlett, Frederic C. 1932. *Remembering: A Study in Experimental and Social Psychology.* Cambridge: Cambridge University Press.

Bateson, Gregory. 1955. A new theory of play and fantasy. *Psychiatric Research Reports* 2:39–51.

Bauer, Patricia J. 1996. What do infants recall of their lives? Memory for specific events by one- to two-year-olds. *American Psychologist* 51:29–41.

Bauer, Patricia J., and Jean Mandler. 1992. Putting the horse before the cart: The use of temporal order in recall of events by one-year-old children. *Developmental Psychology* 28:441–452.

Bauer, Patricia J., and Sandi Saeger Wewerka. 1997. Saying is revealing: Verbal expression of event memory in the transition from infancy to early childhood. In van den Broek, Bauer, and Bourg 1997, 139–168.

Baumeister, Roy F. 2007. Is there anything good about men? Address to American Psychological Association. August 24. Available at http://www.psy.fsu.edu/~baumeistertice/goodaboutmen.htm.

Bedaux, Jan Baptist, and Brett Cooke, eds. 1999. *Sociobiology and the Arts.* Amsterdam: Rodopi.

Bekoff, Marc. 2002a. *Minding Animals: Awareness, Emotions and Heart.* New York: Oxford University Press.

———. 2002b. Virtuous nature. *New Scientist,* 13 July: 34–37.

———. 2007. *The Emotional Lives of Animals.* Novato, Calif.: New World Library.

Bekoff, Marc, ed. 2000. *The Smile of a Dolphin: Remarkable Accounts of Animal Emotions.* New York: Discovery Books.

Bekoff, Marc, and Colin Allen. 2002. The evolution of social play: Interdisciplinary analyses of cognitive processes. In Bekoff, Allen, and Burghardt 2002, 429–435.

Bekoff, Marc, Colin Allen, and Gordon M. Burghardt, eds. 2002. *The Cognitive Animal: Empirical and Theoretical Perspectives on Animal Cognition.* Cambridge, Mass.: Bradford/MIT Press.

Berman, Ruth A., and Dan Isaac Slobin. 1994. *Relating Events in Narrative: A Crosslinguistic Developmental Study.* Hillsdale, N.J.: Lawrence Erlbaum.

Bettinger, Robert L., Robert Boyd, and Peter J. Richerson. 1996. Style, function, and cultural evolutionary processes. In *Darwinian Archaeologies,* ed. S. Shennan. New York: Plenum, 133–166.

Bjorklund, David F., Jane F. Gaultney, and Bradni L. Green. 1993. "I watch, therefore I can do": The development of meta-imitation during the preschool years and the advantage of optimism about one's imitative skills. In *Emerging Themes in Cognitive Development.* Vol. 2: *Competencies,* ed. R. Pasnak and M. L. Howe. New York: Springer Verlag, 79–102.

Bjorklund, David F., and B. L. Green. 1992. The adaptive nature of cognitive immaturity. *American Psychologist* 47:46–54.

Bloom, Howard. 2000. *Global Brain: The Evolution of Mass Mind from the Big Bang to the 21st Century.* New York: Wiley.

Bloom, Paul. 2004. *Descartes' Baby: How the Science of Child Development Explains What Makes Us Human.* New York: Basic Books.

Bloom, Paul, and C. Veres. 1999. The perceived intentionality of groups. *Cognition* 71:B1–B9.

Blythe, Philip W., Peter M. Todd, and Geoffrey Miller. 1999. How motion reveals intention: Categorizing social interactions. In Gigerenzer, Todd, and ABC Research Group 1999, 257–285.

Boden, Margaret A. 1991. *The Creative Mind: Myths and Mechanisms.* 1990. Reprint, New York: Basic Books.

Boehm, Christopher. 1993. Egalitarian society and reverse dominance hierarchy. *Current Anthropology* 34:227–254.

———. 1997. Egalitarian behaviour and the evolution of political intelligence. In *Machiavellian Intelligence II,* ed. R. W. Byrne and A. Whiten. Cambridge: Cambridge University Press, 341–364.

———. 1999. *Hierarchy in the Forest.* Cambridge, Mass.: Harvard University Press.

———. 2000. Conflict and the evolution of social control. In *Evolutionary Origins of Morality: Cross-Disciplinary Perspectives,* ed. L. D. Katz. Bowling Green, Ohio: Imprint Academic, 79–101.

Bonner, John Tyler. 1980. *The Evolution of Culture in Animals.* Princeton: Princeton University Press.

Bordwell, David. 1985. *Narration in the Fiction Film.* Madison: University of Wisconsin Press.

———. 1989. *Making Meaning: Inference and Rhetoric in the Interpretation of Cinema.* Cambridge, Mass.: Harvard University Press.

———. 1996. Convention, construction, and cinematic vision. In Bordwell and Carroll 1996, 87–107.

———. 1997. *On the History of Film Style.* Cambridge, Mass.: Harvard University Press.

———. 2004. Foreword. In *Moving Image Theory: Ecological Considerations,* ed. J. D. Anderson and B. F. Anderson. Carbondale: Southern Illinois University Press.

———. 2005. *Figures Traced in Light: On Cinematic Staging.* Berkeley: University of California Press.

———. 2006. *The Way Hollywood Tells It: Story and Style in Modern Movies.* Berkeley: University of California Press.

———. 2008. *Poetics of Cinema.* New York: Routledge.

Bordwell, David, and Noël Carroll, eds. 1996. *Post-Theory: Reconstructing Film Studies.* Madison: University of Wisconsin Press.

Bortolussi, Marisa, and Peter Dixon. 2003. *Psychonarratology: Foundations for the Empirical Study of Literary Response.* Cambridge: Cambridge University Press.

Bourg, Tammy, Patricia J. Bauer, and Paul van den Broek. 1997. Building the bridges: The development of event comprehension and representation. In van den Broek, Bauer, and Bourg 1997, 385–407.

Bower, Gordon H. 2008. The evolution of a cognitive psychologist: A journey from simple behaviors to complex mental acts. *Annual Review of Psychology* 59:1–27.

Bowlby, John. 1988. *A Secure Base: Parent-Child Attachment and Healthy Human Development.* New York: Basic Books.

Boyd, Brian. 1998. Jane, meet Charles: Literature, evolution and human nature. *Philosophy and Literature* 22:1–30.

———. 1999. Literature and discovery. *Philosophy and Literature* 23:313–333.

———. 2001. The origin of stories: *Horton Hears a Who. Philosophy and Literature* 25:197–214.

———. 2004. Laughter and literature: A play theory of humor. *Philosophy and Literature* 28:1–22.

———. 2005a. Evolutionary theories of art. In Gottschall and Wilson 2005, 147–176.

———. 2005b. Literature and evolution: A biocultural approach. *Philosophy and Literature* 29:1–23.

———. 2006a. Fiction and theory of mind. *Philosophy and Literature* 30:571–581.

———. 2006b. Getting it all wrong: Bioculture critiques cultural critique. *American Scholar* 75 (4):18–30.

———. 2006c. Theory: Dead like a zombie. Review of Patai and Corral 2005 and Levin 2003. *Philosophy and Literature* 30:289–298.

———. 2007a. Fiction and theory of mind: An exchange. *Philosophy and Literature* 31:196–198.

———. 2007b. Getting it all wrong. *American Scholar* 75:156–158.

———. 2008a. Art and evolution: The avant-garde as test case: Spiegelman in *The Narrative Corpse. Philosophy and Literature* 32:31–57.

———. 2008b. Artistic animals: Common and unique features of music and visual art. In *The Visual Animal,* ed. I. North. Adelaide: Centre for Contemporary Art, 15–26.

———. 2008c. The art of literature and the science of literature. *American Scholar* 77:118–127.

———. 2008d. Art as adaptation: A challenge. Comment on Joseph Carroll, "An Evolutionary Paradigm for Literary Study." *Style.* 42:138–143.

———. 2009b. Evolution, art, purpose. *American Scholar.*

Boyd, Brian, Joseph Carroll, and Jonathan Gottschall, eds. In press. *Evolutionary Approaches to Literature and Film: A Reader in Art and Science.* New York: Columbia University Press.

Boyd, Robert, and Peter J. Richerson. 1985. *Culture and the Evolutionary Process.* Chicago: University of Chicago Press.

———. 1992. Punishment allows the evolution of cooperation (or anything else) in sizable groups. *Ethology and Sociobiology* 13:171–195.

Boyer, Pascal. 1994. *The Naturalness of Religious Ideas.* Berkeley: University of California Press.

———. 1996. What makes anthropomorphism natural: Intuitive ontology and cultural representations. *Journal of the Royal Anthropological Institute,* n.s. 2:83–97.

———. 2001. *Religion Explained: The Evolutionary Origins of Religious Thought.* New York: Basic Books.

Brann, Eva. 2002. *Homeric Moments: Clues to Delight in Reading the* Odyssey *and the* Iliad. Philadelphia: Paul Dry.

Braunmuller, A. R. 1992. Henry VIII. In *Which Shakespeare? A User's Guide to Editions,* ed. A. Thompson, T. L. Berger, A. R. Braunmuller, P. Edwards, and L. Potter. Milton Keynes: Open University Press, 78–81.

Britton, Bruce K., and A. D. Pellegrini, eds. 1990. *Narrative Thought and Narrative Language.* Hillsdale, N.J.: Lawrence Erlbaum.

Brosnan, Sarah, and Frans B. M. de Waal. 2003. Monkeys reject unequal pay. *Nature* 425:297 299.

———. 2004. Fair refusal by capuchin monkeys. *Nature* 428:140.

Brothers, Leslie. 1997. *Friday's Footprint: How Society Shapes the Human Mind.* New York: Oxford University Press.

Brown, Andrew. 1999. *The Darwin Wars: The Scientific Battle for the Soul of Man.* London: Simon and Schuster.

Brown, Culum. 2001. Familiarity with the test environment improves escape responses in the crimson spotted rainbowfish, *Melanotaenia duboulayi. Animal Cognition* 4:109–113.

Brown, Donald E. 1991. *Human Universals.* Philadelphia: Temple University Press.

Brown, Steven. 2000a. Evolutionary models of music: From sexual selection to group selection. *Perspectives in Ethology* 13:231–281.

———. 2000b. The "musilanguage" model of music evolution. In Wallin, Merker, and Brown 2000, 271–300.

Brown, Stuart. 1998. Play as an organizing principle: Clinical evidence and personal observations. In *Animal Play: Evolutionary, Comparative, and Ecological Perspectives,* ed. M. Bekoff and J. A. Byers. Cambridge: Cambridge University Press, 243–259.

Brown, Stuart L., and Brian Sutton-Smith. 1995. Concepts of childhood and play: An interview with Brian Sutton-Smith. *ReVision* 17 (4):35–42.

Brownell, Hiram H., Dee Michel, John Powelson, and Howard Gardner. 1983. Surprise but not coherence: Sensitivity to verbal humor in right-hemisphere patients. *Brain and Language* 18:20–27.

Buckner, Randy L. 2007. Prospection and the brain. *Behavioral and Brain Sciences* 30:318–319.

Budgen, Frank. 1960. *James Joyce and the Making of "Ulysses."* Bloomington: Indiana University Press.

Buller, David J. 2005. *Adapting Minds: Evolutionary Psychology and the Persistent Quest for Human Nature.* Cambridge, Mass.: Bradford/MIT Press.

Burke, Séan. 1998. *The Death and Return of the Author: Criticism and Subjectivity in Barthes, Foucault, and Derrida.* 1992. 2d ed. Edinburgh: Edinburgh University Press.

Buss, David M. 1999a. Evolutionary psychology: A new paradigm for psychological science. In *Evolution of the Psyche,* ed. David H. Rosen and Michael C. Luebbert. Westport, Conn.: Praeger, 3–33.

———. 1999b. Human nature and individual difference: The evolution of human personality. In *Handbook of Personality: Theory and Research,* ed. L. A. Pervin and O. P. John. New York: Guilford, 31–56.

———, ed. 2005. *The Handbook of Evolutionary Psychology.* Hoboken, N.J.: Wiley.

Butler, R. A. 1954. Incentive conditions which influence visual exploration. *Journal of Experimental Psychology* 48:17–23.

Byrne, Richard W. 1997. The technical intelligence hypothesis: An additional evolutionary stimulus to intelligence. In Whiten and Byrne 1997, 289–311.

Byrne, Richard W., and Andrew Whiten, eds. 1988. *Machiavellian Intelligence: Social Expertise and the Evolution of Intellect in Monkeys, Apes and Humans.* Oxford: Clarendon.

Cain, A. J. 1964. The perfection of animals. In *Viewpoints in Biology,* ed. J. D. Carthy and C. L. Duddington. London: Butterworths.

Campbell, Anne. 2002. *A Mind of Her Own: The Evolutionary Psychology of Women.* Oxford: Oxford University Press.

———. 2005. Aggression. In Buss 2005, 628–651.

Campbell, Donald. 1965. Variation and selective retention in socio-cultural evolution. In *Social Change in Developing Areas: A Reinterpretation of Evolutionary Theory,* ed. H. R. Barringer, G. I. Blanksten, and R. W. Mack. Cambridge, Mass.: Schenkma, 19–49.

———. 1974. Evolutionary epistemology. In *The Philosophy of Karl R. Popper,* ed. P. A. Schilpp. La Salle, Ill.: Open Court, 413–463.

———. 1983. Two distinct routes beyond kin selection to ultrasociality: Implications for the humanities and social sciences. In *Nature of Prosocial Development: Theories and Strategies,* ed. D. L. Bridgeman. New York: Academic, 11–41.

———. 1987. Blind variation and selective retention in creative thought as in other knowledge processes. 1960. In *Evolutionary Epistemology, Rationality, and the Sociology of Knowledge,* ed. G. Radnitzky and W. W. Bartley III. La Salle, Ill.: Open Court, 91–114.

Caramazza, Alfonso. 2000. The organization of conceptual knowledge in the brain. In Gazzaniga 2000, 1037–66.

Caramazza, Alfonso, and Bradford Z. Mahon. 2003. The organization of conceptual knowledge: The evidence from category-specific semantic deficits. *Trends in Cognitive Science* 7:354–361.

Caramazza, A., and J. R. Shelton. 1998. Domain-specific knowledge systems in the brain: The animate-inanimate distinction. *Journal of Cognitive Neuroscience* 10:1–34.

Carey, John. 2006. *What Good Are the Arts?* Oxford: Oxford University Press.

Carroll, Joseph. 1995. *Evolution and Literary Theory.* Columbia: University of Missouri Press.

———. 1998. Steven Pinker's cheesecake for the mind. *Philosophy and Literature* 22:578–585.

———. 2004. *Literary Darwinism: Evolution, Human Nature and Literature.* New York: Routledge.

———. 2007. Evolutionary approaches to literature and drama. In R. I M. Dunbar and L. Barrett, 2007, 637–648.

Carroll, Joseph, Jonathan Gottschall, John Johnson, and Daniel Kruger.

Under submission. *Graphing Jane Austen: A Quantitative Analysis of Human Nature in British Novels of the Nineteenth Century.*

Carroll, Lewis. [1893] 2006. *Sylvia and Bruno Concluded.* N.p.: Objective Systems.

Carroll, Noël. 1988. *Mystifying Movies: Fads and Fallacies in Contemporary Film Theory.* New York: Columbia University Press.

———. 1996a. Prospects for film theory: A personal assessment. In Bordwell and Carroll 1996, 37–68.

———. 1996b. *Theorizing the Moving Image.* Cambridge: Cambridge Univrsity Press.

———. 1998. *A Philosophy of Mass Art.* Oxford: Clarendon.

———. 1999. *Philosophy of Art: A Contemporary Introduction.* London: Routledge.

Carruthers, Peter, and Andrew Chamberlain, eds. 2000. *Evolution and the Human Mind.* Cambridge: Cambridge University Press.

Cavalli-Sforza, Luigi Luca, Paolo Menozzi, and Alberto Piazza. 1994. *The History and Geography of Human Genes.* Princeton: Princeton University Press.

Chafe, Wallace. 1990. Some things that narratives tell us about the mind. In Britton and Pellegrini 1990, 79–98.

Chagnon, Napoleon. 1997. *Yąnomamö.* 1968. 5th ed. N.p.: Wadsworth.

Chance, Michael, and C. J. Jolly. 1970. *Social Groups of Monkeys, Apes, and Men.* New York: Dutton.

Chance, Michael, and Ray Larsen. 1976. *The Social Structure of Attention.* London: Wiley.

Chauvet, Jean-Marie, Eliette Brunel Deschamps, and Christian Hillaire. 1996. *Chauvet Cave: The Discovery of the World's Oldest Paintings.* London: Thames and Hudson.

Cheney, D. L., and R. M. Seyfarth. 1990. *How Monkeys See the World: Inside the Mind of Another Species.* Chicago: University of Chicago Press.

Chomsky, Noam. 1959. A review of B. F. Skinner's *Verbal Behavior. Language* 35 (1):26–58.

———. 1965. *Cartesian Linguistics.* New York: Harper and Row.

Chwe, Michael Suk-Young. 2001. *Rational Ritual: Culture, Coordination, and Common Knowledge.* Princeton: Princeton University Press.

Clayton, N. S., and A. Dickinson. 1998. Episodic-like memory during cache recovery by scrub jays. *Nature* 395:272–274.

Clutton-Brock, T. H., and G. A. Parker. 1995. Punishment in animal socie-
    ties. *Nature* 373:209–216.

Coe, Kathryn. 2003. *The Ancestress Hypothesis: Visual Art as Adaptation.*
    New Brunswick, N.J.: Rutgers University Press.

Cohen, Charles D. 2004. *The Seuss, the Whole Seuss, and Nothing but the
    Seuss.* New York: Random House.

Collins, John, Ned Hall, and L. A. Paul, eds. 2004. *Causation and
    Counterfactuals.* Cambridge, Mass.: Bradford/MIT Press.

Cooke, Brett. 1995. Microplots: The case of *Swan Lake. Human Nature*
    6:183–196.

Cooke, Brett, and Frederick Turner, eds. 1999. *Biopoetics: Evolutionary Ex-
    plorations in the Arts.* Lexington, Ky.: International Conference on
    the Unity of the Sciences.

Corballis, Michael C. 2002. *From Hand to Mouth: The Origins of Language.*
    Princeton: Princeton University Press.

Corballis, Michael C., and Stephen E. G. Lea, eds. 1999. *The Descent of
    Mind: Psychological Perspectives on Hominid Evolution.* Oxford: Ox-
    ford University Press.

Cosmides, Leda. 1989. The logic of social exchange: Has natural selection
    shaped how humans reason? Studies with the Wason Selection Task.
    Cognition 31:187–276.

Cosmides, Leda, Stanley B. Klein, John Tooby, and S. Chance. 2001. Deci-
    sions and the evolution of multiple memory systems: Using person-
    ality judgments to test the scope hypothesis. Paper presented at the
    annual meeting of the Human Behavior and Evolution Society, Lon-
    don.

Cosmides, Leda, and John Tooby. 1989. Evolutionary psychology and the
    generation of culture, part II. Case study: A computational theory of
    social exchange. *Ethology and Sociobiology* 10:51–97.

———. 1992. Cognitive adaptations for social exchange. In Barkow, Cos-
    mides, and Tooby 1992, 163–228.

———. 1994. Origins of domain specificity: The evolution of functional
    organization. In *Mapping the Mind: Domain Specificity in Cognition
    and Culture,* ed. L. A. Hirschfeld and S. A. Gelman. Cambridge: Cam-
    bridge University Press, 85–116.

———. 1995. Beyond intuition and instinct blindness: Toward an evolu-
    tionarily rigorous cognitive science. In *Cognition on Cognition,* ed. J.

Mehler and S. Franck. Cambridge, Mass.: Bradford/MIT Press, 69–105.

———. 1998. *Evolutionary Psychology: A Primer.* Available at http://www.psych.ucsb.edu/research/cep/primer.html.

———. 2000. Consider the source: The evolution of adaptations for decoupling and metarepresentation. In *Metarepresentations: A Multidisciplinary Perspective,* ed. D. Sperber. Oxford: Oxford University Press, 53–115.

———. 2005. Neurocognitive adaptations designed for social exchange. In Buss 2005, 584–627.

Crawford, Charles, and Dennis L. Krebs, eds. 1998. *Handbook of Evolutionary Psychology: Ideas, Issues, and Applications.* Mahwah, N.J.: Lawrence Erlbaum.

Critical Inquiry. 2004. Critical Inquiry Symposium special issue. *Critical Inquiry* 30.

Cronin, Helen. 1991. *The Ant and the Peacock: Altruism and Sexual Selection from Darwin to Today.* Cambridge: Cambridge University Press.

Cronk, Lee. 1999. *That Complex Whole: Culture and the Evolution of Human Behavior.* Boulder: Westview.

Cronk, Lee, Napoleon Chagnon, and William Irons, eds. 2002. *Adaptation and Human Behavior: An Anthropological Perspective.* New York: Aldine.

Csikszentmihalyi, Mihaly. 1996. *Creativity: Flow and the Psychology of Discovery and Invention.* New York: Harper.

Cummins, Denise Dellarosa. 2005. Dominance, status, and social hierarchies. In Buss 2005, 676–697.

Daly, Martin, and Margo Wilson. 1983. *Sex, Evolution, and Behavior.* Belmont, Calif.: Wadsworth.

———. 1988. *Homicide.* Hawthorne, N.Y.: Aldine de Gruyter.

———. 1998a. The evolutionary social psychology of family violence. In Crawford and Krebs 1998, 431–456.

———. 1998b. *The Truth about Cinderella: A Darwinian View of Parental Love.* London: Weidenfeld and Nicolson.

Damasio, Antonio. 1996. *Descartes' Error: Emotion, Reason and the Human Brain.* 1994. Reprint, London: Papermac.

———. 2000. *The Feeling of What Happens: Body, Emotion, and the Making of Consciousness.* London: Vintage.

———. 2003. *Looking for Spinoza: Joy, Sorrow, and the Feeling Brain.* Orlando, Fla.: Harcourt.

Damasio, Hanna, D. Tranel, T. Grabowski, R. Adolphs, and A. Damasio. 2004. Neural systems behind word and concept retrieval. *Cognition* 92:179–229.

Darwin, Charles. [1859] 2003. *On the Origin of Species by Means of Natural Selection,* ed. Joseph Carroll. Peterborough, Ont.: Broadview.

———. [1871/1874] 2003. *The Descent of Man and Selection in Relation to Sex.* 1874. 2d ed. Reprint, London: Gibson Square Books.

———. [1872] 1998. *The Expression of the Emotions in Man and Animals.* 1872. 3d ed. Ed. Paul Ekman. London: Harper Collins.

Davies, Paul. 1988. *The Cosmic Blueprint: New Discoveries in Nature's Creative Ability to Order the Universe.* New York: Simon and Schuster.

Davies, Stephen. 1997. First art and art's definition. *Southern Journal of Philosophy* 35:19–34.

———. 2000. Non-Western art and art's definition. In *Theories of Art Today,* ed. N. Carroll. Madison: University of Wisconsin Press, 199–216.

Dawkins, Richard. 1983. Universal Darwinism. In *Evolution from Molecules to Men,* ed. D. S. Bendall. Cambridge: Cambridge University Press, 403–425.

———. 1985. Sociobiology: The debate continues. Review of Richard Lewontin, Steven Rose, and Leon Kamin, *Not in Our Genes* (1984). *New Scientist,* 24 January: 59–60.

———. 1989a. *The Extended Phenotype: The Long Reach of the Gene.* 1982. 2d ed. Oxford: Oxford University Press.

———. 1989b. *The Selfish Gene.* 1976. 2d ed. Oxford: Oxford University Press.

———. 1991. *The Blind Watchmaker.* 1986. Reprint, London: Penguin.

———. 1996. *River out of Eden.* 1995. Reprint, London: Phoenix.

———. 1997. *Climbing Mount Improbable.* 1996. Reprint, London: Penguin.

———. 1998. *Unweaving the Rainbow: Science, Delusion, and the Appetite for Wonder.* New York: Houghton Mifflin.

———. 1999. Foreword. In *The Meme Machine,* ed. Susan Blackmore. Oxford: Oxford University Press, vii–xvii.

———. 2003. Introduction. In Darwin [1871/1874] 2003, vii–xxx.

————. 2004. *The Ancestor's Tale: A Pilgrimage to the Dawn of Life.* London: Weidenfeld and Nicolson.

Dawkins, R., and J. R. Krebs. 1978. Animal signals: Information or manipulation? In *Behavioural Ecology: An Evolutionary Approach,* ed. J. R. Krebs and N. B. Davies. Oxford: Blackwell Scientific Publications, 282–309.

Deacon, Terrence. 1997. *The Symbolic Species: The Co-evoluton of Language and the Human Brain.* London: Allen Lane.

Dehaene, Stanislas. 1997. *The Number Sense: How the Mind Creates Mathematics.* London: Penguin.

de Jong, Irene. 1985. Eurykleia and Odysseus' scar: *Odyssey* 19.393–466. *Classical Quarterly* 35:517–518.

————. 1999a. Introduction: Homer and literary criticism. In de Jong 1999b, 3:1–24.

————. 2001. *A Narratological Commentary on the* Odyssey. Cambridge: Cambridge University Press.

————, ed. 1999b. *Homer: Critical Assessments.* 4 vols. London: Routledge.

Deliège, Irène, and John Sloboda. 1997. *Perception and Cognition of Music.* Hove, Sussex: Psychology Press.

Dennett, Daniel C. 1978. *Brainstorms.* Montgomery, Vt.: Bradford.

————. 1993. *Consciousness Explained.* 1991. Reprint, London: Penguin.

————. 1996a. *Darwin's Dangerous Idea: Evolution and the Meanings of Life.* 1995. Reprint, London: Penguin.

————. 1996b. *Kinds of Minds: Towards an Understanding of Consciousness.* London: Weidenfeld and Nicolson.

————. 2006. *Breaking the Spell: Religion as a Natural Phenomenon.* New York: Viking.

de Quervain, Dominique J. F., Urs Fischbacher, Valerie Treyer, Melanie Schellhammer, Ulrich Schnyder, Alfred Buck, and Ernst Fehr. 2004. The neural basis of altruistic punishment. *Science* 305:1254–58.

Dessalles, Jean-Louis. 1998. Altruism, status and the origin of relevance. In Hurford, Studdert-Kennedy, and Knight 1998, 130–147.

de Waal, Frans B. M. 1996. *Good Natured: The Origins of Right and Wrong in Humans and Other Animals.* Cambridge, Mass.: Harvard University Press.

————. 1998. *Chimpanzee Politics: Power and Sex among Apes.* 1982. Rev. ed. Baltimore: Johns Hopkins University Press.

———. 2001a. *The Ape and the Sushi Master: Cultural Reflections of a Primatologist.* New York: Basic Books.

———. 2001b. Man versus ape. *Sunday Times,* 5 August.

———. 2002. Evolutionary psychology: The wheat and the chaff. *Current Directions in Psychological Science* 11:187–191.

———. 2004. Brains and the beast: Can the behaviorists' insistence on distinguishing animal from human cognition be reconciled with evolutionary continuity? *Natural History* 114 (May): 53–57.

———. 2006. Morally evolved: Primate social instincts, human morality, and the rise and fall of "veneer theory." In Frans B. M. de Waal, Robert Wright, Christine M. Korsgaard, Philip Kitcher, and Peter Singer. *Primates and Philosophers: How Morality Evolved,* ed. Stephen Macedo and Josiah Ober. Princeton: Princeton University Press, 1–80.

de Waal, Frans B. M., and Peter L. Tyack, eds. 2003. *Animal Social Complexity: Intelligence, Culture, and Individualized Societies.* Cambridge, Mass.: Harvard University Press.

Diamond, Lisa. 2003. What does sexual orientation orient? A biobehavioral model distinguishing romantic love and sexual desire. *Psychological Review* 110:173–192.

Dickie, George. 1997. Art: Function or procedure—nature or culture? *Journal of Aesthetics and Art Criticism* 55 (1):19–28.

Didion, Joan. 2005. *The Year of Magical Thinking.* London: Fourth Estate.

Dimberg, U., M. Thunberg, and K. Elmehed. 2000. Unconscious facial reactions to emotional facial expressions. *Psychological Science* 11:86–89.

Dissanayake, Ellen. 1988. *What Is Art For?* Seattle: University of Washington Press.

———. 1995. *Homo Aestheticus: Where Art Comes From and Why.* 1992. Reprint, Seattle: University of Washington Press.

———. 2000. *Art and Intimacy: How the Arts Began.* Seattle: University of Washington Press.

Dobzhansky, Theodosius. 1973. Nothing in biology makes sense except in the light of evolution. *American Biology Teacher* 35:125–129.

Doidge, Norman. 2007. *The Brain That Changes Itself: Stories of Personal Triumph from the Frontiers of Brain Science.* New York: Penguin.

Donald, Merlin. 1998. Mimesis and the executive suite: Missing links in language evolution. In Hurford, Studdert-Kennedy, and Knight 1998, 44–67.

———. 1999. Preconditions for the evolution of protolanguages. In Corballis and Lea 1999, 138–155.

Douglas, Kate. 2001. Playing fair. *New Scientist,* 10 March: 38–42.

Duchan, Judith F., Gail A. Bruder, and Lynne E. Hewitt. 1995. *Deixis in Narrative: A Cognitive Science Perspective.* Hillsdale, N.J.: Lawrence Erlbaum.

Dugatkin, Lee A. 2002. Cooperation in animals: An evolutionary overview. *Biology and Philosophy* 17:459–476.

Dunbar, Robin. 1996. *Grooming, Gossip and the Evolution of Language.* London: Faber.

———. 1998a. The social brain hypothesis. *Evolutionary Anthropology* 6:178–190.

———. 1998b. Theory of mind and the evolution of language. In Hurford, Studdert-Kennedy, and Knight 1998, 92–110.

———. 2000. On the origin of the human mind. In Carruthers and Chamberlain 2000, 238–253.

Dunbar, Robin, and Louise Barrett, eds. 2007. *The Oxford Handbook of Evolutionary Psychology.* New York: Oxford University Press.

Dunbar, Robin, Chris Knight, and Camilla Power, eds. 1999. *The Evolution of Culture: An Interdisciplinary View.* New Brunswick, N.J.: Rutgers University Press.

Dunbar, Robin, Anna Marriott, and N. D. C. Duncan. 1997. Human conversational behavior. *Human Nature* 8 (3):31–46.

Dürer, Albrecht. 1971. *Sketchbook of His Journey to the Netherlands, with Excerpts from His Diary.* Trans. P. Troutman. New York: Praeger.

Durkheim, Emile. 1915. *The Elementary Forms of Religious Life.* Trans. Joseph Ward Swain. London: Allen and Unwin.

Dutton, Denis. 1977. Art, behavior, and the anthropologists. *Current Anthropology* 18:387–407.

———. 1993. Tribal art and artifact. *Journal of Aesthetics and Art Criticism* 51 (1):13–21.

———. 1995. Mythologies of tribal art. *African Arts.* Summer: 32–45.

———. 2000. But they don't have our concept of art. In *Theories of Art Today,* ed. Noël Carroll. Madison: University of Wisconsin Press, 217–238.

————. 2009. *The Art Instinct.* New York: Bloomsbury Press.

Easterlin, Nancy. 2004. "Loving ourselves best of all": Ecocriticism and the adapted mind. *Mosaic* 37 (3):1–18.

Edelman, Gerald M. 1987. *Neural Darwinism: The Theory of Neuronal Group Selection.* New York: Basic Books.

Editors of Lingua Franca. 2000. *The Sokal Hoax: The Sham That Shook the Academy.* Lincoln: University of Nebraska Press.

Edwards, A. W. F. 2003. Human genetic diversity: Lewontin's fallacy. *BioEssays* 25:798–801.

Edwards, Mark W. 1987. *Homer: Poet of the Iliad.* Baltimore: Johns Hopkins University Press.

Eibl-Eibesfeldt, Irenäus. 1982. Warfare, man's indoctrinability and group selection. *Zeitschrift für Tierpsychologie* 60:177–198.

————. 1988. The biological foundations of aesthetics. In *Beauty and the Brain: Biological Aspects of Aesthetics,* ed. I. Rentschler, B. Herzberger, and D. Epstein. Basel: Birkhäuser, 29–68.

Eiseley, Loren. 1979. *The Star Thrower.* New York: Harcourt Brace Jovanovich.

Ekman, Paul, ed. 1982. *Emotion in the Human Face.* 1971. Reprint, Cambridge: Cambridge University Press.

Ekman, Paul, and Wallace V. Friesen. 1975. *Unmasking the Face: A Guide to Recognizing Emotions from Facial Clues.* Englewood Cliffs, N.J.: Prentice-Hall.

Ellis, Bruce J., and David F. Bjorklund, eds. 2005. *Origins of the Social Mind: Evolutionary Psychology and Child Development.* New York: Guilford.

Ellis, John M. 2005. Is theory to blame? 1997. In Patai and Corral 2005, 92–109.

Ellis, L. 1995. Dominance and reproductive success among nonhuman animals: A cross-species comparison. *Ethology and Sociobiology* 16:257–333.

Emmott, Catherine. 1997. *Narrative Comprehension: A Discourse Perspective.* Oxford: Clarendon.

Erdrich, Louise. 1990. *Jacklight.* 1984. Reprint, London: Abacus.

————. 2003. *Books and Islands in Ojibwe Country.* Washington, D.C.: National Geographic.

Ermer, Elsa, Scott A. Guerin, Leda Cosmides, John Tooby, and Michael B. Miller. 2007. Theory of mind broad and narrow: Reasoning about social exchange engages ToM areas, precautionary reasoning does not. 2006. In Saxe and Baron-Cohen 2007, 196–219.

Fagen, Robert. 1995. Animal play, games of angels, biology, and the brain. In Pellegrini 1995, 23–44.

Fauconnier, Gilles. 1997. *Mappings in Thought and Language.* Cambridge: Cambridge University Press.

Fauconnier, Gilles, and Mark Turner. 2002. *The Way We Think: Conceptual Blending and the Mind's Hidden Complexities.* New York: Basic Books.

Fehr, Ernst, and Simon Gächter. 2002. Altruistic punishment in humans. *Nature* 415:137–140.

Fehr, Ernst, and Joseph Henrich. 2003. Is strong reciprocity a maladaptation? On the evolutionary foundations of human altruism. In *Genetic and Cultural Evolution of Cooperation,* ed. P. Hammerstein. Cambridge, Mass.: MIT Press, 55–82.

Felson, Nancy, and Laura Slatkin. 2004. Gender and Homeric epic. In Fowler 2004, 91–114.

Finley, Moses. 1979. *The World of Odysseus.* 2d ed. Harmondsworth: Penguin.

Fischler, Ira. 1998. Attention and language. In Parasuraman 1998, 381–399.

Fisher, Helen. 2004. *Why We Love: The Nature and Chemistry of Romantic Love.* New York: Henry Holt.

Fivush, Robyn. 2002. Scripts, schemas, and memory of trauma. In *Representation, Memory, and Development,* ed. N. L. Stein, P. J. Bauer, and M. Rabinowitz. Mahwah, N.J.: Lawrence Erlbaum, 53–74.

Fivush, Robyn, and Catherine A. Haden. 1997. Narrating and representing experience: Preschoolers' developing autobiographical accounts. In van den Broek, Bauer, and Bourg 1997, 169–198.

Flesch, William. 2007. *Comeuppance: Costly Signaling, Altruistic Punishment, and Other Biological Components of Fiction.* Cambridge, Mass.: Harvard University Press.

Fletcher, Charles R., Amy Briggs, and Brian Linzie. 1997. Understanding the causal structure of narrative events. In van den Broek, Bauer, and Bourg 1997, 343–360.

Flinn, Mark V., David C. Geary, and Carol V. Ward. 2005. Ecological dominance, social competition, and coalitionary arms races: Why humans evolved extraordinary intelligence. *Evolution and Human Behavior* 26:10–46.

Flinn, Mark V., and Carol V. Ward. 2005. Ontogeny and evolution of the social child. In Ellis and Bjorklund 2005, 19–44.

Fodor, Jerry A. 1983. *The Modularity of Mind: An Essay on Faculty Psychology.* Cambridge, Mass.: Bradford/MIT Press.

Foley, John Miles. 2004. Epic as genre. In Fowler 2004, 171–187.

Fossey, Dian. 1983. *Gorillas in the Mist.* Boston: Houghton Mifflin.

Fowler, R. L., ed. 2004. *The Cambridge Companion to Homer.* Cambridge: Cambridge University Press.

Fox, Robin. 1972. Alliance and constraint: Sexual selection in the evolution of human kinship systems. In *Sexual Selection and the Evolution of Man, 1871–1971,* ed. B. G. Campbell. London: Heinemann, 282–331.

———. 1995. Sexual conflict in the epics. *Human Nature* 8:135–144.

Frank, Robert H. 1988. *Passions within Reason: The Strategic Role of the Emotions.* New York: Norton.

Franklin, N., and B. Tversky. 1990. Searching imagined environments. *Journal of Experimental Psychology: General* 119:63–76.

Frisch, Karl von. 1950. *Bees: Their Vision, Chemical Senses, and Language.* Ithaca, N.Y.: Cornell University Press.

———. 1967. *The Dance Language and Orientation of Bees.* Cambridge, Mass.: Harvard University Press.

———. 1974. *Animal Architecture.* New York: Harcourt Brace.

Fromm, Harold. 2003. The new Darwinism in the humanities: Part II, back to nature, again. *Hudson Review* 56 (2):315–327.

Frye, Northrop. 1957. *The Anatomy of Criticism: Four Essays.* Princeton: Princeton University Press.

Fukui, N. 2001. Is music the peacock's tail? Paper presented at Human Behavior and Evolution Society conference, London.

Gallagher, Helen L., and Christopher D. Frith. 2003. Functional imaging of "theory of mind." *Trends in Cognitive Sciences* 7 (2):77–83.

Gangestad, Steven W., and Jeffrey A. Simpson, eds. 2007. *The Evolution of Mind: Fundamental Questions and Controversies.* New York: Guilford.

Garcia, John, and F. Ervin. 1968. Gustatory-visceral and telereceptor-cutaneous conditioning: Adaptation in internal and external milieus. *Communications in Behavioral Biology, Part A* 1:389–415.

Garcia, John, F. R. Ervin, and R. A. Koelling. 1966. Learning with prolonged delay in reinforcement. *Psychonomic Science* 5:121–122.

Garcia, John, and R. A. Koelling. 1966. Relation of cue to consequence in avoidance learning. *Psychonomic Science* 5:123–124.

Gaulin, Steven J. C., and Donald H. McBurney. 2001. *Psychology: An Evolutionary Approach.* Upper Saddle River, N.J.: Prentice-Hall.

Gazzaniga, Michael S. 1994. *Nature's Mind: The Biological Roots of Thinking, Emotions, Sexuality, Language, and Intelligence.* 1992. Reprint, Harmondsworth: Penguin.

———. 2008. *Human: The Science behind What Makes Us Unique.* New York: HarperCollins.

———, ed. 2000. *The New Cognitive Neurosciences.* 2d ed. Cambridge, Mass.: Bradford/MIT Press.

Geary, David C. 2005. *The Origin of Mind: Evolution of Brain, Cognition, and General Intelligence.* Washington, D.C.: American Psychological Association.

Geissman, Thomas. 2000. Gibbon songs and human music from an evolutionary perspective. In Wallin, Merker, and Brown 2000, 103–123.

Gelman, Rochel. 1990. First principles organize attention to and learning about relevant data: Number and the animate-inanimate distinction as examples. *Cognitive Science* 14:79–106.

Gergely, György, Zoltán Nádasdy, Gergely Csibra, and Szilvia Biró. 1995. Taking the intentional stance at 12 months of age. *Cognition* 56:165–193.

Gerrig, Richard J. 1993. *Experiencing Narrative Worlds: On the Psychological Activities of Reading.* New Haven: Yale University Press.

Gerstmyer, J. S. 1991. Toward a theory of play as performance: An analysis of videotaped episodes of a toddler's play performance. Ph.D. diss., University of Pennsylvania.

Gervais, Matt, and David Sloan Wilson. 2005. The evolution and functions of laughter and humor: A synthetic approach. *Quarterly Review of Biology* 80:395–430.

Gigerenzer, Gerd, and Reinhard Selten, eds. 2001. *Bounded Rationality: The Adaptive Toolbox.* Cambridge, Mass.: Bradford.

Gigerenzer, Gerd, and Peter M. Todd. 1999. Fast and frugal heuristics: The adaptive toolbox. In Gigerenzer, Todd, and ABC Research Group 1999, 3–34.

Gigerenzer, Gerd, Peter M. Todd, and ABC Research Group, eds. 1999. *Simple Heuristics That Make Us Smart.* New York: Oxford University Press.

Glausiusz, Josie. 2001. The genetic mystery of music. *Discover* 22 (8): 71–75.

Goldberg, Rick. 2002. Costly signaling in the Jewish context. Paper presented at Human Behavior and Evolution Society conference, Rutgers University.

Goldinger, S. D. 1996. Words and voices: Episodic traces in spoken word identification and recognition memory. *Journal of Experimental Psychology: Learning, Memory and Cognition* 22 (5):1166–83.

———. 1998. Echoes of echoes? An episodic theory of lexical access. *Psychological Review* 105:251–279.

Goleman, Daniel. 2006. *Social Intelligence: The New Science of Human Relationships.* New York: Bantam.

Gombrich, Ernst. 1960. *Art and Illusion: A Study in the Psychology of Pictorial Representation.* Princeton: Princeton University Press.

———. 1979. *Ideals and Idols: Essays on Values in History and Art.* Oxford: Phaidon.

———. 1982. *The Image and the Eye.* Oxford: Phaidon.

———. 1991. *Topics of Our Time: Twentieth-Century Issues in Learning and in Art.* London: Phaidon.

———. 1995. *The Story of Art.* 1950. Reprint, London: Phaidon.

Göncü, Artin, and Elisa L. Klein. 2001. Children in play, story, and school: A tribute to Greta G. Fein. In *Children in Play, Story, and School,* ed. A. Göncü and E. L. Klein. New York: Guilford, 3–15.

Goodall, Jane. 1986. *The Chimpanzees of Gombe: Patterns of Behavior.* Cambridge, Mass.: The Belknap Press of Harvard University Press.

———. 1990. *Through a Window.* Boston: Houghton Mifflin.

Goodman, Nelson. 1972. *Problems and Projects.* Indianapolis: Bobbs-Merrill.

Gopnik, Alison, and Janet Wilde Astington. 2000. Children's understanding of representational change and its relation to the understanding of false belief and the appearance-reality distinction. 1988. In *Childhood Cognitive Development: The Essential Readings,* ed. K. Lee. Oxford: Blackwell.

Gopnik, Alison, and Andrew N. Meltzoff. 1997. *Words, Thoughts, and Theories.* Cambridge, Mass.: Bradford/MIT Press.

Gordon, Nakia S., Sharon Burke, Huda Akil, Stanley J. Watson, and Jaak

Panksepp. 2003. Socially induced brain "fertilization": Play promotes brain-derived neurotrophic factor transcription in the amygdala and dorsolateral frontal cortex in juvenile rats. *Neuroscience Letters* (341):17–20.

Gottschall, Jonathan. 2003. The tree of knowledge and Darwinian literary study. *Philosophy and Literature* 27:255–268.

———. 2008a. *Literature, Science, and a New Humanities.* New York: Palgrave.

———. 2008b. *The Rape of Troy: Evolution, Violence, and the World of Homer.* Cambridge: Cambridge University Press.

Gottschall, Jonathan, and David Sloan Wilson, eds. 2005. *The Literary Animal: Evolution and the Nature of Narrative.* Evanston: Northwestern University Press.

Gould, Stephen Jay. 1980. Sociobiology and the theory of natural selection. In *Sociobiology: Beyond Nature/Nurture? Reports, Definitions and Debate,* ed. G. W. Barlow and J. Silverberg. Boulder: Westview, 257–269.

———. 1981. *The Mismeasure of Man.* New York: Norton.

———. 1985. *The Flamingo's Smile: Reflections in Natural History.* New York: Norton.

———. 1991. *Wonderful Life: The Burgess Shale and the Nature of History.* 1989. Reprint, London: Penguin.

———. 1992. *Bully for Brontosaurus: Further Reflections in Natural History.* 1991. Reprint, London: Penguin.

Gould, Stephen Jay, and Richard Lewontin. 1979. The spandrels of San Marco and the Panglossian paradigm: A critique of the adaptationist paradigm. *Proceedings of the Royal Society of London B* 205:581–598.

Gray, Russell D., Megan Heaney, and Scott Fairhall. 2003. Evolutionary psychology and the challenge of adaptive explanation. In *From Mating to Mentality: Evaluating Evolutionary Psychology,* ed. K. Sterelny and J. Fitness. New York: Psychology Press, 247–268.

Green, C. S., and Daphne Bavelier. 2007. Action-video-game alters the spatial resolution of vision. *Psychological Science* 18:88–94.

Greenblatt, Stephen. 1988. *Shakespearean Negotiations: The Circulation of Social Energy in Renaissance England.* Berkeley: University of California Press.

Greenspan, S. I., and S. G. Shanker. 2004. *The First Idea.* Cambridge, Mass.: Da Capo.

Grice, Paul. 1975. Logic and conversation. In *Syntax and Semantics.* Vol. 3: *Speech Acts,* ed. P. Cole and J. Morgan. New York: Academic.

Griffin, Donald R. 2001. *Animal Minds: Beyond Cognition to Consciousness.* 1992. 2d ed. Chicago: University of Chicago Press.

Griffin, Jasper. 1977. The epic cycle and the uniqueness of Homer. *Journal of Hellenic Studies* 97:39–53.

———. 1980. *Homer on Life and Death.* Oxford: Clarendon.

Griffiths, Paul E. 1997. *What Emotions Really Are: The Problem of Psychological Categories.* Chicago: University of Chicago Press.

Groos, K. 1898. *The Play of Animals.* New York: Appleton.

Gunn, D. M. 1970. Thematic composition and Homeric authorship. *Harvard Studies in Classical Philology* 75:1–31.

Haan, Michelle de, and Mark H. Johnson. 2003. Mechanisms and theories of development. In *The Cognitive Neuroscience of Development.* Hove, Sussex: Psychology Press, 1–18.

Haidt, Jonathan. 2006. *The Happiness Hypothesis: Finding Modern Truth in Ancient Wisdom.* New York: Basic Books.

Hamilton, W. D. 1963. The evolution of altruistic behavior. *American Naturalist* 97:354–356.

———. 1964. The genetical evolution of social behaviour. *Journal of Theoretical Biology* 7:1–51.

———. 1975. Gamblers since life began: Barnacles, aphids, elms. *Quarterly Review of Biology* 50:175–180.

Hansen, Brian. 1999. A Prehistory of theatre: A path with six turnings. In Cooke and Turner 1999, 347–365.

Hare, B., J. Call, and M. Tomasello. 2001. Do chimpanzees know what conspecifics know? *Animal Behaviour* 61:1085–89.

Harlow, Harry F., and R. R. Zimmerman. 1958. The development of affectional responses in infant monkeys. *Proceedings of the American Philosophical Society* 102:501–509.

Harris, Judith Rich. 2006. *No Two Alike: Human Nature and Human Individuality.* New York: Norton.

Harris, Paul L. 1990. The child's theory of mind and its cultural context. In *The Causes of Development,* ed. G. Butterworth and P. Bryant. Hemel Hempstead: Harvester Wheatsheaf, 215–237.

———. 1991. The work of the imagination. In *Natural Theories of Mind:*

Evolution, Development and Simulation of Everyday Mindreading, ed. A. Whiten. Oxford: Blackwell, 238–304.

Hatfield, Rab. 2002. *The Wealth of Michelangelo.* Rome: Edizioni di storia e letteratura.

Hauser, Marc. 2000. *Wild Minds: What Animals Really Think.* London: Allen Lane/Penguin.

Hauser, Marc D., and B. Spaulding. 2006. Monkeys generate causal inferences about possible and impossible physical transformations. *Proceedings of the National Academy of Sciences* 103:7181–85.

Haydon, Brian. 1990. Nimrods, piscators, pluckers and planters: The emergence of food production. *Journal of Anthropological Archaeology* 9:31–69.

Headlam Wells, Robin. 2005. *Shakespeare's Humanism.* Cambridge: Cambridge University Press.

Heberlein, A. S., R. Adolphs, D. Tranel, D. Kemmerer, S. Anderson, and Antonio Damasio. 1998. Impaired attribution of social meanings to abstract dynamic visual patterns following damage to the amygdala. *Society for Neuroscience Abstracts* 24:1176.

Heider, Fritz, and Marianne Simmel. 1944. An experimental study of apparent behavior. *American Journal of Psychology* 57:243–259.

Henrich, Joseph, Robert Boyd, Sam Bowles, Herbert Gintis, and Ernst Fehr. 2001. In search of *Homo economicus:* Experiments in 15 small-scale societies. *American Economic Review* 91:73–78.

Henrich, Joseph, and Francisco J. Gil-White. 2001. The evolution of prestige: Freely conferred deference as a mechanism for enhancing the benefits of cultural transmission. *Evolution and Human Behavior* 22:165–196.

Hernadi, Paul. 2001. Literature and evolution. *Substance* 30 (1–2):55–71.

Hewlett, Barry S., and Luigi Luca Cavalli-Sforza. 1986. Cultural transmission among Aka pygmies. *American Anthropologist* 88:922–934.

Heyes, Cecilia. 2003. Four routes of cognitive evolution. *Psychological Review* 110:713–727.

Hilfer, Tony. 2003. *The New Hegemony in Literary Studies: Contradictions in Theory.* Evanston: Northwestern University Press.

Hintzman, D. H. 1986. "Schema abstraction" in a multiple-trace memory model. *Psychological Review* 93:411–428.

Hirstein, William. 2005. *Brain Fiction: Self-Deception and the Riddle of Confabulation.* Cambridge, Mass.: MIT Press.

Hobson, J. Allan. 1994. *The Chemistry of Conscious States.* Boston: Little, Brown.

Hobson, Peter. 2004. *The Cradle of Thought: Exploring the Origins of Thinking.* Oxford: Oxford University Press.

Hoffrage, Ulrich, and Ralph Hertwig. 1999. Hindsight bias: A price worth paying for fast and frugal memory. In Gigerenzer, Todd, and ABC Research Group 1999, 191–208.

Hogan, Patrick Colm. 2003. *The Mind and Its Stories: Narrative Universals and Human Emotion.* Cambridge: Cambridge University Press.

Hogan, Robert. 1983. A socioanalytic theory of personality. In *Nebraska Symposium in Motivation,* ed. M. M. Page. Lincoln: University of Nebraska Press, 55–89.

Hollingham, Richard. 2004. In the realm of your senses. *New Scientist,* 31 January: 40–43.

Homer. 1951. *The Iliad.* Trans. R. Lattimore. Chicago: University of Chicago Press.

———. 1968. *The Odyssey.* Trans. R. Lattimore. 1965. Reprint, New York: Harper.

———. 1988. *The Odyssey.* Trans. R. Fitzgerald. 1961. Reprint, London: Collins Harvill.

———. 1993. *The Odyssey.* Trans. A. Cook. 1967. Reprint, New York: Norton.

———. 1995. *The Odyssey.* Trans. A. T. Murray and G. E. Dimock. 2 vols. 1919. Loeb Classical Library. Cambridge, Mass.: Harvard University Press.

———. 1996. *The Odyssey.* Trans. R. Fagles. New York: Viking.

———. 2000. *Odyssey.* Trans. S. Lombardo. Indianapolis: Hackett.

———. 2003. *The Odyssey.* Trans. Rodney Merrill. Ann Arbor: University of Michigan Press.

Horton, Robin. 1993. *Patterns of Thought in Africa and the West: Essays on Magic, Religion and Science.* Cambridge: Cambridge University Press.

Hrdy, Sarah Blaffer. 1999. *Mother Nature: A History of Mothers, Infants, and Natural Selection.* New York: Pantheon.

Hull, David L. 1988. *Science as a Process: An Evolutionary Account of the So-*

cial and Conceptual Development of Science. Chicago: University of Chicago Press.

Humphrey, Nicholas K. 1976. The social function of intellect. In *Growing Points in Ethology,* ed. P. P. G. Bateson and R. A. Hinde. Cambridge: Cambridge University Press, 303–317.

Hurford, James R., Michael Studdert-Kennedy, and Chris Knight, eds. 1998. *Approaches to the Evolution of Language.* Cambridge: Cambridge University Press.

Huxley, Julian. 1932. *Problems of Relative Growth.* London: Methuen.

Iacoboni, Marco. 2008. *Mirroring People: The New Science of How We Connect with Others.* New York: Farrar, Straus and Giroux.

Irons, William. 2001. Religion as a hard-to-fake sign of commitment. In *Evolution and the Capacity for Commitment,* ed. R. Nesse. New York: Russell Sage Foundation, 292–309.

Jackson, Michael. 1982. *Allegories of the Wilderness: Ethics and Ambiguity in Kuranko Narratives.* Bloomington: Indiana University Press.

Jaynes, Julian. 1976. *The Origin of Consciousness in the Breakdown of the Bicameral Mind.* Boston: Houghton Mifflin.

Jenkins, W. M., M. M. Merzenich, M. T. Ochs, T. Allard, and E. Guic-Robles. 1990. Functional reorganization of primary somatosensory cortex in adult owl monkeys after behaviorally controlled tactile stimulation. *Journal of Neurophysiology* 63 (1):82–104.

Johansson, Gunnar. 1973. Visual perception of biological motion and a model for its analysis. *Perception and Psychophysics* 14:201–211.

Johansson, Gunnar, Sten Sture Bergström, William Epstein, and Gunnar Jansson, eds. 1994. *Perceiving Events and Objects.* Hillsdale, N.J.: Lawrence Erlbaum.

Johnson, Mark H. 1998. Developing an attentive brain. In Parasuraman 1998, 428–443.

Johnson, Ryan T., Joshua A. Burk, and Lee A. Kirkpatrick. 2007. Dominance and prestige as differential predictors of aggression and testosterone levels in men. *Evolution and Human Behavior* 28:345–351.

Johnson, Susan, Amy Booth, and Kirsten O'Hearn. 2001. Inferring the goals of a non-human agent. *Cognitive Development* 16:637–656.

Johnson, Vivien. 2003. *Clifford Possum Tjapaltjarri.* Adelaide: Art Gallery of South Australia.

Johnson, Vivien, and Clifford Possum Tjapaltjarri. 1994. *The Art of Clifford*

Possum Tjapaltjarri. East Roseville, New South Wales: Gordon and Breach Arts International.

Jolly, Alison. 1966. Lemur social behavior and primate intelligence. *Science* 153 (3735):501–506.

———. 1999. *Lucy's Legacy: Sex and Intelligence in Human Evolution*. Cambridge, Mass.: Harvard University Press.

Jones, Owen D., and Timothy H. Goldsmith. 2005. Law and behavioral biology. *Columbia Law Review* 105:405–502.

Jones, Peter. 1991. *Homer: The Odyssey 1 & 2*. Warminster, Eng.: Aris and Phillips.

———. 2003. *Homer's* Iliad: *A Commentary on Three Translations*. Bristol, Eng.: Bristol Classical.

Jung, Wonil Edward. 2003. The inner eye theory of laughter: Mindreader signals cooperator value. *Evolutionary Psychology* 1:214–253.

Kaplan, Joan A., Hiram H. Brownell, Janet R. Jacobs, and Howard Gardner. 1990. The effects of right hemisphere damage on the pragmatic interpretation of conversational remarks. *Brain and Language* 38:315–333.

Katz, Leonard D., ed. 2000. *Evolutionary Origins of Morality: Cross-Disciplinary Perspectives*. Bowling Green, Ohio: Imprint Academic.

Keesing, R. M. 1981. *Cultural Anthropology: A Contemporary Perspective*. New York: Holt, Rinehart and Winston.

Keller, Heidi, Axel Schölmerich, and Irenaus Eibl-Eibesfeldt. 1988. Communication patterns in adult-human interactions in western and non-western cultures. *Journal of Cross-Cultural Psychology* 19:427–445.

Kendrick, Keith. 2004. Here's looking at ewe. *New Scientist*, 12 June: 48–49.

Kenrick, Douglas T., Edward K. Sadalla, and Richard C. Keele. 1998. Evolutionary cognitive psychology: The missing heart of cognitive science. In Crawford and Krebs 1998, 485–514.

Kinderman, Peter, Robin Dunbar, and Richard P. Bentall. 1998. Theory of mind deficits and causal attributions. *British Journal of Psychology* 89:191–204.

King, James E., Duane M. Rumbaugh, and E. Sue Savage-Rumbaugh. 1999. Perception of personality traits and semantic learning in evolving hominids. In Corballis and Lea 1999, 98–116.

Kingdon, Jonathan. 2003. *Lowly Origin: Where, When, and Why Our First Ancestors Stood Up*. Princeton: Princeton University Press.

Kirk, G. S. 1993. *Heroic Age and Heroic Poetry in Homer: The* Odyssey, ed. A. Cook. 1962. Reprint, New York: Norton, 284–294.

Kirkpatrick, Lee A. 2005. *Attachment, Evolution, and the Psychology of Religion.* New York: Guilford.

Klein, Stanley B., Leda Cosmides, Cynthia E. Gangi, Betsy Jackson, and Kristi Costabile. In press. Evolution and episodic memory: An analysis and demonstration of a social function of episodic recollection. *Social Cognition.*

Klein, Stanley B., Leda Cosmides, John Tooby, and Sarah Chance. 2002. Decisions and the evolution of memory: Multiple systems, multiple functions. *Psychological Review* 109:306–329.

Klin, Ami, David Schultz, and Donald Cohen. 2000. Theory of mind in action: Developmental perspectives on social neuroscience. In Baron-Cohen, Tager-Flusberg, and Cohen 2000, 357–488.

Kniffin, Kevin M., and David Sloan Wilson. 2005. Utilities of gossip across organizational levels: Multilevel selection, free-riders, and teams. *Human Nature* 16 (3):278–292.

Knight, Chris. 1998. Ritual/speech coevolution: A solution to the problem of deception. In Hurford, Studdert-Kennedy, and Knight 1998, 68–91.
———. 1999. Sex and language as pretend-play. In Dunbar, Knight, and Power 1999, 228–247.

Knox, Bernard. 1996. Introduction. In Homer 1996, 3–64.

Kobayashi, H., and D. Koshima. 2001. Unique morphology of the human eye and its adaptive meaning: Comparative studies on external morphology of the primate eye. *Journal of Human Evolution* 40:419–435.

Kohlberg, S. 1969. Stage and sequence: The cognitive-developmental approach to socialization. In *Handbook of Socialization Theory and Research,* ed. D. A. Goslin. New York: Rand McNally, 347–380.

Kohn, Marek, and Steve Mithen. 1999. Handaxes: Products of sexual selection? *Antiquity* 73:518–526.

Krebs, Dennis. 2000. The evolution of moral dispositions in the human species. In *Evolutionary Perspectives on Human Reproductive Behavior,* ed. D. LeCroy and P. Moller. New York: New York Academy of Sciences, 132–148.

Krebs, J. R., and Richard Dawkins. 1984. Animal signaling: Mind-reading and manipulation. In *Behavioral Ecology,* ed. J. R. Krebs and N. B. Davies. Sunderland, Mass.: Sinauer Associates, 380–402.

Kugiumtzakis, G. 1988. Neonatal imitation in the intersubjective companion space. In *Intersubjective Communication and Emotion in Early Ontogeny,* ed. S. Braten. Cambridge: Cambridge University Press, 63–88.

Kurzban, Robert, and Martie Haselton. 2006. Making hay out of straw? Real and imagined controversies in evolutionary psychology. In Barkow 2006, 149–161.

Kurzban, Robert, John Tooby, and Leda Cosmides. 2001. Can race be erased? Coalitional computation and social categorization. *Proceedings of the National Academy of Sciences* 98:15387–92.

Labov, William. 1972. *Language in the Inner City: Studies in the Black English Vernacular.* Philadelphia: University of Pennsylvania Press.

Lakoff, George, and Mark Johnson. 1999. *Philosophy in the Flesh.* New York: Basic Books.

————. 2003. *Metaphors We Live By.* 1980. 2d ed. Chicago: University of Chicago Press.

Lane, Richard. 1989. *Hokusai: Life and Work.* New York: Dutton.

Langellier, Kristin M., and Eric E. Peterson. 1992. Spinstorying: An analysis of women storytelling. In *Performance, Culture, and Identity,* ed. E. C. Fine and J. H. Speer. Westport, Conn.: Praeger, 157–179.

Law. 2004. Law and the Brain special issue. *Philosophical Transactions of the Royal Society: Biological Sciences* 1697.

Lea, Stephen E. G. 1999. The background to hominid intelligence. In Corballis and Lea 1999, 16–39.

Leach, E. R. 1954. *Political Systems of Highland Burma: A Study of Kachin Social Structure.* Cambridge, Mass.: Harvard University Press.

Leary, M. R., and D. L. Downs. 1995. Interpersonal functions of the self-esteem motive: The self-esteem system as a sociometer. In *Efficacy, Agency, and Self-Esteem,* ed. M. H. Kernis. New York: Plenum, 123–144.

LeDoux, Joseph. 1998. *The Emotional Brain: The Mysterious Underpinnings of Emotional Life.* 1996. Reprint, New York: Touchstone.

Lehnert, Wendy. 1994. Cognition, computers, and car bombs: How Yale prepared me for the 1990s. In *Beliefs, Reasoning, and Decision Making: Psycho-Logic in Honor of Bob Abelson,* ed. R. C. Schank and E. Langer. Hillsdale, N.J.: Lawrence Erlbaum, 143–174.

Lehrer, Jonah. 2008. *Proust Was a Neuroscientist.* Boston: Houghton Mifflin.

Lenain, Thierry. 1997. *Monkey Painting.* 1990. Trans. Caroline Beamish. London: Reaktion.

Leslie, Alan. 1987. Pretence and representation: The origins of "theory of mind." *Psychological Review* 94:412–426.

———. 1994. Pretending and believing: Issues in the theory of ToM. *Cognition* 50:211–238.

Levin, Richard. 1971. *The Multiple Plot in English Renaissance Drama.* Chicago: University of Chicago Press.

———. 1979. *New Readings vs. Old Plays: Recent Trends in the Reinterpretation of English Renaissance Drama.* Chicago: University of Chicago Press.

———. 2003. *Looking for an Argument: Critical Encounters with the New Approaches to the Criticism of Shakespeare and His Contemporaries.* Madison, N.J.: Fairleigh Dickinson University Press.

Levitin, Daniel J. 2006. *This Is Your Brain on Music: The Science of a Human Obsession.* New York: Dutton.

Lewis, David. 1973. *Counterfactuals.* Cambridge, Mass.: Harvard University Press.

Lewis-Williams, David. 2000. *Stories That Float from Afar: Ancestral Folklore of the San of South Africa.* Cape Town: David Philip.

———. 2002. *The Mind in the Cave: Consciousness and the Origins of Art.* London: Thames and Hudson.

Lewontin, Richard. 1972. The apportionment of human diversity. *Evolutionary Biology* 6:391–398.

Livingston, Paisley. 1988. *Literary Knowledge: Humanistic Inquiry and the Philosophy of Science.* Ithaca, N.Y.: Cornell University Press.

———. 2005. *Art and Intention: A Philosophical Study.* Oxford: Clarendon.

Long, Debra L., Brian J. Oppy, and Mark R. Seely. 1997. A "global-coherence" view of event comprehension: Inferential processing as question answering. In van den Broek, Bauer, and Bourg 1997, 361–384.

Lorenz, Konrad. 1971. Part and parcel in animal and human societies. 1950. In Lorenz, *Studies in Animal and Human Behavior.* Trans. Robert Martin. Vol. 2. Cambridge, Mass.: Harvard University Press, 115–195.

———. 1978. *Behind the Mirror: A Search for a Natural History of Human Knowledge.* Trans. R. Taylor. New York: Harcourt Brace Jovanovich.

Lowe, N. J. 2000. *The Classical Plot and the Invention of Western Narrative.* Cambridge: Cambridge University Press.

Lukács, Georg. 1971. *The Theory of the Novel.* 1920. Reprint, Cambridge, Mass.: MIT Press.

MacDonald, Ruth K. 1988. *Dr. Seuss.* Boston: Twayne.

Maehler, H. 1999. The singer in the *Odyssey.* 1963. In de Jong 1999b, 4:6–20.

Maltz, D., and R. Borker. 1982. A cultural approach to male-female miscommunication. In *Language and Social Identity,* ed. J. Gumperz. New York: Cambridge University Press, 195–216.

Marker, Peter. 2000. Origins of music and speech. In Wallin, Merker, and Brown 2000, 31–48.

Marr, David. 1982. *Vision: A Computational Investigation into the Human Representation and Processing of Visual Information.* San Francisco: Freeman.

Marten, Ken, Karim Shariff, Suchi Psarakos, and Don J. White. 1996. Ring bubbles of dolphins. *Scientific American,* August: 64–69.

Martindale, Colin. 1990. *The Clockwork Muse: The Predictability of Artistic Change.* New York: Basic Books.

Marzluff, John M., Bernd Heinrich, and Colleen S. Marzluff. 1996. Raven roosts are mobile information centres. *Animal Behaviour* 51:89–103.

Mather, J. A., and R. C. Anderson. 1999. Exploration, play and habituation in octopuses *(Octopus dolfeini). Journal of Comparative Psychology* 113 (3):333–338.

Matsuzawa, Tetsuro. 2003. Koshima monkeys and Bossou chimpanzees: Long-term research on culture in nonhuman primates. In de Waal and Tyack 2003, 374–387.

Maynard Smith, John. 1982a. *Evolution and the Theory of Games.* Cambridge: Cambridge University Press.

———. 1982b. Introduction. In *Current Problems in Sociobiology,* ed. King's College Sociobiology Group. Cambridge: Cambridge University Press, 1–3.

Maynard Smith, John, and Eörs Szathmáry. 1995. *The Major Transitions in Evolution.* Oxford: Oxford University Press.

Mayr, Ernst. 1961. Cause and effect in biology. *Science* 134:1501–6.

McCloud, Scott. 1993. *Understanding Comics.* Northampton, Mass.: Kitchen Sink.

McEwan, Ian. 2004. *Enduring Love.* 1997. Reprint, London: Vintage.

McGrew, W. C. 2003. Ten dispatches from the chimpanzee cultural wars. In de Waal and Tyack 2003, 419–439.

McNeill, William H. 1995. *Keeping Together in Time: Dance and Drill in Human History.* Cambridge, Mass.: Harvard University Press.

Mead, Margaret. 1932. An investigation of the thought of primitive children with special reference to animism. *Royal Anthropological Institute* 62:173–190.

Meltzoff, Andrew N. 1988. Imitation, objects, tools, and the rudiments of language in human ontogeny. *Human Evolution* 3:45–64.

———. 1996. The human infant as imitative generalist: A 20-year progress report on infant imitation with implications for comparative psychology. In *Social Learning in Animals: The Roots of Culture,* ed. Cecilia M. Heyes and Bennett G. Galef Jr. San Diego: Academic, 347–370.

Meltzoff, Andrew N., and M. Keith Moore. 1995. Infants' understanding of people and things: From body imitation to folk psychology. In *The Body and the Self,* ed. J. L. Bermúdez, A. Marcel, and N. Eilan. Cambridge, Mass.: Bradford/MIT Press, 43–70.

Menand, Louis. 2002. Cat people: What Dr. Seuss really taught us. *New Yorker,* 23–30 December, 148–154.

———. 2005. Dangers within and without. *Profession,* 10–17.

Merzenich, M. M., P. Tallal, B. Peterson, S. Miller, and W. M. Jenkins. 1999. Some neurological principles relevant to the origins of—and the cortical plasticity-based remediation of—developmental language impairments. In *Neural Plasticity: Building a Bridge from the Laboratory to the Clinic,* ed. J. Grafman and Y. Christen. Berlin: Springer Verlag, 168–187.

Michod, R. E., M. F. Wojciechowski, and M. A. Hoelzer. 1988. DNA repair and the evolution of transformation in the bacterium *Bacillus subtilis. Genetics* 118:31–39.

Michotte, Albert. 1963. *The Perception of Causality.* London: Methuen.

Midgley, Mary. 1995. *Beast and Man: The Roots of Human Nature.* 1979. 2d ed. London: Routledge.

Miller, Geoffrey. 1999. Sexual selection for cultural displays. In Dunbar, Knight, and Power 1999, 71–91.

———. 2000a. Evolution of human music through sexual selection. In Wallin, Merker, and Brown. 2000, 329–360.

———. 2000b. *The Mating Mind: How Sexual Choice Shaped the Evolution of Human Nature.* New York: Doubleday.

Miller, Laura. 2005. Far from Narnia. *New Yorker,* 26 December, 54–75.

Minear, Richard H. 1999. *Dr. Seuss Goes to War: The World War II Editorial Cartoons of Theodor Seuss Geisel.* New York: New Press.

Mineka, Susan, M. Davidson, M. Cook, and R. Keir. 1984. Observational conditioning of snake fear in rhesus monkeys. *Journal of Abnormal Psychology* 93:355–372.

Mock, Douglas W. 2004. *More than Kin and Less than Kind: The Evolution of Family Conflict.* Cambridge, Mass.: The Belknap Press of Harvard University Press.

Morgan, Judith, and Neil Morgan. 1995. *Dr. Seuss and Mr. Geisel.* New York: Random House.

Morris, Desmond. 1962. *The Biology of Art.* London: Methuen.

Morris, Ian, and Barry Powell, eds. 1997. *A New Companion to Homer.* Leiden: Brill.

Morris, Simon Conway. 2003. *Life's Solution: Inevitable Humans in a Lonely Universe.* Cambridge: Cambridge University Press.

Most, G. W. 1999. The structure and function of Odysseus' *apologoi.* In de Jong 1999b, 4:486–503.

Murphy, Dominic, and Stephen Stich. 2000. Darwin in the madhouse: evolutionary psychology and the classification of mental disorders. In Carruthers and Chamberlain 2000, 62–92.

Nabokov, Vladimir. 1962. *Pale Fire.* New York: Putnam.

Nel, Philip. 2004. *Dr. Seuss: American Icon.* New York: Continuum.

Nelson, Katherine. 1996. *Language in Cognitive Development: The Emergence of the Mediated Mind.* New York: Cambridge University Press.

———. 1997. Event representations then, now, and next. In van den Broek, Bauer, and Bourg 1997, 1–26.

Nelson, Katherine, and G. Ross. 1980. The generalities and specifics of long-term memory in infants and young children. In *New Directions for Child Development—Children's Memory,* ed. M. Perlmutter. New York: Plenum, 87–110.

Nettle, Daniel. 1999. Language variation and the evolution of societies. In Dunbar, Knight, and Power 1999, 214–217.

———. 2005. The wheel of fire and the mating game: Explaining the origins of tragedy and comedy. *Journal of Cultural and Evolutionary Psychology* 3:39–56.

New, Joshua, Leda Cosmides, and John Tooby. 2007. Category-specific attention for animals reflects ancestral priorities, not expertise. *Pro-*

ceedings of the National Academy of Sciences of the United States of America 104: 16598–603.

Nilsson, D.-E., and S. Pelger. 1994. A pessimistic estimate of the time required for an eye to evolve. *Proceedings of the Royal Society of London B* 256:53–58.

Nisbett, Richard F., and Dov Cohen. 1996. *Culture of Honor: The Psychology of Violence in the South.* New York: HarperCollins.

Noble, Jason. 2000. Cooperation, competition and the evolution of prelinguistic communication. In *The Evolutionary Emergence of Language: Social Function and the Origins of Linguistic Form,* ed. C. Knight. Cambridge: Cambridge University Press, 40–61.

Nordlund, Marcus. 2007. *Shakespeare and the Nature of Love: Literature, Culture, and Evolution.* Evanston: Northwestern University Press.

Novitz, David. 1996. Disputes about art. *Journal of Aesthetics and Art Criticism* 54:153–163.

Nuñez, M. 2001. "Cheating" as facilitating intentional context for the False Belief Task. Paper presented at Human Behavior and Evolution Society conference, London.

Nyreröd, Marie. 2004. *Ingmar Bergman—3 dokumentärer om film, teater, Färö och livet.* SVT Nöje.

Oakley, Barbara. 2007. *Evil Genes: Why Rome Fell, Hitler Rose, Enron Failed, and My Sister Stole My Mother's Boyfriend.* Amherst, N.Y.: Prometheus Books.

Oatley, Keith. 1999. Why fiction may be twice as true as fact: Fiction as cognitive and emotional simulation. *Review of General Psychology* 3:101–117.

Ohman, A., and Susan Mineka. 2001. Fears, phobias and preparedness: Toward an evolved module of fear and fear learning. *Psychological Review* 108:483–522.

Olson, S. Douglas. 1995. *Blood and Iron: Stories and Storytelling in Homer's Odyssey.* Leiden: Brill.

Ostrom, E., R. Gardner, and J. M. Walker. 1994. *Rules, Games, and Common-Pool Resources.* Ann Arbor: University of Michigan Press.

Packer, C. 1977. Reciprocal altruism in *Papilio anubis. Nature* 265:441–443.

Panksepp, Jaak. 1998. *Affective Neuroscience: The Foundations of Human and Animal Emotions.* New York: Oxford University Press.

Parasuraman, Raja, ed. 1998. *The Attentive Brain.* Cambridge, Mass.: Bradford/MIT Press.

Patai, Daphne, and Will H. Corral, eds. 2005. *Theory's Empire: An Anthology of Dissent.* New York: Columbia University Press.

Pellegrini, Anthony D., ed. 1995. *The Future of Play Theory: A Multidisciplinary Inquiry into the Contributions of Brian Sutton-Smith.* Albany: State University of New York Press.

Pellis, Sergio M. 2002. Keeping in touch: Play fighting and social knowledge. In Bekoff, Allen, and Burghardt 2002, 421–427.

Perner, Josef. 1991. *Understanding the Representational Mind.* Cambridge, Mass.: MIT Press.

Peterson, Carole, and Alyssa McCabe. 1983. *Developmental Psycholinguistics: Three Ways of Looking at a Child's Narrative.* New York: Plenum.

Petyarre, Kathleen. 2001. *Kathleen Petyarre: Genius of Place.* Kent Town, South Africa: Wakefield.

Pinker, Steven. 1995. *The Language Instinct: The New Science of Language and Mind.* 1994. Reprint, London: Penguin.

———. 1997. *How the Mind Works.* New York: Norton.

———. 1999. *Words and Rules: The Ingredients of Language.* New York: Basic Books.

———. 2002. *The Blank Slate: The Modern Denial of Human Nature.* New York: Viking.

———. 2005. Foreword. In Buss 2005, xi–xvi.

———. 2007a. *The Stuff of Thought: Language as a Window into Human Nature.* London: Allen Lane.

———. 2007b. Toward a consilient study of literature. *Philosophy and Literature* 31:162–178.

Plotkin, Henry. 1994. *Darwin Machines and the Nature of Knowledge.* Cambridge, Mass.: Harvard University Press, 1993. Reprint, Harmondsworth: Penguin.

———. 1997. *Evolution in Mind: An Introduction to Evolutionary Psychology.* Cambridge, Mass.: Harvard University Press.

Pollock, Lina A. 1983. *Forgotten Children: Parent-Child Relations from 1500–1900.* Cambridge: Cambridge University Press.

Popper, Karl R. 1976. *Unended Quest: An Intellectual Autobiography.* Glasgow: Fontana.

———. 1999. *All Life Is Problem-Solving*. Trans. P. Camiller. London: Routledge.

Povinelli, Daniel J., and T. M. Preuss. 1995. Theory of mind: Evolutionary history of a cognitive specialization. *Trends in Neuroscience* 18:418–424.

Powell, Barry. 1991. *Homer and the Origin of the Greek Alphabet*. Cambridge: Cambridge University Press.

Premack, David, and Ann James Premack. 1995. Afterword. In Sperber, Premack, and Premack 1995, 651–654.

———. 2003. *Original Intelligence: Unlocking the Mystery of Who We Are*. New York: McGraw-Hill.

Premack, David, and Guy Woodruff. 1978. Does the chimpanzee have a theory of mind? *Behavioral and Brain Sciences* 1:515–526.

Provine, Robert R. 2000. *Laughter: A Scientific Investigation*. London: Faber and Faber.

Pulvermüller, F., H. Huss, F. Kherif, F. M. P. Martin, O. Hauk, and Y. Shtyrov. 2006. Motor cortex maps articulatory features of speech sounds. *Proceedings of the National Academy of Sciences of the United States of America* 103:7865–70.

Pusey, Anne, Jennifer Williams, and Jane Goodall. 1997. The influence of dominance rank on the reproductive success of female chimpanzees. *Science* 277:828–830.

Pylyshyn, Z. W. 1978. When is attribution of beliefs justified? *Behavior and Brain Science* 1:592–593.

Quine, Willard Van Orman. 1960. *Word and Object*. Cambridge, Mass.: Harvard University Press.

Rabkin, Eric S. 1999. The descent of fantasy. In Cooke and Turner 1999, 47–57.

Radcliffe Richards, Janet. 2000. *Human Nature after Darwin: A Philosophical Introduction*. London: Routledge.

Rakoczy, Hannes, Michael Tomasello, and Tricia Striano. 2005. How children turn objects into symbols: A cultural learning account. In *Symbol Use and Symbolic Representation: Developmental and Comparative Perspectives*, ed. L. L. Namy. Mahwah, N.J.: Lawrence Erlbaum, 69–97.

Ramachandran, V. S. 2001. Mirror neurons and imitation learning as the driving force behind "the great leap forward" in human evolution. Available at www.edge.org/documents/archive/edge69.html.

Ramachandran, V. S., and Sandra Blakeslee. 1998. *Phantoms in the Brain: Human Nature and the Architecture of the Mind.* London: Fourth Estate.

Reeve, Hudson Kern. 1998. Acting for the good of others: Kinship and reciprocity with some new twists. In Crawford and Krebs 1998, 43–85.

Renner, M. J., and M. R. Rosenzweig. 1987. *Enriched and Impoverished Environments.* New York: Springer Verlag.

Richerson, Peter J., and Robert Boyd. 2005. *Not by Genes Alone: How Culture Transformed Human Evolution.* Chicago: University of Chicago Press.

Ridley, Matt. 2003. *Nature via Nurture: Genes, Experience, and What Makes Us Human.* New York: HarperCollins.

Root-Bernstein, Robert. 1989. *Discovering.* Cambridge, Mass.: Harvard University Press.

———. 1996. The sciences and arts share a common creative aesthetic. In *The Elusive Synthesis: Aesthetics and Science,* ed. A. I. Tauber. Dordrecht: Kluwer, 49–82.

———. 2000. Art advances science. *Nature* 407 (14 September): 134.

———. 2003. The art of innovation: Polymaths and universality of the creative process. In *International Handbook on Innovation,* ed. Larisa Shavinina. Amsterdam: Elsevier, 267–278.

Rosch, Eleanor. 1973. On the internal structure of perceptual and semantic categories. In *Cognitive Development and the Acquisition of Language,* ed. T. Moore. New York: Academic, 111–144.

———. 1975. Cognitive representations of semantic categories. *Journal of Experimental Psychology* 104:192–233.

Rosch, Eleanor, and C. Mervis. 1975. Family resemblances: Studies in the internal structure of categories. *Cognitive Psychology* 7:573–605.

Rosser, James C., Paul J. Lynch, Laurie Cuddihy, Douglas A. Gentile, Jonathan Klonsky, and Ronald Merrell. 2007. The impact of video games on training surgeons in the 21st century. *Archives of Surgery (JAMA)* 142:181–186.

Rubin, K. H., G. G. Fein, and B. Vandenberg. 1983. Play. In *Handbook of Child Psychology,* ed. P. H. Mussen. New York: Wiley, 693–774.

Rutherford, R. B. 1999. The philosophy of the *Odyssey.* 1986. In Jong 1999b, 4:271–298.

Sacks, Oliver. 1995. *An Anthropologist on Mars.* New York: Knopf.

Sale, W. M. 1987. The formularity of the place phrases in the *Iliad. Transactions of the American Philological Association* 117:21–50.

———. 1989. The Trojans, statistics, and Milman Parry. *Greek, Roman and Byzantine Studies* 30:341–410.

Santos, Laurie R., Marc D. Hauser, and Elizabeth Spelke. 2002. Domain-specific knowledge in human children and nonhuman primates: Artifacts and foods. In Bekoff, Allen, and Burghardt 2002, 205–215.

Sapolsky, Robert. 2006. The 2% difference. *Discover* 27:42–45.

Sawyer, R. Keith. 1997. *Pretend Play as Improvisation: Conversation in the Preschool Classroom*. Mahwah, N.J.: Lawrence Erlbaum.

Saxe, Rebecca. 2004. Reading your mind: How our brains help us understand other people. *Boston Review,* February/March.

Saxe, Rebecca, and Simon Baron-Cohen. 2007. *Theory of Mind.* Special issue of *Social Neuroscience.* 2006. Hove: Psychology Press.

Scalise Sugiyama, Michelle. 2001a. Food, foragers, and folklore: The role of narrative in human subsistence. *Evolution and Human Behavior* 22:221–240.

———. 2001b. Narrative theory and function: Why evolution matters. *Philosophy and Literature* 25:233–250.

———. 2004. Predation, narration, and adaptation: "Little Red Riding Hood" revisited. *Interdisciplinary Literary Studies* 5:108–127.

———. 2005. Reverse engineering narrative: Evidence of special design. In Gottschall and Wilson 2005, 177–196.

Schacter, Daniel L., and Donna Rose Addis. 2007a. The cognitive neuroscience of constructive memory: Remembering the past and imagining the future. *Philosophical Transactions of the Royal Society B* 362:773–786.

———. 2007b. On the constructive episodic simulation of past and future events. *Behavioral and Brain Sciences* 30:331–332.

Schacter, Daniel L., Donna Rose Addis, and Randy L. Buckner. 2008. Episodic simulation of future events: Concepts, data, and applications. *Annals of the New York Academy of Sciences* 1124:39–60.

Schaller, Susan. 1991. *A Man without Words.* New York: Summit Books.

Schank, Roger C. 1995. *Tell Me a Story: Narrative and Intelligence.* 1990. Reprint, Evanston: Northwestern University Press.

Schank, Roger C., and Robert P. Abelson. 1977. *Scripts, Plans, Goals, and Understanding: An Inquiry into Human Knowledge Structures.* Hillsdale, N.J.: Lawrence Erlbaum.

Schank, Roger C., and Peter G. Childers. 1984. *The Cognitive Computer: On*

Language, Learning, and Artificial Intelligence. Reading, Mass.: Addison-Wesley.

Schiller, Friedrich. [1795] 1967. Fourteenth letter. In *Letters on the Aesthetic Education of Man,* ed. E. H. Wilkinson and L. A. Willoughby. Oxford: Oxford University Press.

Schnider, Armin, Nathalie Valenza, Stephanie Morand, and Christoph M. Michel. 2002. Early cortical distinction between memories that pertain to ongoing reality and memories that don't. *Cerebral Cortex* 12 (1):54–61.

Scodel, Ruth. 2002. *Listening to Homer: Tradition, Narrative, and Audience.* Ann Arbor: University of Michigan Press.

———. 2004. The story-teller and his audience. In Fowler 2004, 45–55.

Scott, A. O. 2000. Sense and nonsense. *New York Times Magazine,* 26 November. Available at http://www.nytimes.com/library/magazine/home/20001126mag-seuss.html.

Sebeok, Thomas A. 1979. Prefigurements of art. *Semiotica* 27:3–73.

Segerstråle, Ullica. 2000. *Defenders of the Truth: The Battle for Science in the Sociobiology Debate and Beyond.* Oxford: Oxford University Press.

Sensu, Atsushi, Mark H. Johnson, and Gergely Csibra. 2007. The development and neural basis of gaze perception. In Saxe and Baron-Cohen 2007, 220–234.

Seuss, Dr. 1951. Horton and the Kwuggerbug. *Redbook* (January): 46–47.

———. 1957. *The Cat in the Hat.* New York: Random House.

———. 1989. *And to Think That I Saw It on Mulberry Street.* 1938. Reprint, New York: Random House.

———. 1995. *The Secret Art of Dr. Seuss.* New York: Random House.

———. 1998. *Horton Hears a Who!* 1954. Reprint, New York: Collins.

Seyfarth, R. M., D. L. Cheney, and P. Marler. 1980. Vervet monkey alarm calls: Evidence for predator classification and semantic communication. *Animal Behavior* 28:1070–94.

Shaw, George Bernard. 1921. *Back to Methuselah.* London: Constable.

Sheff, David. 1987. Seuss on wry. *Parenting Magazine,* February: 52–57.

Shklovsky, Viktor. 1990. *Theory of Prose.* 1925. Trans. Benjamin Sher. Elmwood Park, Ill.: Dalkey Archive.

Sidanius, Jim, and Felicia Pratto. 1999. *Social Dominance: An Intergroup Theory of Social Hierarchy and Oppression.* Cambridge: Cambridge University Press.

Sidney, Sir Philip. [1595] 1966. *A Defence of Poetry,* ed. J. A. Van Dorsten. Oxford: Oxford University Press.

Silk, Michael. 2004. The *Odyssey* and its explorations. In Fowler 2004, 31–44.

Simonton, Dean Keith. 1999. *Origins of Genius: Darwinian Perspectives on Creativity.* New York: Oxford University Press.

Singer, Peter. 1999. *A Darwinian Left: Politics, Evolution and Cooperation.* London: Weidenfeld and Nicolson.

Sirois, Sylvain, and Iain Jackson. 2007. Social cognition in infancy: A critical review of research on higher order abilities. *European Journal of Developmental Psychology* 4:46–64.

Siviy, Steven. 1998. Neurobiological substrates of play behavior: Glimpses into the structure of mammalian playfulness. In *Animal Play: Evolutionary, Comparative and Ecological Perspectives,* ed. M. Bekoff and J. A. Byers. New York: Cambridge University Press, 221–242.

Slingerland, Edward. 2008. *What Science Has to Offer the Humanities.* Cambridge: Cambridge University Press.

Sloboda, John. 2003. Power of music. *New Scientist,* 29 November: 38–42.

Smail, David Lord. 2007. *On Deep History and the Brain.* Berkeley: University of California Press.

Smith, Peter K. 1982. Does play matter? Functional and evolutionary aspects of animal and human play. *Behavioral and Brain Sciences* 5:139–184.

———. 2005. Play: Types and functions in human development. In Ellis and Bjorklund 2005, 271–291.

———, ed. 1984. *Play in Animals and Humans.* New York: Oxford University Press.

Smuts, Barbara. 2000. Battle of the sexes. In Bekoff 2000, 92–95.

Snell, Bruno. 1975. *Die Entdeckung des Geistes: Studien zur Entstehung des europäischen Denkens bei den Griechen.* 1946. Reprint, Göttingen: Vandenhoeck und Ruprecht.

Sodian, Beate. 1986. *Wissen durch Denken? Über den naiven Empirismus im Denken von Vorschulkindern.* Münster: Aschendorf.

Sokal, Alan D. 1996. Transgressing the boundaries: Towards a transformative hermeneutics of quantum gravity. *Social Text* 46/47 (Spring/Summer):217–252.

———. 2008. *Beyond the Hoax: Science, Philosophy and Culture.* Oxford: Oxford University Press.

Sokal, Alan D., and Jean Bricmont. 1998. *Fashionable Nonsense: Postmodern Intellectuals' Abuse of Modern Science.* New York: Picador.

Solso, Robert L. 1994. *Cognition and the Visual Arts.* Cambridge, Mass.: Bradford/MIT Press.

———. 2003. *The Psychology of Art and the Evolution of the Conscious Brain.* Cambridge, Mass.: MIT Press.

Sosis, Richard. 2000. Religion and intragroup cooperation: Preliminary results of a comparative analysis of utopian communities. *Cross-Cultural Research* 34:70–87.

———. 2004. The adaptive value of religious ritual. *American Scientist* 92:166–172.

Sosis, Richard, and E. Bressler. 2003. Cooperation and commune longevity: A test of the costly signaling theory of religion. *Cross-Cultural Research* 37:211–239.

Spelke, Elizabeth. 1995. Initial knowledge: Six suggestions. In *Cognition on Cognition,* ed. J. Mehler and S. Franck. Cambridge, Mass.: Bradford/MIT Press, 433–447.

Sperber, Dan. 1996. *Explaining Culture: A Naturalistic Approach.* Oxford: Blackwell.

———. 2000. Metarepresentations in an evolutionary perspective. In *Metarepresentations: A Multidisciplinary Perspective,* ed. D. Sperber. Oxford: Oxford University Press, 117–138.

———. 2006. An evolutionary perspective on testimony and argumentation. In *Biological and Cultural Bases of Human Inference,* ed. R. Viale, D. Andler, and L. A. Hirschfeld. Mahwah, N.J.: Lawrence Erlbaum, 177–189.

Sperber, Dan, David Premack, and Ann James Premack, eds. 1995. *Causal Cognition: A Multidisciplinary Debate.* Oxford: Clarendon.

Sperber, Dan, and Deirdre Wilson. 1988. *Relevance: Communication and Cognition.* Oxford: Blackwell.

Spiegelman, Art. 1986. *Maus: A Survivor's Tale.* Vol. 1: *My Father Bleeds History.* New York: Pantheon.

———. 1991. *Maus: A Survivor's Tale.* Vol. 2: *And Here My Troubles Began.* New York: Pantheon.

―――. 1999. *Comix, Essays, Graphics and Scraps: From Maus to Now to Maus to Now.* New York: Raw.

Špinka, Marek, Ruth C. Newberry, and Marc Bekoff. 2001. Mammalian play: Training for the unexpected. *Quarterly Journal of Biology* 76 (2):141–168.

Spitzer, Leo. 1948. *Linguistics and Literary History.* Princeton: Princeton University Press.

Stanford, W. B. 1999. The untypical hero. 1963. In de Jong 1999b, 4:190–205.

Stein, Nancy L., and Maria D. Liwag. 1997. Children's understanding, evaluation, and memory for emotional events. In van den Broek, Bauer, and Bourg 1997, 199–235.

Steinfels, Peter. 1995. Beliefs. *New York Times,* 19 August: 9.

Steptoe, Andrew. 1996. Mozart's personality and creativity. In *Wolfgang Amadè Mozart: Essays on His Life and Music,* ed. S. Sadie. Oxford: Clarendon, 21–34.

Sterelny, Kim. 2003. *Thought in a Hostile World: The Evolution of Cognition.* Oxford: Blackwell.

Stern, Daniel. 1977. *The First Relationship: Mother and Infant.* Cambridge, Mass.: Harvard University Press.

―――. 1985. *The Interpersonal World of the Infant: A View from Psychoanalysis and Developmental Psychology.* New York: Basic Books.

Sternberg, Meir. 1978. *Expositional Modes and Temporal Ordering in Fiction.* Baltimore: Johns Hopkins University Press.

―――. 2003a. Universals of narrative and their cognitivist fortunes (I). *Poetics Today* 24 (2):297–395.

―――. 2003b. Universals of narrative and their cognitivist fortunes (II). *Poetics Today* 24 (3):517–638.

Stofflet, Mary. 1986. Introduction. In *Dr. Seuss from Then till Now,* ed. San Diego Museum of Art. New York: Random House, 8–69.

Storey, Robert. 1996. *Mimesis and the Human Animal: On the Biogenetic Foundations of Literary Representation.* Evanston: Northwestern University Press.

Strawson, Galen. 2004. Tales of the unexpected. *The Guardian,* 10 January.

Suddendorf, Thomas. 1999. The rise of the metamind. In Corballis and Lea 1999, 218–260.

Suddendorf, Thomas, and Michael C. Corballis. 2007. The evolution of foresight: What is mental time travel, and is it unique to humans? *Behavioral and Brain Sciences* 30:299–351.

Suddendorf, Thomas, and Andrew Whiten. 2001. Mental evolution and development: Evidence for secondary representation in children, great apes, and other animals. *Psychological Bulletin* 127:629–650.

Sugiyama, Y. 1972. Social characteristics and socialization of wild chimpanzees. In *Primate Socialization,* ed. F. E. Poirier. New York: Random House, 145–163.

Sutton-Smith, Brian. 1981. *The Folk-Stories of Children.* Philadelphia: University of Pennsylvania Press.

———. 1986. Children's fiction-making. In *Narrative Psychology: The Storied Nature of Human Conduct,* ed. T. R. Sarbin. New York: Praeger, 67–90.

———. 1997. *The Ambiguity of Play.* Cambridge, Mass.: Harvard University Press.

———. 2001. Emotional breaches in play and narrative. In *Children in Play, Story, and School,* ed. A. Göncü and E. L. Klein. New York: Guilford, 161–176.

Symons, Donald. 1979. *The Evolution of Human Sexuality.* New York: Oxford University Press.

———. 1992. On the use and misuse of Darwinism in the study of human behavior. In Barkow, Cosmides, and Tooby 1992, 137–159.

Taylor, Marjorie. 1999. *Imaginary Companions and the Children Who Create Them.* New York: Oxford University Press.

Thayer, Bradley A. 2004. *Darwin and International Relations: On the Evolutionary Origins of War and Ethnic Conflict.* Lexington: University Press of Kentucky.

Thieme, H. 1997. Lower Paleolithic hunting spears from Germany. *Nature* 385:807–810.

Thiessen, Del, and Yoko Umezawa. 1998. The sociobiology of everyday life: A new look at a very old novel. *Human Nature* 9:293–320.

Thompson, D'Arcy. 1942. *On Growth and Form.* Cambridge: Cambridge University Press.

Thornhill, Randy, and Craig Palmer. 2000. *A Natural History of Rape: Biological Bases of Sexual Coercion.* Cambridge, Mass.: MIT Press.

Tinbergen, Niko. 1963. On aims and methods in ethology. *Zeitschrift für Tierpsychologie* 20:410–433.

Tomasello, Michael. 1995. Joint attention as social cognition. In *Joint Attention: Its Origins and Role in Development,* ed. C. Moore and P. Dunham. Hillsdale, N.J.: Lawrence Erlbaum, 103–130.

———. 1998. Introduction: A cognitive-functional perspective on language structure. In *The New Psychology of Language: Cognitive and Functional Approaches to Language Structure,* ed. M. Tomasello. Vol. 1. Mahwah, N.J.: Lawrence Erlbaum, vii–xxiii.

———. 2003. Introduction: Some surprises for psychologists. In *The New Psychology of Language: Cognitive and Functional Approaches to Language Structure,* ed. M. Tomasello. Vol. 2. Mahwah, N.J.: Lawrence Erlbaum, 1–14.

Tomasello, Michael, and Josep Call. 1997. *Primate Cognition.* New York: Oxford University Press.

Tomasello, M., M. Carpenter, J. Call, T. Behne, and H. Moll. 2005. Understanding and sharing intentions: The origins of cultural cognition. *Behavioral and Brain Sciences* 28:675–735.

Tooby, John, and Leda Cosmides. 1992. The psychological foundations of culture. In Barkow, Cosmides, and Tooby 1992, 19–36.

———. 1996. Friendship and the Banker's Paradox: Other pathways to the evolution of adaptations for altruism. *Proceedings of the British Academy* 88:119–143.

———. 2001. Does beauty build adapted minds? Toward an evolutionary theory of aesthetics, fiction and the arts. *Substance* 30 (1–2):6–27.

Tooby, John, and Irwin DeVore. 1987. The reconstruction of hominid behavioral evolution through strategic modeling. In *The Evolution of Human Behavior: Primate Models,* ed. W. G. Kinzey. Albany: State University of New York Press, 183–237.

Trabasso, Tom, and Nancy L. Stein. 1997. Narrating, representing, and remembering event sequences. In van den Broek 1997, 237–270.

Tracy, Stephen V. 1990. *The Story of the* Odyssey. Princeton: Princeton University Press.

Travis, Lisa L. 1997. Goal-based organization of event memory in toddlers. In van den Broek 1997, 111–138.

Trehub, Sandra. 2000. Human processing predispositions and musical universals. In Wallin, Merker, and Brown 2000, 427–448.

Trevarthen, Colwyn. 1979. Communication and cooperation in early infancy: A description of primary intersubjectivity. In *Before Speech:*

The Beginning of Human Communication, ed. M. Bullowa. Cambridge: Cambridge University Press, 321–347.

———. 1993. The function of emotions in early communication and development. In *New Perspectives in Early Communicative Development,* ed. Jacqueline Nadel and Luigi Camaioni. London: Routledge, 48–81.

Trivers, Robert. 1971. The evolution of reciprocal altruism. *Quarterly Review of Biology* 46:35–57.

———. 1972. Parental investment and sexual selection. In *Sexual Selection and the Descent of Man: 1871–1971,* ed. B. Campbell. Chicago: Aldine, 163–179.

———. 1985. *Social Evolution.* Menlo Park, Calif.: Benjamin/Cummings.

———. 2000. The elements of a scientific theory of self-deception. In *Evolutionary Perspectives on Human Reproductive Behavior,* ed. D. LeCroy and P. Moller. New York: New York Academy of Sciences, 114–131.

———. 2002a. *Natural Selection and Social Theory: Selected Papers of Robert Trivers.* New York: Oxford University Press.

———. 2002b. Parent-offspring conflict. 1974. In Trivers 2002a, 123–153.

Tucker, M., and R. Ellis. 1998. On the relations between seen objects and components of potential actions. *Journal of Experimental Psychology: Human Perception and Performance* 24:830–846.

Tulving, Endel. 1985. Memory and consciousness. *Canadian Journal of Psychology* 26:1–12.

Turner, Frank. 1997. The Homeric question. In *A New Companion to Homer,* ed. I. Morris and B. Powell. Leiden: Brill, 123–145.

Turner, Frederick, and Ernst Pöppel. 1988. Metered poetry, the brain, and time. In *Beauty and the Brain: Biological Aspects of Aesthetics,* ed. I. Rentschler, B. Hertzberger, and D. Epstein. Basel: Birkhäuser, 71–90.

Turner, Mark. 1996. *The Literary Mind.* Oxford: Oxford University Press.

van den Broek, Paul. 1997. Discovering the cement of the universe: The development of event comprehension from childhood to adulthood. In van den Broek, Bauer, and Bourg 1997, 321–342.

van den Broek, Paul W., Patricia J. Bauer, and Tammy Bourg, eds. 1997. *Developmental Spans in Event Comprehension and Representation: Bridging Fictional and Actual Events.* Mahwah, N.J.: Lawrence Erlbaum.

van Hooff, Jan A. R. A. M., and Signe Preuschoft. 2003. Laughter and smiling: The intertwining of nature and culture. In de Waal and Tyack 2003, 260–287.

van Wees, H. J. 1999. Introduction: Homer and early Greece. In de Jong 1999b, 2:1–32.

Verhagen, Aric. 2007. Construal and perspectivization. In *Cognitive Linguistics*, ed. Dirk Geergerts and Hubert Cuyckens. Oxford: Oxford University Press, 48–81.

———. 2008a. Intersubjectivity and the architecture of the language system. In *The Shared Mind: Perspectives on Intersubjectivity*, ed. J. Zlatev, T. P. Racine, C. Sinha, and E. Itkonen. Amsterdam: John Benjamins, 307–331.

———. 2008b. Intersubjectivity and explanation in linguistics—a reply to Hinzen and Van Lambalgen. *Cognitive Linguistics* 19:125–143.

Vickers, Brian. 1993. *Appropriating Shakespeare: Contemporary Critical Quarrels*. New Haven: Yale University Press.

Vinden, Penelope G., and Janet Wilde Astington. 2000. Culture and understanding other minds. In Baron-Cohen, Tager-Flusberg, and Cohen 2000, 503–519.

Waelti, P., A. Dickinson, and Wolfram Schultz. 2001. Dopamine responses comply with basic assumptions of formal learning theory. *Nature* 412:43–48.

Wallin, Nils L., Björn Merker, and Steven Brown, eds. 2000. *The Origins of Music*. Cambridge, Mass.: Bradford/MIT Press.

Walsh, Thomas R., and Rodney Merrill. 2002. The *Odyssey:* The tradition, the singer, the performance. In Homer 2003, 1–53.

Ward, P., and Amotz Zahavi. 1973. The importance of certain assemblages of birds as "information centres" for food finding. *Ibis* 115 (4):517–534.

Wason, Peter. 1966. Reasoning. In *New Horizons in Psychology*, ed. B. M. Foss. Harmondsworth: Penguin.

Weiss, Paul. 1955. Beauty and the beast: Life and the rule of order. *Scientific Monthly* 81 (December): 286–299.

Wellman, Henry W. 1990. *The Child's Theory of Mind*. Cambridge, Mass.: MIT Press.

Wellman, Henry W., and Susan A. Gelman. 1998. Knowledge acquisition in foundational domains. In *Handbook of Child Psychology*. Vol. 2: *Cog-*

nition, Perception, and Language, ed. D. Kuhn and R. Siegler. New York: Wiley, 523–573.

Wellman, Henry W., and Kristin H. Lagattuta. 2000. Developing understandings of mind. In Baron-Cohen, Tager-Flusberg, and Cohen 2000, 21–49.

Wells, H. G. 2007. *The Country of the Blind and Other Selected Stories,* ed. Patrick Parrinder. London: Penguin.

Wells, Randall S. 2003. Dolphin social complexity: Lessons from long-term study and life history. In de Waal and Tyack 2003, 32–56.

White, Don J. 2001. Mystery of the silver rings. Available at http://earthtrust.org/delrings.html.

Whiten, Andrew. 1997. The Machiavellian mindreader. In Whiten and Byrne 1997, 144–173.

Whiten, Andrew, and Richard W. Byrne, eds. 1997. *Machiavellian Intelligence II: Extensions and Evaluations.* Cambridge: Cambridge University Press.

Wilkinson, G. C. 1984. Reciprocal food sharing in the vampire bat. *Nature* 308:181–184.

Williams, Bernard. 1993. *Shame and Necessity.* Berkeley: University of California Press.

Williams, George C. 1966. *Adaptation and Natural Selection.* Princeton: Princeton University Press.

———. 1992. *Natural Selection: Domains, Levels, and Challenges.* New York: Oxford University Press.

Wilson, David Sloan. 1994. Adaptive genetic variation and human evolutionary psychology. *Ethology and Sociobiology* 15:219–235.

———. 2002. *Darwin's Cathedral: Evolution, Religion, and the Nature of Society.* Chicago: University of Chicago Press.

———. 2005. Evolutionary social constructivism. In Gottschall and Wilson 2005, 20–37.

———. 2007a. *Evolution for Everyone: How Darwin's Theory Can Change the Way We Think about Our Lives.* New York: Delacorte.

———. 2007b. Foreword. In Oakley 2007.

Wilson, David Sloan, David C. Near, and Ralph R. Miller. 1996. Machiavellianism: A synthesis of the evolutionary and psychological literature. *Psychological Bulletin* 119:285–299.

Wilson, David Sloan, and Edward O. Wilson. 2007. Rethinking the theoretical foundation of sociobiology. *Quarterly Review of Biology* 82 (4):328–348.

Wilson, Edward O. 1975/2000. *Sociobiology: The New Synthesis.* 25th Anniversary Edition, 2000. Cambridge, Mass.: The Belknap Press of Harvard University Press.

———. 1999. *Consilience: The Unity of Knowledge.* 1998. Reprint, New York: Vintage.

Wilson, Margo, Martin Daly, and J. E. Scheib. 1997. Femicide: An evolutionary psychological perspective. In *Feminism and Evolutionary Biology: Boundaries, Intersections, and Frontiers,* ed. P. A. Gowaty. New York: Chapman and Hall.

Winkielman, Piotr, Jamin Halberstadt, Tedra Fazendeiro, and Steve Catty. 2006. Prototypes are attractive because they are easy on the mind. *Psychological Science* 17:799–806.

Wittgenstein, Ludwig. 1980. *Remarks on the Philosophy of Psychology.* 2 vols. Trans. G. E. M. Anscombe. Ed. G. E. M. Anscombe and G. H. von Wright. Chicago: University of Chicago Press.

Wolf, Maryanne. 2007. *Proust and the Squid: The Story and Science of the Reading Brain.* New York: HarperCollins.

Wolpert, Lewis. 1993. *The Unnatural Nature of Science.* 1992. Reprint, London: Faber.

Woodhouse, William. 1930. *The Composition of Homer's* Odyssey. Oxford: Clarendon.

Worden, Robert. 1998. The evolution of language from social intelligence. In Hurford, Studdert-Kennedy, and Knight 1998, 148–166.

Wright, Robert. 1996. *The Moral Animal: Evolutionary Psychology and Everyday Life.* London: Abacus.

———. 2000. *Nonzero: The Logic of Human Destiny.* Boston: Little, Brown.

Xu, Jiang, Stefan Kemeny, Grace Park, Carlo Frattali, and Allen Braun. 2005. Language in context: Emergent features of word, sentence, and narrative comprehension. *NeuroImage* 25:1002–15.

Zahavi, Amotz, and Avishag Zahavi. 1997. *The Handicap Principle: A Missing Piece of Darwin's Puzzle.* Oxford: Oxford University Press.

Zuberbühler, Klaus, and Richard W. Byrne. 2006. Social cognition. *Current Biology* 16:R786–790.

Zunshine, Lisa. 2006. *Why We Read Fiction: Theory of Mind and the Novel.* Columbus: Ohio State University Press.

———. 2007. Fiction and theory of mind: An exchange. *Philosophy and Literature* 31:189–195.

INDEX

Page numbers in italic refer to illustrations.